LTE SECURITY

LTE SECURITY

Dan Forsberg
Security Consultant, ISECure.fi Oy, Finland

Günther Horn
Senior Security Specialist, Nokia Siemens Networks, Germany

Wolf-Dietrich Moeller
Senior Security Specialist, Nokia Siemens Networks, Germany

Valtteri Niemi
Nokia Fellow, Nokia Corporation, Switzerland

A John Wiley and Sons, Ltd., Publication

This edition first published 2010
© 2010 John Wiley & Sons Ltd

Library of Congress Cataloging-in-Publication Data

LTE security/Dan Forsberg, Günther Horn. . . [et al.].
    p. cm.
  Includes bibliographical references and index.
  ISBN 978-0-470-66103-1 (hardback)
  1. Long-Term Evolution (Telecommunications)   2. Global system for mobile communications.   I. Forsberg, Dan.
II. Horn, Günther.
  TK5103.48325.L74 2010
  621.3845′6–dc22

                                                                                    2010022116

A catalogue record for this book is available from the British Library.

Print ISBN: 9780470661031 (hb)
ePDF ISBN: 9780470973288
oBook ISBN: 9780470973271

Set in 10/12pt Times by Aptara Inc., New Delhi, India

Printed in the UK by CPI Antony Rowe Chippenham and Eastbourne

Contents

Foreword

The early to mid 1980s saw the commercial opening across Europe of public-access mobile communications systems. These cellular systems all used analogue technology, but outside of the Nordic countries no attempt was made to standardize the systems – so the technology adopted differed from country to country. Unfortunately, one thing they did have in common was a total absence of adequate security features, which made them open to abuse by criminals, journalists and all manner of opportunists. User's calls could be eavesdropped on the air using readily available and comparatively inexpensive interception devices, and there were celebrated cases of journalistic invasion of privacy. A well-known example was the "squidgy" tapes, where mobile telephone calls between members of the British royal family were recorded. Mobile telephone operators and their customers became very concerned.

The operators also had another problem with serious financial consequences. When a mobile phone attempted to connect to a network the only check made on authenticity was to see that the telephone number and the phone's identity correctly corresponded. These numbers could be intercepted on the air and programmed to new phones creating clones of the original. Clones were used by criminals to run up huge charges for calls which had nothing to do with the legitimate owner. Cloning became very widespread, with criminals placing their "cloning" equipment in cars parked at airports to capture the numbers from business people announcing their arrival back home to their families. It represented a serious financial problem for operators who ended up covering the charges themselves. The problems caused by lack of security in European analogue systems were a significant factor in accelerating the creation and adoption of GSM.

GSM is a standard for digital mobile communications, designed originally for Europe but now adopted all over the world. Being an international standard it brings economy of scale and competition, and it enables users to roam across borders from one network to another. Being digital it brings transmission efficiency and flexibility, and enables the use of advanced cryptographic security. The security problems of the original analogue systems are addressed in GSM by encryption on the air interface of user traffic, in particular voice calls, and authentication by network operators of their customers on an individual basis whenever they attempt to connect to a network, irrespective of where that network may be. From both a technical and a regulatory perspective the use of cryptography in GSM was ground-breaking. Initially manufacturers and operators feared it would add too much complexity to the system, and security agencies were concerned that it may be abused by criminals and terror organizations. The legitimate fears and concerns constrained what was possible, especially with the encryption algorithm, which was designed against a philosophy of "minimum strength to provide adequate security". Despite this, and the continuing efforts of organized hackers, eavesdropping on the air of GSM calls protected using the original cipher has still to be demonstrated in a real deployment, and with a stronger cipher already available in the wings, any future success will be largely pointless. This doesn't mean that GSM is free from security weaknesses – the ability to attack it using false base stations is very real.

GSM is the first in an evolving family of technologies for mobile communications. The second member of the family is 3G (or UMTS as it is often referred to in Europe) and the third, and most recent, is LTE (EPS to give it its proper title which is used throughout the main body of this book). With each technology evolution the security features have been enhanced to address learning from its predecessor, as well as to accommodate any changes in system architecture or services. The underlying GSM security architecture has proved to be extremely robust, and consequently has remained largely unchanged with the evolving technology family. It has also been adapted for use in other communications systems, including WLAN, IMS and HTTP. It is characterized by authentication data and encryption key generation being confined to a user's home network authentication center and personal SIM, the two elements where all user-specific static security data is held. Only dynamic and user session-specific security data goes outside these domains.

3G sees the addition to the GSM security features of user authentication of the access network – to complement user authentication by the network, integrity protection of signalling and the prevention of authentication replay. Start and termination of ciphering is moved from the base station further into the network. Of course the false base station attack is countered. A new suite of cryptographic algorithms based on algorithms open to public scrutiny and analysis is introduced, and changes of regulation governing the export of equipment with cryptographic functionality make their adoption easier for most parts of the world.

LTE heralds the first technology in the family that is entirely packet-switched – so voice security has to be addressed in an entirely different way from GSM and 3G. LTE is a much flatter architecture, with fewer network elements, and is entirely IP-based. Functionality, including security functionality, is migrated to the edge of the network, including encryption functionality which is moved to the edge of the radio network, having been moved from the base station to the radio network controller in the evolution from GSM to 3G. While maintaining compatibility with the security architecture developed for GSM and evolved for 3G, the security functionality has been significantly adapted, enhanced and extended to accommodate the changes that LTE represents, as well as security enhancements motivated by practical experience with 3G. Much of this plays back into 3G itself as new security challenges arise with the advent of femto cells – low-cost end nodes in exposed environments that are not necessarily under the control of the operator of the network to which they are attached.

The book takes the reader through the evolution of security across three generations of mobile, focusing with clarity and rigor on the security of LTE. It is co-authored by a team who continue to be at the heart of the working group in 3GPP responsible for defining the LTE security standards. Their knowledge, expertise and enthusiasm for the subject shines through.

Professor Michael Walker
Chairman of the ETSI Board

Acknowledgements

This book presents the results of research and specification work by many people over an extended period. Our thanks therefore go to all those who helped make LTE possible through their hard work. In particular, we thank the people working in 3GPP, the standardization body that publishes the LTE specifications, and, especially, the delegates to the 3GPP security working group, SA3, with whom we were working to produce the LTE security specifications over the past years.

We would also like to express our gratitude to our colleagues at Nokia and Nokia Siemens Networks for our longstanding fruitful collaboration. We are particularly indebted to Wolfgang Bücker, Devaki Chandramouli, Jan-Erik Ekberg, Silke Holtmanns, Jan Kåll, Raimund Kausl, Christian Markwart, Kaisa Nyberg, Martin Öttl, Jukka Ranta, Manfred Schäfer, Peter Schneider, Hans-Jürgen Schwarzbauer, José Manuel Tapia Pérez, Janne Tervonen, Robert Zaus and Dajiang Zhang who helped us improve the book through their invaluable comments.

Finally, we would like to thank the editing team at Wiley whose great work turned our manuscript into a coherent book.

The authors welcome any comments or suggestions for improvements.

Copyright Acknowledgements

The authors would like to include additional thanks and full copyright acknowledgement as requested by the following copyright holders in this book.

© **2009, 3GPP™**. TSs and TRs are the property of ARIB, ATIS CCSA, ETSI, TTA and TTC who jointly own the copyright in them. They are subject to further modifications and are therefore provided here 'as is' for information purposes only. Further use is strictly prohibited.

© **2010, 3GPP™**. TSs and TRs are the property of ARIB, ATIS CCSA, ETSI, TTA and TTC who jointly own the copyright in them. They are subject to further modifications and are therefore provided here 'as is' for information purposes only. Further use is strictly prohibited.

© **2010, Nokia Corporation** – for permission to reproduce the Nokia Corporation UE icon within Figures 2.1, 3.1, 3.2, 3.3, 6.1, 6.2, 6.3, 7.1 and 14.1.

Please see the individual figure captions for copyright notices throughout the book.

1

Overview of the Book

Mobile telecommunications systems have evolved in a stepwise manner. A new cellular radio technology has been designed once per decade. Analogue radio technology was dominant in the 1980s and paved the way for the phenomenal success of cellular systems. The dominant second-generation system GSM was introduced in the early 1990s, while the most successful third-generation system 3G – also known as UMTS, especially in Europe – was brought into use in the first years of the first decade of the new millennium.

At the time of writing, the fourth generation of mobile telecommunications systems is about to be introduced. Its new radio technology is best known under the acronym 'LTE' (Long Term Evolution). The complete system is named 'SAE/LTE', where 'SAE' (System Architecture Evolution) stands for the entire system, which allows combining access using the new, high-bandwidth technology LTE with access using the legacy technologies such as GSM, 3G and HRPD. The technical term for the SAE/LTE system is Evolved Packet System (EPS), and we shall be using this term consistently in the book. The brand name of the new system has been chosen to be LTE, and that is the reason why the title of the book is *LTE Security*.

With the pervasiveness of telecommunications in our everyday lives, telecommunications security has also moved more and more to the forefront of attention. Security is needed to ensure that the system is properly functioning and to prevent misuse. Security includes measures such as encryption and authentication, which are required to guarantee the user's privacy as well as ensuring revenue for the mobile network operator.

The book will address the security architecture for EPS. This is based on elements of the security architectures for GSM and 3G, but it needed a major redesign effort owing to the significantly increased complexity, and new architectural and business requirements. The book will present the requirements and their motivation and then explain in detail the security mechanisms employed to meet these requirements.

To achieve global relevance, a communication system requires world-wide interoperability that is easiest to achieve by means of standardization. The standardized part of the system guarantees that the entities in the system are able to communicate with each other even if they are controlled by different mobile network operators or manufactured by different vendors. There are also many parts in the system where interoperability does not play a role, such as the internal structure of the network entities. It is better not to standardize

LTE Security   Dan Forsberg, Günther Horn, Wolf-Dietrich Moeller, and Valtteri Niemi
© 2010 John Wiley & Sons, Ltd

wherever it is not necessary because then new technologies can be introduced more rapidly and differentiation is possible among operators as well as among manufacturers, thus encouraging healthy competition.

As an example in the area of security, communication between the mobile device and the radio network is protected by encrypting the messages. It is important that we standardize how the encryption is done and which encryption keys are used, otherwise the receiving end could not do the reverse operation and recover the original content of the message. On the other hand, both communicating parties have to store the encryption keys in such a way that no outsider can get access to them. From the security point of view, it is important that this be done properly but we do not have to standardize how it is done, thus leaving room for the introduction of better protection techniques without the burden of standardizing them first. The emphasis of our book is on the standardized parts of EPS security, but we include some of the other aspects as well.

The authors feel that there will be interest in industry and academia in the technical details of SAE/LTE security for quite some time to come. The specifications generated by standardization bodies only describe *how* to implement the system (and this only to the extent required for interoperability), but almost never inform readers about *why* things are done the way they are. Furthermore, specifications tend to be readable only by a small group of experts and lack the context of the broader picture. This book is meant to fill this gap by providing first-hand information from insiders who participated in decisively shaping SAE/LTE security in the relevant standardization body, 3GPP, and can therefore explain the rationale for the design decisions in this area.

The book is based on versions of 3GPP specifications from March 2010 but corrections approved by June 2010 were still taken into account. New features will surely be added into these specifications in later versions and there will most probably also be further corrections to the existing security functionality. For the obvious reason of timing, these additions cannot be addressed in this book.

The book is intended for telecommunications engineers in research, development and technical sales and their managers as well as engineering students who are familiar with architectures of mobile telecommunications systems and interested in the security aspects of these systems. The book will also be of interest to security experts who are looking for examples of the use of security mechanisms in practical systems. Both readers from industry and from academia should be able to benefit from the book. The book is probably most beneficial to advanced readers, with subchapters providing sufficient detail so that the book can also be useful as a handbook for specialists. It can also be used as textbook material for an advanced course, and especially the introductory parts of each chapter, when combined, give a nice overall introduction to the subject.

The book is organized as follows. Chapter 2 gives the necessary background information on cellular systems, relevant security concepts, standardization matters and so on. As explained earlier, LTE security relies heavily on security concepts introduced for the predecessor systems. Therefore, and also to make the book more self-contained, Chapters 3, 4 and 5 are devoted to security in legacy systems, including GSM and 3G, and security aspects of cellular–WLAN interworking.

Chapter 6 provides an overall picture of the EPS security architecture. The next four chapters provide detailed information about the core functionalities in the security architecture. Chapter 7 is devoted to authentication and key agreement which constitute the cornerstones for

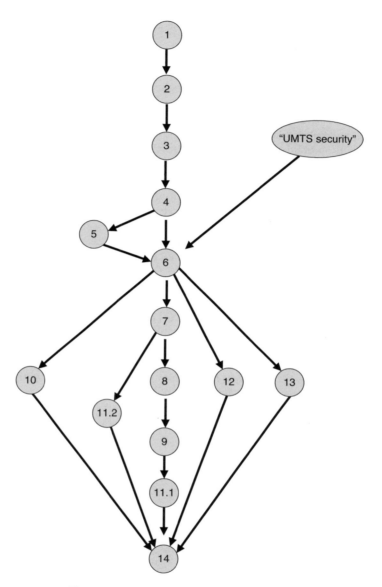

Figure 1.1   Major dependencies among chapters

the whole security architecture. Chapter 8 shows how user data and signalling data is protected in the system, including protecting confidentiality and integrity of the data. A very characteristic feature in cellular communication is the possibility of handing over the communication from one base station to another. Security for handovers and other mobility issues is handled in Chapter 9. Another cornerstone of the security architecture is the set of cryptographic algorithms that are used in the protection mechanisms. The algorithms used in EPS security are introduced in Chapter 10.

In the design of EPS, it has been taken into account already from the beginning how interworking with access technologies that are not defined by 3GPP is arranged. Also, interworking with legacy 3GPP systems has been designed into the EPS system. These two areas are discussed in detail in Chapter 11.

The EPS system is exclusively packet-based; there are no circuit-switched elements in it. This implies, in particular, that voice services have to be provided on top of IP packets. The security for such a solution is explained in Chapter 12.

Partially independently of the introduction of EPS, 3GPP has specified solutions that enable the deployment of base stations covering very small areas, such as in private homes. This type of base station may serve restricted sets of customers (e.g. people living in a house), but open usage in hotspots or remote areas is also envisaged. These home base stations are also planned for 3G access, not only for LTE access. Such a new type of base station may be placed in a potentially vulnerable environment not controlled by the network operator and therefore many new security measures are needed, compared to conventional base stations. These are presented in detail in Chapter 13.

Finally, Chapter 14 contains a discussion of both near-term and far-term future challenges in the area of securing mobile communications.

Many of the chapters depend on earlier ones, as can be seen from the above descriptions. However, it is possible to read some chapters without reading first all of the preceding ones. Also, if the reader has prior knowledge of GSM and 3G systems and their security features, the first four chapters can be skipped. This kind of knowledge could have been obtained, for example, by reading the book *UMTS Security* [Niemi and Nyberg 2003]. The major dependencies among the chapters of the book are illustrated in Figure 1.1.

2

Background

2.1 Evolution of Cellular Systems

Mobile communications were originally introduced for military applications. The concept of a cellular network was taken into commercial use much later, near the beginning of the 1980s, in the form of the Advanced Mobile Phone System (AMPS) in the USA and in the form of the Nordic Mobile Telephone system (NMT) in northern Europe. These first-generation cellular systems were based on analogue technologies. Simultaneous access by many users in the same cell was provided by the Frequency Division Multiple Access (FDMA) technique. Handovers between different cells were already possible in these systems and a typical use case was a phone call from a car.

The second generation of mobile systems (2G) was introduced roughly a decade later, at the beginning of the 1990s. The dominant 2G technology has been the Global System for Mobile (GSM) communications, with more than three and a half billion users worldwide at the time of writing. The second generation introduced digital information transmission on the radio interface between the mobile phone and the base station. The multiple access technology is Time Division Multiple Access (TDMA).

The second generation provided an increased capacity of the network (owing to more efficient use of radio resources), better speech quality (from digital coding techniques) and a natural possibility for communicating data. Furthermore, it was possible to use new types of security feature, compared to analogue systems.

Again roughly one decade later, the third-generation technologies (3G) were introduced at the beginning of the twenty-first century. Although GSM had become a phenomenal success story already at that point, there were also other successful 2G systems, both in Asia and in North America. One of the leading ideas for 3G was to ensure fully global roaming: to make it possible for the user to use the mobile system services anywhere in the world. A collaborative effort of standards bodies from Europe, Asia and North America developed the first truly global cellular technologies in the 3rd Generation Partnership Project (3GPP). At the time of writing, there are almost half a billion 3G subscriptions in the world.

The third generation provided a big increase in data rates, up to 2 megabits per second (Mbps) in the first version of the system that was specified in Release 99 of 3GPP. The multiple-access technology is Wideband Code Division Multiple Access (WCDMA).

LTE Security   Dan Forsberg, Günther Horn, Wolf-Dietrich Moeller, and Valtteri Niemi
© 2010 John Wiley & Sons, Ltd

Both GSM and 3G systems were divided into two different domains, based on the underlying switching technology. The circuit-switched (CS) domain is mainly intended for carrying voice and short messages while the packet-switched (PS) domain is mainly used for carrying data traffic.

One more decade passed and the time was ripe for taking another major step forward. In 3GPP the development work was done under the names of 'Long Term Evolution' (LTE) of radio technologies and 'System Architecture Evolution' (SAE). Both names emphasized the evolutionary nature of this step, but the end result is in many respects a brand new system, both from the radio perspective and from the system perspective. The new system is called Evolved Packet System (EPS) and its most important component, the new radio network, is called Evolved Universal Terrestrial Radio Access Network (E-UTRAN).

The EPS contains only a packet-switched domain. It offers a big increase in data rates, up to more than 100 Mbps. The multiple-access technology is again based on FDMA, namely Orthogonal Frequency Division Multiple Access (OFDMA) for the downlink traffic (from the network to the terminal) and Single Carrier FDMA (SC-FDMA) for the uplink traffic (from the terminal to the network).

2.1.1 Third-generation Network Architecture

In this section we give a brief overview of the 3GPP network architecture. A more thorough description of the 3G architecture can be found elsewhere [Kaaranen *et al.* 2005].

A simplified picture of the 3GPP Release 99 system is given in Figure 2.1.

The network model consists of three main parts, all of which are visible in Figure 2.1. The part closest to the user is the terminal that is also called the User Equipment (UE). The UE has a radio connection to the Radio Access Network (RAN), which itself is connected to the Core Network (CN). The core network takes care of coordination of the whole system.

The core network contains the PS domain and the CS domain. The former is an evolution of the GPRS domain of the GSM system and its most important network elements are the Serving GPRS Support Node (SGSN) and the Gateway GPRS Support Node (GGSN). The CS domain is an evolution from the original circuit-switched GSM network with the Mobile Switching Centre (MSC) as its most important component.

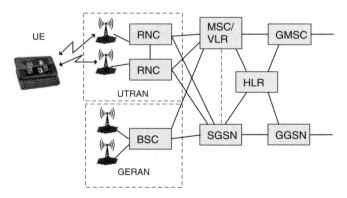

Figure 2.1 The 3G system

In addition to the various network elements, the architecture defines also interfaces or, more correctly, reference points between these elements. Furthermore, protocols define how different elements are able to communicate over the interfaces. Protocols involving the UE are grouped into two main strata: the Access Stratum (AS) contains protocols that are run between the UE and the access network, while the Non-Access Stratum (NAS) contains protocols between the UE and the core network. In addition to these two, there are many protocols that are run between different network elements.

The core network is further divided into the home network and the serving network. The home network contains all the static information about the subscribers, including the static security information. The serving network handles the communication to the UE (via the access network). If the user is roaming, then the home and the serving network are controlled by different mobile network operators.

2.1.2 Important Elements of the 3G Architecture

The user equipment consists of two parts: the Mobile Equipment (ME) and the Universal Subscriber Identity Module (USIM). The ME is typically a mobile device that contains the radio functionality and all the protocols that are needed for communications with the network. It also contains the user interface, including a display and a keypad. The USIM is an application that is run inside a smart card called Universal Integrated Circuit Card (UICC) [TS31.101]. The USIM contains all the operator-dependent data about the subscriber, including the permanent security information.

There are two types of radio access network in the 3G system. The Universal Terrestrial Radio Access Network (UTRAN) is based on WCDMA technology, and the GSM/EDGE Radio Access Network (GERAN) is an evolution of GSM technology.

The radio access network contains two types of element. The base station (BS) is the termination point of the radio interface on the network side, and it is called Node B in the case of UTRAN and Base Transceiver Station (BTS) in GERAN. The base station is connected to the controlling unit of the RAN, which is the Radio Network Controller (RNC) in UTRAN or the Base Station Controller (BSC) of GERAN.

In the core network, the most important element in the circuit-switched domain is the switching element MSC that is typically integrated with a Visitor Location Register (VLR) that contains a database of the users currently in the location area controlled by the MSC. The Gateway MSC (GMSC) takes care of connections to external networks, an example being the Public Switched Telephone Network (PSTN). In the packet-switched domain, the role of MSC/VLR is taken by the SGSN, while the GGSN takes care of connecting to IP services within the operator network and to the outside world, such as the Internet.

The static subscriber information is maintained in the Home Location Register (HLR). It is typically integrated with the Authentication Centre (AuC) that maintains the permanent security information related to subscribers. The AuC also creates temporary authentication and security data that can be used for security features in the serving network, such as authentication of the subscriber and encryption of the user traffic.

In addition to the elements mentioned here and illustrated in Figure 2.1, there are many other components in the 3G architecture, an example being the Short Message Service Centre (SMSC) that supports storing and forwarding of short messages.

2.1.3   Functions and Protocols in the 3GPP System

The main functionalities in the 3GPP system are:

* Communication Management (CM) for user connections, such as call handling and session management;
* Mobility Management (MM) covering procedures related to user mobility, as well as important security features;
* Radio Resource Management (RRM) covering, for example, power control for radio connections, control of handovers and system load.

The CM functions are located in the non-access stratum while RRM functions are located in the access stratum. The MM functions are taken care of by both the core network and the radio access network.

The division into user plane and control plane (also called signalling plane) defines an important partition among the protocols. User-plane protocols deal, as the name indicates, with the transport of user data and other directly user-related information, such as speech. Control-plane protocols are needed to ensure correct system functionality by transferring necessary control information between elements in the system.

In a telecommunication system, in addition to the user and control planes, there is also a management plane that, for example, keeps all elements of the system in operation. Usually, there is less need for standardization in the management plane than there is for the user plane and the control plane.

The most important protocols for the Internet are Internet Protocol (IP), User Datagram Protocol (UDP) and Transmission Control Protocol (TCP). In the wireless environment there is a natural reason to favour UDP over TCP: fading and temporary loss of coverage make it difficult to maintain reliable transmission of packets on a continuous basis. There is also a 3GPP specific protocol that is run on top of UDP/IP. This is the GPRS Tunnelling Protocol (GTP). It has been optimized for data transfer in the backbone of the PS domain.

The interworking of the different types of protocol can be illustrated by a typical use case: a user receiving a phone call. First the network pages for the user. Paging is an MM procedure; the network has to know in which geographical area the user could be found. After the user has successfully received the paging message, the radio connection is established by RRM procedures. When the radio connection exists, an authentication procedure may follow, and this belongs again to the MM. Next the actual call set-up (CM procedure) occurs during which the user may be informed about who is calling. During the call there may be many further signalling procedures, such as for handovers. At the end of the call, the call is first released by a CM procedure and after that the radio connection is released by the RRM.

2.1.4   The EPS System

The goals of the EPS are [TS22.278]:

* higher data rates;
* lower latency;

- high level of security;
- enhanced quality of service (QoS);
- support for different access systems with mobility and service continuity between them;
- support for access system selection;
- capabilities for interworking with legacy systems.

The main means to achieve these goals are:

- the new radio interface and the new RAN based on it (E-UTRAN);
- a flat IP-based architecture that has only two network elements on the user plane (evolved NodeB and Serving Gateway).

Figure 2.2 (adapted from [TS23.401]) illustrates the EPS network architecture in a case where the UE is not roaming into a different network than where it has its subscription. Note that the legacy radio access networks UTRAN and GERAN are included in the system together with the legacy core network element SGSN.

The new core network element is called Mobility Management Entity (MME). The HLR of the original GSM and 3G architecture is extended to the Home Subscriber Server (HSS). The core network element for user-plane handling is called Serving Gateway (S-GW). The PDN Gateway (PDN GW) handles the traffic towards packet data networks. It is also possible that S-GW and PDN GW are co-located. The core network of the EPS is called Evolved Packet Core (EPC).

The architecture of E-UTRAN is depicted in Figure 2.3 (see also [TS36.300]). The base station eNB is the only type of network element in E-UTRAN. On the other hand, there is an interface between two eNBs facilitating fast handovers between different base stations.

2.2 Basic Security Concepts

It is not easy to define 'security' even though people tend to understand quite well what is meant by it. Protection methods against malicious actions lie at the core of security. There is also a clear distinction between security, on one hand, and fault-tolerance and robustness, on the other.

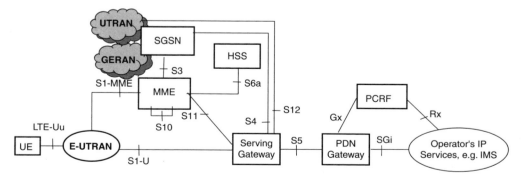

Figure 2.2 The EPS architecture (non-roaming case). Adapted with permission from © 2010, 3GPP™

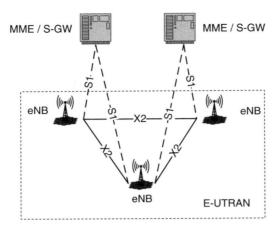

Figure 2.3 The E-UTRAN architecture

Many aspects of security are relevant for a communication system. There are physical security aspects and information security aspects. The former include issues such as locked rooms, safes and guards: all these are needed when operating a large-scale network. Another property that belongs to the area of physical security is tamper-resistance. Smart cards play a major role in the system we describe in this book, and tamper-resistance is a key property of smart cards. Sometimes guaranteed tampering evidence is a sufficient protection method against physical intrusion: if tampering can be detected quickly enough, corrupted elements can be cut out of the network before too much damage is caused.

Biometric protection mechanisms are examples of methods between physical security and information security. For example, checking of fingerprints assumes both sophisticated measurement instruments and a sophisticated information system to support the use of these instruments as access control devices.

In this book we concentrate mainly on aspects belonging to the broad category of information security. In particular, we put focus on communication security. But physical security is also important for EPS security and will be covered to some extent as well.

2.2.1 Information Security

In the context of information security, the following areas can be studied fairly independently of each other:

- *System security*. An example is trying to ensure that the system does not contain any weak parts. Attackers typically try to find a point weak enough to be broken.
- *Application security*. Banking over the Internet, for example, typically uses security mechanisms that are tailored to meet the application-specific requirements.
- *Protocol security*. Communicating parties are, for example, able to achieve security goals by executing well-defined communication steps in a certain well-defined order.

- *Platform security*. The network elements and mobile terminals depend on the correct functionality of the operating system that controls them. Physical security, too, has an important role in platform security.
- *Security primitives*. These are the basic building blocks on top of which all protection mechanisms are built. Typical examples are cryptographic algorithms, but also items like a protected memory can be seen as a security primitive (thus bringing physical security also into the picture).

In this book we put the main emphasis on system security, protocol security and security primitives. Platform security is covered only briefly, and application security is seen as more or less orthogonal to the purposes of this book.

In the design of a practical security system there are always tight constraints. The cost of implementing protection mechanisms must be balanced with the amount of risk mitigated by these mechanisms. The usability of the system must not suffer because of security. These trade-offs depend also on the intended use of the system: in a military system, for example, trade-offs between security, cost and usability are done on a different basis from in a public or a general-purpose communication system.

2.2.2   Design Principles

The design process of a security system contains typically the following phases:

- *Threat analysis*. The intention is to list all possible threats against the system, regardless of the difficulty and cost of carrying out an attack to materialize a particular threat.
- *Risk analysis*. The weight of each threat is measured quantitatively or, at least, in relation to other threats. Estimates are needed for both the probability of various attacks and the potential gain for the attacker and/or damage to the attacked side caused by them.
- *Requirements capture*. Based on the earlier phases, it is now decided what kind of protection is required for the system.
- *Design phase*. The actual protection mechanisms are designed in order to meet the requirements. Existing building blocks, such as security protocols or primitives, are identified, possibly new mechanisms are created, and a security architecture is built. Here the constraints have to be taken into account, and it is possible that not all requirements can be met. This may cause a need to re-visit earlier phases, especially the risk analysis.
- *Security analysis*. An evaluation of the results is carried out independently of the previous phase. Usually, automatic verification tools can be used only for parts of a security analysis. There are often holes in the security system that can be revealed only by using creative methods.
- *Reaction phase*. While planning of the system management and operation can be seen as part of the mechanism design phase, reaction to all unexpected security breaches cannot be planned beforehand. In the reaction phase it is vital that the original design of the system is flexible enough and allows enhancements; it is useful to have a certain amount of safety margin in the mechanisms. These margins tend to be useful in cases where new attack methodologies appear faster than expected.

We have listed here only the phases that can be considered part of the design process. In addition, implementation and testing are also important in building a secure system.

One factor that affects several phases is the fact that often the security system is part of a much larger system that is under design at the same time. This has been the case for EPS specification work also. An iterative approach is needed because the general system architecture and requirements are changing in parallel to the security design. Although these iterations seem to slow down the process, it is important that the security for the system be designed at the same time as the system itself is designed. Trying to add security to an existing completed system typically leads to impractical and inefficient solutions.

2.2.3 *Communication Security Features*

Although security as an abstract concept is hard to define, its ingredients or features are typically easier to grasp in definitions. In the following we list the most important features in communication security:

- *Authenticity*. In a classical scenario where parties A and B are communicating over some channel, both typically want to begin with identifying each other. Authentication is the process of verifying the identities.
- *Confidentiality*. In the same classical scenario, parties A and B may want to limit the intelligibility of the communication just to the two parties themselves, to keep the communication confidential.
- *Integrity*. If all messages sent by the party A are identical to the ones received by the party B, and vice versa, then integrity of the communication has been preserved. Sometimes the property that the message is indeed sent by A is called 'proof-of-origin', while the term 'integrity' is restricted to the property that the message is not altered on the way.
- *Non-repudiation*. It is often useful for the receiving party B to store a message received from the sending party A. Now non-repudiation of the message means that A cannot later deny having sent it.
- *Availability*. This is an underlying assumption for the classical scenario of A and B communicating with each other. The communication channel must be available for parties A and B.

Typical attacks and attackers against these features are as follows:

- Authentication – an imposter tries to masquerade as one of the communicating parties.
- Confidentiality – an eavesdropper tries to get information about at least some parts of the communication.
- Integrity – a third party tries to modify, insert or delete messages in the communication channel.
- Non-repudiation – it may sometimes give a benefit for the sender of a certain message if he can later deny sending of it. For example, the message may relate to a financial transaction, or a commitment to buy or sell something.
- Availability – a Denial of Service (DoS) attack tries to prevent access to the communication channel, at least for some of the communicating parties.

The main emphasis in this book is on the first three features: authenticity, confidentiality and integrity. The whole point of introducing LTE and EPS is to improve the availability of the cellular access channel. The non-repudiation feature is still of less importance in EPS networks; it is much more relevant for the application layer.

2.3 Basic Cryptographic Concepts

Cryptology is sometimes defined as the art and science of secret writing. The possibility to apply cryptology for protecting the confidentiality of communications is obvious. Additionally, it has been found that similar techniques can be successfully applied to provide many other security features, such as for authentication.

Cryptology consists of two parts:

- *cryptography* – designing systems based on secret writing techniques;
- *cryptanalysis* – analysing cryptographic systems and trying to find weaknesses in them.

The twofold nature of cryptology reflects a more general characteristic in security. As explained earlier, it is very difficult to find testing methods that can be applied to reliably assess whether a designed system is secure. The reason for this is that the true test for a system begins when it is deployed in real life. Then attackers may appear who use whatever ways they can find to break the system. What makes the situation even more difficult is that these real-life attackers typically try to hide their actions and methods as far as possible. Cryptanalysis (and security analysis more widely) tries to anticipate what attackers might do and is constantly searching for novel ways of attacking systems. In this manner, cryptanalysis (and security analysis) contributes indirectly to achieving a better security level.

The role of cryptanalysis in modelling attackers is a complex issue. It is perfectly fine to find weaknesses in systems that are still under design and not deployed in practice. This is because then it is still easy and relatively cheap to take corrective action. However, when the system is already in wide use the role of cryptanalysis may become controversial. A clever attack found by a researcher may be reproduced by a real-life attacker who would not have invented it by himself. In this case, the attack found by the researcher seems to cause a decrease in the level of security rather than an increase.

One obvious solution to this dilemma is to keep the cryptanalytic result confidential until corrective action has been done to remove any real-life vulnerabilities. After these vulnerabilities have been removed, publishing the results helps to avoid similar vulnerabilities in future implementations. Note that there are similar debates on how to handle vulnerabilities discovered in, for example, operating systems and browsers. There seems to be no general agreement on the appropriate handling of vulnerabilities in the security community.

Another solution to the problem is to be secretive even in the design phase. If real-life attackers do not know what kind of cryptographic algorithms are in use in the real-life systems it is difficult for them to apply any cryptanalytic results in their attacks. In fact, this approach was widely used until the 1970s. Before that time, academic published results in cryptology were scarce, and their potential relation to real-life systems was not known in public. The big disadvantage of the secretive approach, sometimes called 'security by obscurity', is that feedback from practical experience to academic research is completely missing, which slows down progress on the academic side.

As long as cryptography is used in closed and tightly controlled environments, such as for military communications or protecting databases of financial institutions, there is no need to open up the used systems to academic cryptanalysis. But the situation changes when cryptographic applications are used in commercial systems involving consumers. First, there could be potential attackers among the users of the system and, therefore, the design of the system could leak out to the public through various reverse-engineering efforts. Second, it is harder to build trust in the system among bona fide users if no information is given about how the system has been secured. This trend towards usage of cryptology in more open environments is one reason for the boom in public cryptologic research since the 1970s.

Another, perhaps bigger, reason was the introduction of novel, mathematically intriguing cryptologic concepts, most notably the public key cryptography [Diffie and Hellman 1976].

2.3.1   Cryptographic Functions

Let us next present formal definitions of some central cryptographic notions.

- *Plaintext space P* is a subset of the set of all bit strings (denoted by $\{0,1\}^*$); we assume here, for simplicity, that everything is coded in binary.
- *Cryptotext* (or *Ciphertext*) *space C* is also a subset of $\{0,1\}^*$.
- *Key space K* is also a subset of $\{0,1\}^*$. Often $K = \{0,1\}^k$ where k is a fixed security parameter.
- *Encryption* function is $E: P \times K \to C$.
- *Decryption* function is $D : C \times K \to P$.
- *Cryptosystem* consists of all of the above, i.e $(P;C;K;E;D)$.
- *Symmetric* encryption is defined by $D(E(p, k), k) = p$.
- *Asymmetric* encryption is defined by $D(E(p, k_1), k_2) = $ p, where keys k_1 and k_2 are not identical, and moreover k_2 cannot be derived easily from k_1.

Modern cryptography is based on mathematical functions that are non-trivial from the point of view of computational complexity. This means that either the function as such is complex to compute or the function can only be computed once a certain piece of information – a key – is available. Randomness is another fundamental notion in modern cryptography. A pseudorandom generator is an algorithm that takes a truly random bit string as an input (called a 'seed') and expands it into a (much) longer bit string that is infeasible to be distinguished from a truly random bit string of the same length.

One important function type is a one-way function. Roughly speaking, a function has the one-way property if

- it is easy to compute $f(x)$, if x is given; but
- for a given y, it is infeasible to find any x with $f(x) = y$.

A more accurate definition could be given using terminology from complexity theory [Menezes *et al.* 1996], but we do not need it for the purposes of this book.

Another important function type is a trapdoor function. It is similar to the one-way function with one important difference: if a certain piece of information (a secret key) is known then

Table 2.1 Basic cryptographic function types – I

| | FUNCTION: | |
	Easy (with public key)	Easy with secret key
INVERSE:		
Easy (with public key)	Non-cryptographic function	Digital signature
Easy with secret key	Asymmetric encryption	Symmetric encryption
Infeasible	One-way function	Message authentication code

it becomes easy to find x with $f(x) = y$, given y. Trapdoor functions are used in public key cryptography, for example for digital signatures.

One of the simplest examples of a function used in practice as a one-way function is the multiplication of natural numbers. Given two integers n and m it is easy to compute their product nm but no efficient algorithm is known to compute the inverse operation, determining factors of an integer when the integer becomes large enough. This is the case, in particular, if the integer to be factored is a product of two large prime numbers.

The basic cryptographic function types are listed in Table 2.1. This categorization of the function types should be seen as illustrative; the exact definitions of these function types can be found elsewhere [Menezes *et al.* 1996].

We use the following notations in Table 2.1:

- Easy (with public key): easy to compute (but possibly requiring knowledge of a public key);
- Infeasible: infeasible to compute;
- Easy with secret key: feasible to compute if and only if the secret key is known;
- FUNCTION: given x, find $f(x)$;
- INVERSE: given y, find x such that $f(x) = y$.

In Table 2.2 we focus on the bottom right corner of Table 2.1, on keyless or symmetric key algorithms. We have also added one more dimension which is often useful in practice: whether the length (in bits) of x (and respectively of $f(x)$) is *fixed* or *variable*. Again, the table as such

Table 2.2 Basic cryptographic function types – II

| | FUNCTION/INVERSE: | | |
	Easy/infeasible	Easy with secret key/infeasible	Easy with secret key/easy with secret key
LENGTH OF INPUT/OUTPUT:			
Variable/fixed	One-way hash function	Message authentication code	(esoteric case)
Fixed/variable	Pseudorandom generator	Key stream generator for stream cipher	(esoteric case)
Fixed/fixed (and the same)	One-way permutation	Keyed one-way permutation	Block cipher

does not give exact definitions of these cryptographic terms, and exact definitions are given elsewhere [Menezes *et al.* 1996].

Some cases in the table are marked as esoteric: they are not used as widely as the others. One-way permutation is a one-way function that is also a one-to-one (i.e. bijective) mapping.

2.3.2 Securing Systems with Cryptographic Methods

Using good cryptographic functions does not alone guarantee that a communication system is secure. In addition to the issues with policies and configuration, the structure of the system has to be carefully designed.

One basic principle of using cryptographic functions for securing a system is that the system must remain secure even if the functions and the structure are made publicly available; that is, providing 'security by obscurity' is not deemed acceptable (see section 2.2). Only the randomly generated keys are assumed to be kept secret.

One issue in using cryptography is the management of secret keys. Most cryptographic protection methods rely on the concept of a key and these keys themselves have to be protected; whoever has access to the keys can also remove the protection. This leads to a 'chicken-and-egg' situation: in order to be able to communicate securely we first have to communicate securely certain pieces of information, the keys. Fortunately, it is easier to plan the distribution and exchange of the keys than the communication of arbitrary information that is unpredictable as regards volume, timing and so on. Still, the number of entities that need access to keys is typically of the same order of magnitude as the number of entities in the whole system.

In the following subsections we take a brief look at the various cryptographic primitives that can be used as building blocks for the basic security features listed in section 2.2.3. Let us begin by listing the most popular cryptographic primitives for each communication security feature:

- authentication: challenge–response protocols;
- confidentiality: encryption; (also called ciphering);
- integrity: message authentication codes;
- non-repudiation: digital signatures;
- availability: client puzzles – this method is not, however, in wide use yet and we do not explore it any further in this book.

2.3.3 Symmetric Encryption Methods

Symmetric encryption methods are divided into two main classes: *block* ciphers and *stream* ciphers. In a block cipher, a fixed-length plaintext block is transformed into a cryptotext block of the same length using a key (usually also of fixed length). Thus, for any fixed key the block cipher is a bijection:

$$c = E(p, k); p = D(c, k) = D(E(p, k), k).$$

The dominant block cipher in the past was Data Encryption Standard (DES); its block length is 64 bits and the key length is 56 bits. A newer general-purpose cipher, Advanced Encryption Standard (AES), has a block length of 128 bits and a (minimum) key length of 128 bits.

Usually a block cipher becomes stronger if it is iterated several times. In the design of block ciphers, iteration is used also inside the block cipher. These iterations are called 'rounds'. There is a trade-off here, since adding more rounds increases both security and processing time.

There are a few other classical design principles. For example, each plaintext bit and each key bit should affect each ciphertext bit (diffusion); and the relation between plaintext bits and ciphertext bits should be as complex as possible (confusion).

As block ciphers operate with fixed size words, we encounter a practical problem: how to encrypt messages longer than one block? A straightforward solution is called Electronic Code Book (ECB) mode: a message is divided into blocks and each block is encrypted independently. This mode has a major weakness: identical plaintext blocks result in identical ciphertext blocks!

There exist several different modes that avoid this weakness by introducing additional changing input, such as by using earlier created ciphertext or a counter. Special modes can also be created for using a block cipher for purposes other than encryption, for example as a one-way function or a pseudorandom generator.

Next we discuss stream ciphers. The idea of a stream cipher is based on a simple but yet absolutely secure cipher called the one-time pad. Assume the key k is as long as the plaintext p. Then we define $c = p$ xor k.

The one-time pad cannot be broken. Indeed, any ciphertext may be decrypted to any plaintext (with some valid key). On the other hand, the one-time pad has one major weakness: secure transport or storage of the key becomes as demanding a task as secure transport or storage of the plaintext itself. But still it is advantageous that the transport or storage of the key can be done in convenient time prior to the need of using the key.

In a stream cipher, the long random key of the one-time pad is replaced by a pseudorandom sequence. In other words, we start with a fixed size key (a seed) and generate a mask bit stream m (sometimes called a key stream) that is as long as the plaintext. Then $c = p$ xor m.

Usually there is an additional input (e.g. a counter value) which is used together with the key as a seed. Then the same key can be used for encrypting several messages independently of each other. This holds assuming that the additional input changes for every instance, as when the counter value is increased for every new message. One main advantage of a stream cipher is the fact that the mask bit stream can be generated in advance, even before the plaintext is known. This helps in avoiding delays in the communication.

Another advantage is that the number of erroneous bits in the ciphered message introduced by a noisy channel equals the number of erroneous bits in the recovered plaintext; whereas, for a block cipher, one bit error in a ciphered block typically renders the entire block of recovered plaintext unintelligible. This is a reason why stream ciphers are often used for channels with relatively high bit error rates, such as radio channels.

2.3.4   Hash Functions

Now we take a closer look at a type of cryptographic function that does not require any key, namely a hash function. A one-way hash function h has the properties:

- *compression*: $h(x)$ has a fixed length (e.g. 160 bits) while x may be of any length;
- $h(x)$ is easy to compute.

For some purposes it is important that the hash function fulfils further conditions:

- *2nd-preimage resistance*: for a given x, it is infeasible to find any other x_0 different from x such that $h(x) = h(x_0)$;
- *collision resistance*: it is infeasible to find any two distinct x and x_0 such that $h(x) = h(x_0)$.

It is possible to build a specific mode that converts a block cipher into a hash function, but usually tailor-made hash functions require less computation than block ciphers, and they can be implemented more efficiently. Two hash functions of this type are MD5 (128-bit hash) and SHA-1 (160-bit hash), but many collisions have been found for the former and there is evidence that collisions can be generated for the latter as well. At the time of writing, the most popular hash function is SHA-256 (256-bit hash) but there is also an ongoing competition, run by the National Institute of Standards and Technology (NIST), for finding a new standard hash function, called SHA-3 [NIST].

Hash functions are used in many ways in applications. One important use case is as a message digest: a variable length message can have a unique fixed length representation. Of course, the representation cannot be truly unique, but collision resistance implies that it is infeasible to find two messages (of any length) with the same message digest.

It is also possible to design the computation of a hash function around a secret key. Then we speak of message authentication codes (MACs). These keyed hash functions have typically a somewhat shorter output than (keyless) hash functions. The reason for this is the following generic attack against keyless hash functions. Anybody can compute a large table of inputs and corresponding hash output values (also called message digests). If the length of the output is n then it follows from the birthday paradox [Menezes *et al.* 1996] that approximately $2^{n/2}$ hash outputs need to be computed before a collision is found.

A similar approach does not work for keyed hash functions as the key is known only to authorised parties and the table of hash values is different for each possible secret key.

There are three different strategies in the design of a message authentication code: either direct design, or use of a block cipher or keyless hash functions as building blocks. The HMAC construction is an example of the third strategy. If k is the key and x is the input, then the MAC value is obtained by double hashing:

$$\text{HMAC}(x, k) = h((k \text{ xor } opad)|h((k \text{ xor } ipad)|x)),$$

where vertical bars are used to denote concatenation, and *opad* and *ipad* are just constant values used for padding purposes. The result is often truncated to create a shorter MAC value (e.g. by extracting the first 96 bits from a total of 160 bits).

The basic use case of message authentication codes in information security is to ensure the integrity of a message: we append a MAC to each message transferred over an insecure channel. If the receiving party knows the secret key then it can compute the MAC as well in order to check that the message sent and the message received are indeed identical. Note that the requirement of collision resistance is not crucial for the use of a keyed hash function as a MAC.

2.3.5 Public-key Cryptography and PKI

We now look at the basic notions of public key (asymmetric) cryptography; a deeper treatment of the subject can be found elsewhere [Menezes *et al.* 1996]. The idea of public key encryption is simple: we use different keys for encryption and decryption and it is infeasible to derive the decryption key from the encryption key.

If such encryption and decryption functions are available, then the encryption key can be made public, so it is possible for one party to communicate with many other parties (who do not need to mutually trust each other) using the same key. It is important to note that it is not sufficient that the encryption key be made publicly available; in addition, authenticity of the public key has to be guaranteed.

The setting is reversed for a digital signature. It is infeasible to derive the signing key (used to compute the signature function key) from the verifying key (used to compute the inverse of the signature function). The verifying key can be public, so many people can independently verify the same signature. Usually it is actually the message digest that gets digitally signed. As long as the used hash function is collision resistant, signing the message digest is equivalent to signing the message itself.

Some main benefits of public-key cryptography are:

- easier key management for very large systems, especially those with many-to-many relationships;
- the possibility to use digital signatures, and as a consequence, the possibility for non-repudiation;
- the possibility for any entity to authenticate another entity without online connection to any central trusted third party (but typically an off-line connection to a trusted third party is needed to enable entities to verify the public keys of other entities).

The dominant technique for guaranteeing authenticity of the public key is to use a public key infrastructure (PKI). The central concept of a PKI is a *certificate*: user identity and public key are signed by the Certification Authority (CA). It is assumed that it can be verified by other means that the public key of the CA itself is authentic; for example it could be installed in a computing device at the time of manufacture, or it could be downloaded in a physically secured trusted environment.

The Registration Authority (RA) verifies the user identity, often physically, and delivers the certificate to the correct user. Sometimes a user's private key gets compromised (e.g. stolen), and then the certificate must be revoked. This is often done by including revoked certificates into a certificate revocation list (CRL) that is signed by the CA.

In principle, anybody with a certificate is able to create more certificates by signing identities and public keys of others. Typically CAs are arranged in a hierarchical fashion, and each layer in the hierarchy has a certificate signed by the immediate upper-layer CA except for the root CA on the top of the hierarchy, which either has only a self-signed certificate or does not have any certificate at all. Verifying a certificate of a leaf entity involves verifying all certificates of the nodes between the leaf node and the root node. We speak then of 'certificate chains'.

2.3.6   Cryptanalysis

Here we present the basic concepts of cryptanalysis. A classification of attackers can be done, for instance, as follows.

- *A passive attacker* only monitors the communication and tries to break confidentiality.
- *An active attacker* also adds, deletes and modifies messages. He tries to break also other security features in addition to confidentiality.

The following attack models (against encryption) can be identified.

- *Ciphertext only* – the attacker sees only ciphertext and tries to find the key or at least the corresponding plaintext.
- *Known plaintext* – the attacker knows also the plaintext and tries to find the decryption key.
- *Chosen plaintext* – the attacker can choose the plaintext and gets also the corresponding ciphertext (and again tries to find the decryption key).
- *Adaptive chosen plaintext* – the plaintexts to be chosen may depend on previously observed ciphertexts.
- *Chosen ciphertext* – the attacker chooses the ciphertext and gets the plaintext.
- *Adaptive chosen ciphertext* – the ciphertexts to be chosen may depend on previously observed plaintexts.

Only the first two models are available for a passive attacker. Various chosen plaintext and ciphertext scenarios can also be practical attack models. An example is the case where the user has full access to a tamper-resistant cryptographic module and tries to discover the key inside the module.

A similar classification of attack models applies for attacks against authentication and integrity protection. The simplest attack type that applies even in the ciphertext only model is exhaustive search of all keys. If we have a reasonable amount of ciphertext available, there is typically only one key that decrypts the ciphertext into a meaningful plaintext.

The differential cryptanalysis method is an example of a modern method in the chosen plaintext scenario. It is carried out by choosing a big number of pairs of plaintexts with a pre-determined difference. Analysis is done by studying the corresponding differences in the ciphertexts. Another method, requiring only the known plaintext scenario, that has been applied successfully for many block ciphers and stream ciphers is the linear cryptanalysis method. It is based on analysis of the correlation between plaintext and ciphertext bits.

Recently another attack model has gained a lot of popularity:

- *Related key attack* – the attacker is able to ask that the key be changed and in such a way that the relation between the old key and the new key is pre-determined and chosen by the attacker.

This attack scenario is very optimistic from the attacker's point of view, because only under special circumstances is there a real chance for the attacker to try to change keys somehow. The related key scenario is often viewed as a theoretical tool that can be used in the analysis of ciphers and their structures.

The following attack model takes a wider approach and it is often relevant in practical settings:

- *Side channel attack* – the attacker is able to utilize information about the physical implementation of the cryptosystem.

For instance, the attacker could measure time and power consumption related to execution of the cryptographic algorithm and deduce useful information about the key, possibly by using statistical methods. The attacker may also be able to induce controlled faults in the execution of the algorithm, such as by heat treatment or electric shocks.

2.4   Introduction to LTE Standardization

By the end of the last century it had become evident in Japan that the regional second-generation system PDC (Personal Digital Cellular) was no longer going to provide good enough service for the huge market. Therefore, two Japanese standards organizations, the Association of Radio Industries and Businesses [ARIB] and the Telecommunication Technology Committee [TTC] were already in quite an advanced state in creating detailed specifications for a third-generation technology, especially for the radio network part. In parallel with the Japanese activities there were ongoing efforts in the European Telecommunications Standards Institute [ETSI] to prepare the way for a third-generation cellular technology, called Universal Mobile Telecommunications System (UMTS).

In 1998, five standards development organizations (SDOs) decided to combine their efforts to accelerate the work and guarantee global interoperability. The organizations, ETSI from Europe, ARIB and TTC from Japan, the Alliance for Telecommunications Industry Solutions [ATIS] from North America, and the Telecommunications Technology Association [TTA] from South Korea formed the 3rd Generation Partnership Project (3GPP). A little bit later, a sixth partner, the China Communications Standards Association [CCSA], joined the project. The six SDOs are the organizational partners of 3GPP. Each organizational partner has its own individual members, including terminal and infrastructure manufacturers, mobile network operators and telecommunications regulators.

The dream of getting all of the third-generation development work under one project did not, however, become true. In the United States a lot of work had been done for a system called cdma2000® that had evolved from one of the North American second-generation cellular systems. Driven by the Telecommunications Industry Association [TIA], another project was started, called the 3GPP2. At the same time, the International Telecommunication Union [ITU], a sub-organization of the United Nations, changed their original target of creating one single International Mobile Telecommunications-2000 (IMT-2000) standard to creating a family of third-generation standards instead.

The co-operation between the 3GPP partners quickly began to work very well, and a large number of specifications were in a stable state at the end of 1999. In March 2000, the first release, Release 1999 of the 3GPP specification set, was declared 'frozen'. Nonetheless, after that date many corrections were needed in most specifications, a process that is unavoidable in a project of this scale. After Release 99, 3GPP continued by creating more releases: Release 4 was frozen in 2001, Release 5 in 2002, Release 6 in 2005 and Release 7 in 2007.

The first release that covers LTE and EPS is Release 8. It was frozen in 2008 and the most recent release at the time of writing, Release 9, was frozen at end of 2009 with a few explicitly stated exceptions.

As early as the first 3GPP release it was understood that the cycle of one year might sometimes be too short to add significant features to each new release, so 3GPP stopped the practice of naming releases after calendar years.

2.4.1   Working Procedures in 3GPP

The 3GPP formally is just a co-operation project between the partners, the regional SDOs. Therefore, 3GPP produces specifications and they become standards only after each regional partner has approved them. The specification work in 3GPP follows a three-stage model.

- Requirements for new services are defined in stage 1 specifications.
- Stage 2 specifications contain functional architectures that meet the requirements, including description of functional entities and information flows between them.
- In stage 3 specifications, the functional entities are mapped to physical entities, and bit-level descriptions of protocols between the entities are defined.

In addition to these specifications, there are also test specifications that are typically completed some time later (as they are also needed later).

The 3GPP specification work is carried out in working groups. Above the working groups there is another layer, called technical specification groups (TSGs), where the specifications created in working groups are approved. There are four TSGs: Service and System Aspects (SA), Core Network and Terminals (CT), Radio Access Networks (RAN) and GSM EDGE Radio Access Networks (GERAN).

Typically different working groups carry out different stages for the same features. For example, one working group (called SA Working Group 1) concentrates purely on requirements (i.e. stage 1), another one (called SA WG2) creates system architecture specifications (i.e. stage 2), a third one (called CT WG1) creates stage 3 specifications for protocols between the core network and terminal, and so on.

Specifying different stages requires somewhat different types of expertise and skills. This is one reason for the approach of distributing specification work to several working groups. Another reason is that the distributed approach allows more efficient use of time: stage 1 groups are able to start work on the next release at the same time as other working groups are still busy with the previous release.

Figure 2.4 illustrates how work in different stages is typically scheduled. The time unit in the figure is a quarter year. This follows from the fact that new specifications and change requests to old specifications are approved in TSG plenary meetings which are arranged four times a year. An indicative interval between two consecutive releases is 15 months in the figure. This has roughly been the time interval between the recent 3GPP releases; the exact

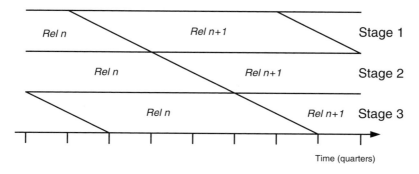

Figure 2.4 Time distribution of work in releases and stages

point of freezing a release is always decided case by case because the optimal timing depends on many factors, some of which stem from the business environment.

As can be seen from Figure 2.4, it is often the case that a working group is doing a significant amount of specification work for two different releases simultaneously. Indeed, there are typically corrections needed to the previous release while work on the next release has already started.

The figure also illustrates that the work in different stages for the same release occurs partially in parallel. Indeed, stage 2 groups do not wait until stage 1 work has been completed, and the same holds between stages 2 and 3. This overlap is useful because typically different working groups need to consult each other in order to guarantee consistency between different stages and specifications. However, it can also be seen from the same figure that there never is a point in time where all stages would work full time for the same release.

The work in 3GPP is contribution driven; individual members send delegates to the working group meetings in order to progress the specification work. Looking at it from another angle, if no member has an interest in progressing work on a particular specification, that specification is never completed. The higher layer body, the TSG, has to approve starting a new work item in a working group. Later the TSG approves the specification resulting from the work item and the change requests to it.

The change request procedure is a formal tool for handling corrections to approved specifications. The correction process is handled by approving each individual change request separately. Each change is also documented explicitly and the documentation includes, in addition to the change itself, a reason for the change, and a brief summary of the change.

In addition to technical specifications (TSs), working groups create also technical reports (TRs). These are informative documents without normative status. Implementers could completely ignore these documents and still build equipment fully compliant with the 3GPP specifications, and the resulting standards. Typically, technical reports are used for (at least) two essentially different purposes:

- for carrying out feasibility studies of features and mechanisms that could later be specified in normative documents (in case the results of the feasibility study are encouraging);
- for analyzing features and adding guidelines and background information that is useful for implementation, deployment and/or operation purposes.

Technical reports of the first type are typically done before corresponding technical specifications are created. Reports of the second type are rather written after the corresponding specification, or at least a draft version, is available.

Technical reports intended for 3GPP internal use have a number of the form 'xx.8xx', while TRs intended for wider distribution have a number of the form 'xx.9xx'.

For security, both types of reports are useful. Often several different approaches could be taken in securing a certain feature. A feasibility study is useful in making each of these approaches explicit and describing them to a similar level of detail. This helps in comparing these approaches with each other and assessing whether they really reach the security goals that are claimed. Of course, similar reasons explain why feasibility studies are also useful for non-security features. For the other type of report, there area specific reasons why they are useful for security purposes.

Analyses of how security features meet the requirements and how the chosen countermeasures address threats cannot be made obsolete by, for example, carrying out field trials and pilots. For security features (e.g. a security protocol), the fact that it can be run efficiently in practice is only a necessary condition for its adoption. It is not a sufficient condition because even a more important condition is that the feature cannot be circumvented.

Guidelines, instructions and clarifications are of special importance in security because specifications often leave some details to be decided during implementation, deployment or operation. It is important that these decisions be made with the understanding of the purpose and characteristic of the security mechanism in question. It is possible to completely undermine a security feature by poor choices of how and when it is in use.

The technical specifications contain normative text using, special wording with reserved words (see section 2.5.2). The normative text may still contain different optional elements: functionalities that may be supported optionally. Sometimes there are also features that are mandatory to support but still optional to use.

As mentioned above, the work in different groups on the same topic requires a fair amount of coordination and communication between working groups. A typical instrument for such purposes is a liaison statement that is sent from one working group to another. Similar liaisons exist also between 3GPP and other organizations, such as the Open Mobile Alliance [OMA] and the GSM Association [GSMA].

In Table 2.3 we show the division of 3GPP specifications into different series. From the point of view of our book, the two security series 33 and 35, and the 23 series, the 24 series and the 36 series, are the most important ones. All 3GPP technical specifications and technical reports are publicly available [3GPP]. Some specifications of cryptographic algorithms in the 35 series need to pass export control first, which introduces a delay in their publication.

Table 2.3   Specification numbering in 3GPP series

Number of series	Subject of series
21	Requirements
22	Service aspects (stage 1)
23	Technical realization (i.e. architectural aspects) (stage 2)
24	Signalling protocols (stage 3) (UE – network)
25	Radio aspects
26	Codecs
27	Data
28	Signalling protocols (Radio network – Core network)
29	Signalling protocols (intra-fixed network)
30	Programme management
31	USIM, IC Cards
32	O&M and Charging
33	Security aspects
34	UE and USIM test specifications
35	Security algorithms
36	LTE and LTE-Advanced radio technology
37	Multiple radio access technology aspects

2.5 Notes on Terminology and Specification Language

2.5.1 Terminology

The reader may have noticed that key terms occurring in this book are sometimes used with different meanings in specifications, technical journal papers, marketing announcements, or in the media at large. Conversely, different terms are sometimes used for the same concept. This section is intended to introduce the key terms as used in this book and clarify how these terms may be used in other forms of publications. The list of terms below is ordered such that it makes sense to read it from start to end. We provide a complete list of abbreviations in alphabetical order towards the back of the book. All 3GPP abbreviations are gathered in [TS21.905]. An arrow followed by an acronym (e.g. → E-UTRAN) indicates that an explanation appears further down the list.

LTE: This book is entitled *LTE Security*. We chose this title as 'LTE' has become a widely known brand name for the 3GPP-defined successor technology of third-generation mobile systems. LTE stands for Long Term Evolution and originally denoted a work item in 3GPP aimed at developing a successor to the third-generation radio technology. Gradually it came to denote first the new radio technology itself, then also encompassed the radio access network (→ E-UTRAN), and is now also used for the entire system succeeding third-generation mobile systems (→ SAE, → EPS) including also the evolved core network (→ EPC), as a quick search for the term LTE on the 3GPP home page [3GPP] will reveal. Only few 3GPP specifications actually use the term LTE, apart from on their cover page, and if they do the term refers to the radio part but never the entire system. Security specifications do not use this term at all. We therefore use the term LTE in the overview parts of the book, but not in the detailed technical parts, so as to make it easier for the reader to use the specifications together with our book without terminological confusion. ETSI has registered 'LTE' as a trademark for the benefit of the 3GPP Partners.

E-UTRAN: E-UTRAN is the Evolved Universal Terrestrial Radio Access Network using the LTE radio technology. The E-UTRAN consists of the network of LTE base stations (eNB). The term E-UTRAN is widely used in the specifications and in the detailed technical parts of this book.

SAE: System Architecture Evolution. The term has made a similar, though not quite as successful, career as the term LTE. Like LTE, it originally denoted a work item in 3GPP on Radio Access Technologies (→ RAT). The aim was to develop a 'framework for an evolution or migration of the third generation mobile system to a higher-data-rate, lower-latency, packet-optimized system that supports multiple RATs' [3GPP 2006]. Combined with LTE it became SAE/LTE, and now often denotes the entire system, encompassing terminals, radio access networks and core networks. It is not commonly used in 3GPP specifications where rather the term → EPS is preferred. It has not become a trademark either. We therefore do not use it in this book any further.

EPS: Evolved Packet System. EPS has the same meaning as SAE/LTE. The term EPS is widely used in 3GPP specifications and in this book.

EPC: Evolved Packet Core. EPC is the core network part of the EPS. The term EPC is widely used in 3GPP specifications and in this book.

RAT: Radio Access Technology. When a terminal moves from a network using one RAT to another network using a different RAT, one often speaks of inter-RAT mobility.

RAN: Radio Access Network. It encompasses the base stations (i.e. eNBs in E-UTRAN, NBs in →UTRAN, and BTSs in →GERAN) and the base station controllers (i.e. RNCs in UTRAN, and BSCs in GERAN). There are no base station controllers in E-UTRAN.

UTRAN: Universal Terrestrial Radio Access Network. The third-generation radio access network encompassing UMTS radio technology.

GERAN: GSM EDGE Radio Access Network. The second-generation radio access network encompassing GSM radio technology and its enhancement, EDGE.

UE: User Equipment. There is a fine distinction between User Equipment and Mobile Equipment (→ME), which is, however, important for security. The UE is the combination of the ME and the →UICC.

ME: Mobile Equipment. The ME is the terminal device without the UICC.

UICC: Universal Integrated Circuit Card. The UICC is a smart card platform, on which applications such as the → USIM reside.

USIM: Universal Subscriber Identity Module. This application on the UICC holds the security parameters and functions that are used in authentication and key agreement in 3G and EPS.

SIM: Subscriber Identity Module. In newer implementations it means a SIM application on the UICC. In older implementations it means the SIM functionality together with the smart card platform.

2.5.2 *Specification Language*

Clear rules for the use of verbal forms in specifications are essential so that the reader can distinguish mandatory requirements from other provisions where there is a certain freedom of choice. 3GPP therefore defined the use of a few key words. The most important key words are: "shall" (meaning "is to", "is required to"), "should" (meaning "it is recommended that", "ought to"), and "may" (meaning "is permitted", "is allowed") – for more details see [TR21.801]. We use this specification language in the detailed parts of this book.

3

GSM Security

3.1 Principles of GSM Security

The goal of security design for the GSM system was that it had to be as good as that of wireline systems. Additionally it was required that security mechanisms should not have a negative impact on the usability of the system.

These goals were clearly reached, and it can be argued that GSM has even better security than wireline systems. On the other hand, it is also clear that there is room for improvement in GSM security. This is, of course, generally true for any system that has been in wide use for a long time. Attack methods and equipment evolve over time, and there should be corresponding improvements in protection methods. Some enhancements in GSM security have been made over the years but the basic structures have remained.

It is always difficult to introduce radical changes into a system that is in wide use, and there is a key learning point in this: security design for a new system should provide adequate protection against *contemporary* attack techniques and include an additional security margin.

The most important security features in the GSM system are:

- subscriber authentication;
- encryption at the radio interface for confidentiality of communication;
- use of temporary identities for identity confidentiality.

All these features were carried over to the third-generation security architecture and later to the EPS security architecture.

As GSM became more and more successful it also became a preferred target for fraudsters. This drew attention to the shortcomings of GSM security. The properties of GSM security that received most criticism are listed below.

- Active attacks are possible in principle. This refers to somebody who has obtained the required equipment to masquerade as a legitimate network element towards the terminal (Figure 3.1).
- Sensitive control data such as keys to be used for radio interface encryption may be sent between different networks without protection.

LTE Security   Dan Forsberg, Günther Horn, Wolf-Dietrich Moeller, and Valtteri Niemi
© 2010 John Wiley & Sons, Ltd

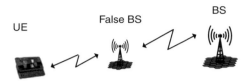

Figure 3.1 Active attack

- Some essential parts of the security architecture were kept confidential (e.g. the cryptographic algorithms), which does not create trust in them in the long run.
- Keys used for the radio interface are short enough to eventually become vulnerable to a exhaustive search attack where the attacker tries all the possible keys until one makes a match.

All these limitations were known at the time when GSM security was designed. However, they were left in because it was estimated that the severity of the threats did not justify the added cost of addressing the limitations. At the time of designing the 3G security architecture around a decade later, a similar comparison between cost and security led to the conclusion that these limitations should be removed for 3G mobile networks.

In the following sections we take a brief look at the most important GSM security features.

3.2   The Role of the SIM

The GSM technology is still the dominant global cellular standard. GSM also contains the world's largest security system: at the time of writing, there are more than four billion actively used security elements in this one single coherent system. The cornerstone of GSM security is the Subscriber Identity Module (SIM) that contains the subscriber identity (IMSI) and an associated 128-bit permanent key (Ki). As will be explained shortly, GSM subscriber authentication is done by a cryptographic challenge–response protocol based on the permanent key. There is also a key generation mechanism integrated with the authentication protocol. Both of these cryptography-based mechanisms are implemented inside the SIM on a smart card.

In the original GSM specifications and until Release 4 of 3GPP specifications, the physical smart card itself was also called SIM, so the specifications use the term 'a SIM card'. In the more recent releases, the setting is such that the smart card itself is called Universal IC Card (UICC) and SIM is an application running in the UICC. This setting allows for other applications that may be run on the same platform.

The SIM is by far the most successful smart card ever. In 2009, volumes of delivered SIM cards exceeded three billion, contributing to approximately 75% of the total smart card market [EUROSMART]. Altogether, the total weight of SIM cards shipped so far equals the weight of more than 400 blue whales [Vedder 2010]!

All smart cards share two fundamental properties that are the reason for the huge success of these tiny devices. From a security point of view, the most important property is tamper-resistance. It requires very sophisticated equipment to tamper with a smart card physically in such a manner that it is possible to find out what is inside. Of course, the device can be disassembled, but it is difficult to gather useful information during the process. Smart cards typically implement various physical protection mechanisms against less intrusive attacks, such as shielding against viewing with an electron microscope.

The other main property of smart cards is portability. In the case of a SIM, this property makes it possible for the GSM subscriber to move a SIM from one terminal device to another, either temporarily or permanently. It also makes it possible to do so-called 'plastic' roaming – to travel to another country with only a SIM card in the wallet, and rent or borrow a mobile phone from the target country. The popularity of this option has nowadays decreased because terminal devices are so small themselves and they often support many frequency bands. Another portability use case related to this travel scenario is that the subscriber replaces his normal SIM with a local pre-paid SIM in order to reduce roaming costs.

Although a SIM card is tamper-resistant, it is still possible to break into an individual card if sophisticated enough machinery is used in a high-tech laboratory. However, the GSM security architecture is built in such a way that it is easy to lock the broken card out of the system as soon as it is observed that tampering may have happened. As a general rule, it is important that systems utilizing small and cheap security elements like smart cards never include global secrets in these devices. If there are only secrets that are applicable to one single device and a single subscriber then the gain of breaking into the device is dramatically reduced.

One of the main motivations for breaking into a SIM card is a type of fraud called 'SIM cloning'. If the permanent key Ki of a subscriber leaks out it is possible to create 'clones' of the broken SIM card. Using these clones it is possible to make many calls simultaneously while using the same subscription. Fortunately, this kind of attack is easily detectable from the network side (one single subscriber is suddenly in many different locations at the same time), and the broken SIM and its clones can be shut out of the network.

In an attack like SIM cloning the attacker typically is the owner of the SIM card himself. In principle, it is also possible for an outsider to perform, for instance, a so-called 'lunch time attack' against an innocent victim: if an attacker gets hold of the victim's SIM card for a long enough time, he can create a copy of the SIM. If the original SIM is destroyed in the tampering, it does not bother the attacker; he can simply replace the original SIM by a copy. If the only objective of the attacker is eavesdropping on calls made by the victim then the attack is difficult to detect. On the other hand, if the attacker begins to make calls on behalf of the victim, the network has much better chances to notice that something irregular is going on. Note here that, assuming the attacker has a chance to make a 'lunch-time attack', there are many other, easier ways to obtain the potential for eavesdropping, such as by using traditional 'bugs'.

3.3   Mechanisms of GSM Security

We shall now discuss briefly the most important security features in GSM and GPRS.

3.3.1   Subscriber Authentication in GSM

There exists a permanent, shared secret key Ki for each subscriber. This parmanent key is stored in two locations (Figure 3.2):

- in the subscriber's SIM card;
- in the Authentication Centre.

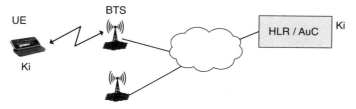

Figure 3.2 GSM system

The key Ki is never moved from either of these two locations. Authentication of the subscriber is done by checking that the subscriber has access to Ki. This can be achieved by challenging the subscriber by sending a random 128-bit string RAND to the terminal. The terminal has to respond by computing a one-way function with inputs of RAND and the key Ki, and returning the 32-bit output SRES to the network. Inside the terminal, the computation of this one-way function, denoted by A3, happens in the SIM card.

During the authentication procedure, a temporary session key Kc is generated as an output of another one-way function A8. The input parameters for A8 are the same as for A3: Ki and RAND. The session key Kc is subsequently used to encrypt communication on the radio interface.

The serving network does not have direct access to the permanent key Ki, so it cannot perform the authentication alone. Instead, all relevant parameters – the authentication triplet (RAND, SRES, Kc) – are sent to the serving network element MSC/VLR (or SGSN in the case of GPRS) from the Authentication Centre. The process of identification, authentication and cipher key generation is depicted in Figure 3.3.

3.3.2 GSM Encryption

As explained above, a secret session key Kc is generated as a by-product of the authentication procedure and used to encrypt all communication between the terminal and the base transceiver station (BTS), called simply base station in the following, including phone calls and signalling. When the next authentication occurs, the key Kc is also changed at the same time.

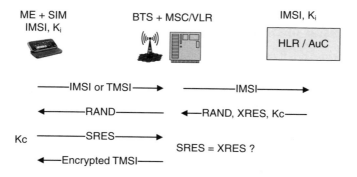

Figure 3.3 Identification and authentication of a subscriber

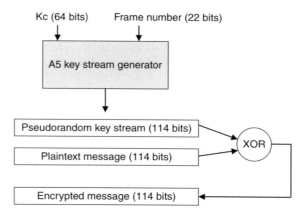

Figure 3.4 GSM encryption

The GSM encryption algorithm is called A5, and Figure 3.4 describes its high-level structure. More details about A5 are given in section 3.4.

Figure 3.4 gives the parameter lengths in their original form. The length of the plain message (similarly, length of key stream and encrypted message) has been made longer for GSM EDGE. The key Kc is also longer in the case of the most recent version of the A5 algorithm, A5/4, that was introduced in Release 9. The key for this algorithm is called Kc_{128} and it has 128 bits that are derived from the 3G keys CK and IK by a key derivation function.

3.3.3 GPRS Encryption

When the packet-switched domain of GSM – that is the General Packet Radio Service (GPRS) – was designed there was a chance to move the termination point of encryption deeper into the network, from the base station to the SGSN. This move implied also that the encryption function is applied at a higher communication layer in GPRS. In (circuit-switched) GSM the encryption is done at the physical layer while in GPRS the encryption is done at the Logical Link Control (LLC) layer. The encryption algorithm structure is very similar to that of A5 but there are a couple of differences also: instead of the frame number the 32-bit LLC counter parameter is used as input, and the output pseudorandom key stream needs to be of variable length. The encryption algorithm is called GEA (GPRS Encryption Algorithm) and is further discussed in section 3.4.

3.3.4 Subscriber Identity Confidentiality

The permanent identity of the subscriber, the International Mobile Subscriber Identity (IMSI), could in principle be tracked by eavesdroppers on the radio interface. For a description of the structure of the IMSI, see Chapter 7. To protect against such an attack, the occasions of sending the permanent identity are limited to necessary cases. Instead of always using the IMSI for identification, another identity, the Temporary Mobile Subscriber Identity (TMSI), is in use whenever it exists.

There is a separate temporary identity, called the Packet TMSI (P-TMSI), in use for GPRS. It is allocated independently of the TMSI by the packet core network element SGSN, but the allocation follows the same principles as the allocation of the TMSI. A similar mechanism is used also in both 3G and in EPS. It is described in more detail in sections 4.2 and 7.1.

3.4   GSM Cryptographic Algorithms

As explained earlier in this chapter, the SIM runs an authentication protocol based on the permanent key Ki and derives also a 64-bit temporary key Kc. This latter key is used in the radio interface encryption between the mobile terminal and the base station. The associated cryptographic algorithms are called A3 (subscriber authentication), A8 (key generation) and A5 (radio encryption).

The GSM system is modular in the sense that it is possible to replace any algorithm by another, probably more modern, algorithm without affecting the rest of the security system, as long as the algorithms share the same input–output structure. Therefore, the symbols A3, A8 and A5 rather refer to families of algorithms than to individual algorithms. The internal structures of the algorithms in a family may be totally different.

For the radio interface encryption, three different stream ciphers A5/1, A5/2 and A5/3 have been standardized so far for 64-bit keys. In addition, the term A5/0 is used for the case where no encryption is applied. In Release 9 of the 3GPP specifications, there is also a variant A5/4 that uses 128-bit keys. The introduction of A5/4 into the system is more cumbersome than the introduction of a new 64-bit key variant because changing the length of temporary keys affects all the interfaces in the system that carry these keys. Additionally, there is an impact on storing these keys. In the roaming case, the key travels from the home network Authentication Centre (AuC) first to the visited network, and inside the visited network from the core network to the radio access network and finally, inside the radio access network, into the base station. Thus, a large number of interfaces and network elements are affected by a change of the key length.

There exists a very efficient attack against the A5/2 algorithm [Barkan *et al.* 2003]. This attack also constitutes a threat to the other algorithms because the key generation in GSM is agnostic to the algorithm. This means an active attacker could try to fool the user to start A5/2 ciphering with the same key that is valid with another A5 algorithm from the point of view of the genuine network. As a countermeasure, 3GPP and the GSM Association (GSMA) launched a process that removed A5/2 completely from mobile terminals.

The situation is even more fragmented for the algorithms A3 and A8. This is a consequence of the fact that these algorithms need not be standardized. The algorithms A3 and A8 are executed in two places, in the SIM card on the user side, and in the Authentication Centre on the network side. Because both the SIM and the AuC are controlled by the same operator, it is possible for each mobile network operator to use their own proprietary algorithms. On the other hand, 3GPP has created, as an example, public algorithm specifications for A3 and A8 [TS55.205].

The history behind the A5 algorithms reflects general developments of mass-market use of cryptography. At the time when GSM was created, there were strict controls on the export of any products that contained cryptography. These types of restriction were in use for most countries, including the European countries that were heavily involved in the specification work of GSM inside ETSI. Cryptography was included in the list of dual-use goods and was basically comparable to items like guns. Similar to many other technologies, cryptographic protocols can be useful also for somebody with malicious intentions.

When the algorithm A5/1 was designed, one obvious requirement was that it had to comply with the export restrictions to the countries in the target market. At that time the target market was mainly seen as Europe in the original phases. As the success of GSM soon became obvious, it turned out that another weaker algorithm was needed in order to get global export licences for GSM terminals and network equipment. The algorithm A5/2 was designed for that purpose. At the end of the 1990s, export restrictions were harmonized between many countries and a multilateral export control regime was created under the name of 'Wassenaar arrangement' [Wassenaar]. At the end of 2009, 40 countries had endorsed the arrangement, including Australia, Russia, Korea, Japan, the USA and many European countries. In many ways, the creation of the Wassenaar arrangement meant also less strict controls. For example, mass-market products like mobile phones were essentially exempted from obtaining export licences.

The UMTS cryptographic algorithms were designed after liberalization of export control restrictions. Therefore, it was possible to introduce 128-bit key encryption algorithm UEA1 based on the block cipher KASUMI [TS35.202]. The algorithm A5/3 was created as a 64-bit key adaptation of UEA1 [TS55.216].

The specifications of A5/1 and A5/2 are kept confidential and are provided only to parties who need to know them for implementation or deployment purposes. However, both algorithms need to be implemented in all GSM terminals (although not anymore for A5/2) and therefore reverse engineering of the algorithms was naturally feasible. There are many cryptanalytic results also against A5/1, and certain time–memory trade-offs are possible. There have been several attempts to collect the vast amount of data that is needed for breaking A5/1. The most recent results on this front have been published [Nohl and Paget 2009], and a GSM Association press release [GSMA 2009] assessed the state of these types of attack study.

The KASUMI algorithm and UEA1 have been publicly available from the beginning of 3G and there are also cryptanalytic results against them. An efficient attack on stand-alone KASUMI has been found in a 'related key' and 'chosen plaintext' setting [Dunkelmann *et al.* 2010] (see section 2.3.6 for discussion of cryptanalytic attack scenarios). This theoretical attack gives valuable cryptanalytic information about KASUMI, but the related key scenario does not apply to A5/3 as used in the GSM.

The packet-switched domain of the GSM system is called GPRS (General Packet Radio Service). The most notable difference between GPRS security and the original GSM security is the following: the A5 encryption algorithm that resides in the physical layer is replaced by the GPRS encryption algorithm (GEA) and encryption is moved to the logical link layer of the radio network. Compared to A5, the changes in the cryptographic algorithm input–output structure are small but the change of layer is more important: the protection is extended further into the network, from the mobile terminal all the way to the core network. So far, three different stream ciphers, GEA1, GEA2 and GEA3, have been standardized for 64-bit keys, the last one being the only one with specifications publicly available [TS55.216]. The GEA3 algorithm uses the same KASUMI-based key stream generator as UEA1, and is therefore very similar to A5/3. Similarly to A5/4, starting from Release 9 there is also the GEA4 algorithm that uses 128-bit keys.

4

Third-generation Security (UMTS)

4.1 Principles of Third-generation Security

The design work for 3G security was based on the practical experiences with GSM security and, to a lesser extent, experiences with the security of other second-generation cellular systems. Before 3GPP was created in 1998, there was a subgroup of the ETSI SMG 10 working group that did preliminary work for UMTS security, but the actual design work was done in the 3GPP security working group SA3. Principles of 3G security, together with design objectives for security work, have been documented [TS33.120].

The major principles for 3G security are:

- it builds on those elements of 2G security that have proven to be both robust and needed;
- it addresses and corrects real and perceived weaknesses in 2G security;
- it adds new security features to address security needs of all new 3G services.

The first two principles were given priority in the beginning of the design work, whereas the third principle became the most important for later releases of 3GPP where more and more features have been added to the 3GPP system.

4.1.1 Elements of GSM Security Carried Over to 3G

Here we list the security features and design principles that were identified as worth retaining in 3G systems. In most areas, further development was done for 3G security. The elements of 2G security considerably strengthened for 3G are as follows.

- *Subscriber Authentication.* This was extended to become mutual authentication between subscribers and the system. Protocols and algorithms were also enhanced. Note that 3G security uses the term "user authentication" rather than "subscriber authentication".
- *Radio interface encryption.* Encryption was extended to cover more than just the radio interface between the terminal and the base station. The strength of the encryption was greatly enhanced by a much longer key size and a publicly verifiable algorithm design.

LTE Security   Dan Forsberg, Günther Horn, Wolf-Dietrich Moeller, and Valtteri Niemi
© 2010 John Wiley & Sons, Ltd

- *SIM as a removable and tamper-resistant security module.* The SIM card was (gradually) replaced by the UICC but its role as a cornerstone of the security architecture remained. Functionality was greatly enhanced for the UICC and the USIM application inside it, compared to the SIM. Related to this, the SIM application toolkit security features were enhanced for the USIM application toolkit.

The elements of GSM security that were eventually seen as adequate also for the 3G environment more or less as they existed already in GSM were as follows.

- *Subscriber identity confidentiality on the radio interface.* The mechanism based on temporary identities provides protection only against passive attackers. Lots of effort was spent on designing a protection also against active attackers, but in the end it turned out that a full protection would require too costly investment. Note that 3G security uses the term "user identity confidentiality" rather than "subscriber identity confidentiality".
- *Transparency for the user.* For the most important security features, like the ones listed above, the user does not have to do anything to get them into operation. The global and pervasive presence of 3G systems emphasizes the importance of this principle.

4.1.2   Weaknesses in GSM Security

Following from the second main principle expressed earlier in this chapter, it was important to explicitly list the weaknesses that were considered to be real at the time when design work for 3G security was started. Of course, in parallel with the 3G security design work, much effort was devoted to mitigating these weaknesses also in the GSM environment.

At the time of writing, more than a decade after the work was started, it is interesting to compare how well the 3G security systems address the listed weaknesses. Partly for that reason, we include all items from the original list (see [TS33.120] for full formulations of these items) in the following.

- Active attacks by 'false networks' are possible. The feature of mutual authentication, in combination with the mandatory integrity protection for signalling, addresses this weakness.
- Encryption keys and credentials for authentication are transmitted adding "in cleartext" is essential for understanding the text between and within networks. In order to address this weakness, network domain security features were added to 3G systems but only in later releases of the 3GPP specifications.
- Encryption does not extend far enough towards the network. In 3G the encryption is run between the user equipment and the Radio Network Controller (RNC) entity.
- Encryption is not used in some networks. From the technical and specification points of view, it would be easy to remove this weakness: just drop all unencrypted calls and sessions. However, this is a regulatory rather than a technical matter, and at the time of writing there still exist big networks that do not regularly use encryption.
- Data integrity is not provided. Protection for signalling data integrity was added from the first release of 3GPP specifications.
- The International Mobile Equipment Identity (IMEI) is an unsecured identity and should be treated as such. Adding an independent authentication system for mobile equipments, in addition to the subscriber authentication system, would have been too costly. Therefore, the

IMEI was kept as an unsecured identity from the network point of view. However, measures to prevent tampering with the IMEI implementation on the mobile equipment itself have been improved.

- Fraud and lawful interception were not considered in the design phase of GSM security. This was changed for 3GPP work, as lawful interception specifications have been developed in parallel with other specifications. Similarly, a fraud information gathering system and support for immediate service termination were provided already in early releases of 3GPP.
- The home network does not know (nor control) whether and how the serving network authenticates roaming subscribers. Mandatory integrity protection addresses the 'whether' part, since integrity protection cannot be started without keys, and obtaining keys requires authentication. Some effort was spent on trying to address also the 'how' part, but in the end it was decided that the 'minimal trust' principle does not justify introduction of a new mechanism for this type of home control.
- There is no flexibility to upgrade security functionality over time. Certain elements to support flexibility and future-proofing have been included in the 3G systems. For instance, there is a secure negotiation mechanism for encryption algorithms, which enables effective introduction of new algorithms and removal of deprecated ones. On the other hand, the authentication and key agreement protocol is more or less hard-wired to the system; only cryptographic algorithms used inside it may be upgraded.

Overall, it appears that the 2G weaknesses have been addressed well in 3G systems, but there is room for improvement in some items. These lessons from the past have also been helping in the design of the LTE and EPS security functions.

4.1.3   Higher Level Objectives

Apart from the fairly concrete design principles stemming from experiences with 2G systems, there was also a list of principles and objectives that helped in meeting the third main principle: securing all new 3G services. For instance, the 3G security was designed to ensure the following.

- All information related to a user is adequately protected.
- Resources and services in the networks are adequately protected.
- Standardized security features are available world-wide, and in particular there is at least one encryption algorithm that can be exported world-wide.
- Security features are adequately standardized to support world-wide interoperability and roaming.
- Protection for 3G subscribers is better than that provided by fixed and mobile (including GSM) systems (of that time).
- 3GPP security mechanisms can be extended as required by new threats and services.

4.2   Third-generation Security Mechanisms

4.2.1   Authentication and Key Agreement

The 3G authentication and key agreement protocol, UMTS AKA, was originally designed for use in the circuit-switched and packet-switched domains of 3G. Variations of it have since been used in a range of other settings. In this chapter we describe UMTS AKA, and refer the reader to the cited literature for other uses of AKA. These include:

- The Internet Engineering Task Force (IETF) specified HTTP Digest AKA in two versions [RFC3310 and RFC4169]. HTTP Digest AKA uses the UMTS AKA functions to dynamically generate passwords for HTTP Digest [RFC2617].
- 3GPP specified IMS AKA, which uses HTTP Digest AKA, embedded in SIP messages, to set up IPsec associations between an IMS UE and a SIP proxy, the P-CSCF [TS33.203] – see Chapter 12.
- 3GPP also uses HTTP Digest AKA in its Generic Bootstrapping Architecture (GBA) for authentication between the UE and a key distribution server, the Bootstrapping Server Function (BSF) [TS33.220].
- 3GPP enhanced UMTS AKA to EPS AKA for authentication across LTE – see Chapter 7. The main enhancement consists in providing a binding of the agreed keys to the name of the serving network.
- IETF specified EAP-AKA [RFC4187], which embeds the UMTS AKA functions in the Extensible Authentication Protocol (EAP) framework [RFC3748]; in this way it makes UMTS AKA functions usable over a range of link layer technologies, including WLAN. EAP-AKA' [RFC5448] extends EAP-AKA by providing a binding of the agreed keys to the name of the access network.

3GPP uses EAP-AKA to provide authentication for subscribers connected across WLAN, and similar access networks, to the Internet and 3GPP core networks – see section 11.2.

3GPP uses EAP-AKA' to provide authentication for subscribers connected across trusted non-3GPP access networks to the Evolved Packet Core of EPS – see Chapter 11.

As we give a detailed description of the EPS AKA protocol in section 7.2, and the elements of UMTS AKA adopted in EPS AKA are clearly distinguished there from the elements, which are new to EPS AKA, we do not present UMTS AKA in this section in full detail. We rather give an overview and discuss the significant enhancements that UMTS AKA provides over the GSM authentication and key agreement protocol.

The 3GPP Security Working Group WG SA3, which designed the UMTS AKA protocol, based its design on a threat and risk analysis of the GSM authentication and key agreement protocol GSM AKA described in Chapter 3. Some of these weaknesses of GSM AKA have already been discussed in Chapter 3. The threat and risk analysis showed that, while GSM security is still adequate today to prevent widespread technical fraud, a determined attacker with significant resources could attack GSM security using so-called false base stations, which are not under the control of a licensed operator. False base stations were not considered a possibility for an attacker when GSM was designed. We quote from a chapter written by Professor Michael Walker [Hillebrand 2001], who was heavily involved in the design of GSM security:

One of the key assumptions when GSM security was designed was that the system would not be subject to 'active' attacks, where the attacker could interfere with the operation of the system or impersonate one or more entities in the system. This assumption was made because it was believed such attacks, which would require the attacker to effectively have their own base station, would be too expensive compared to other methods of attacking GSM as a whole (e.g. wiretapping the fixed links or even just bugging the target).

This view was clearly no longer tenable at the time 3G security was designed; so, protection measures against attacks using false base stations had to be included as part of the 3G security architecture.

3G security includes two such protection measures: integrity protection of signalling, and prevention of replay of authentication messages. The combination of these two measures can effectively mitigate false base station attacks.

Integrity protection is described in section 4.2.3 below. Replay of authentication messages is prevented in UMTS AKA by including a sequence number controlled by the USIM in the challenge sent to the mobile station, and protecting the challenge, together with the sequence number, with a message authentication code. Adding this feature is the main enhancement of UMTS AKA over GSM AKA.

Overview of UMTS AKA

In order to make this book self-contained we include a high-level description of UMTS AKA here, which is similar to a description given elsewhere [Kaaranen *et al.* 2005]. The detailed specification of the protocol can be found in clause 6.3 of [TS33.102].

There are three entities involved in the authentication mechanism of the UMTS system:

- the home network, sometimes called Home Environment (HE);
- the serving network (SN);
- the terminal, more specifically USIM (in a smart card, the UICC).

The basic idea is that the serving network checks the subscriber's identity (as in GSM) by a challenge–response technique, while the terminal checks that the SN has been authorized by the home network to do so. The latter part is a new feature in UMTS (compared to GSM) and it enables the terminal to check that it is connected to a legitimate network.

The mutual authentication protocol itself does not prevent the scenario where an attacker puts up a false base station, but it guarantees (in combination with the other security mechanisms) that the active attacker cannot get any real benefit out of the situation. The only possible gain for the attacker is to be able to disturb the connection, but clearly no protocol methods exist that can circumvent this type of attack completely. For instance, an attacker can implement a malicious action of this kind by radio jamming.

The cornerstone of the authentication mechanism is a permanent key K that is shared between the USIM of the user and the home network database. This is a permanent secret with a length of 128 bits. The key K is never transferred out from the two locations. For instance, the user has no knowledge of his permanent key.

Together with mutual authentication, keys for encryption and integrity are derived. These are temporary keys with the same length of 128 bits as the permanent key K. New keys are derived from K during every authentication event. It is a basic principle in cryptography to limit the use of a permanent key to a minimum and instead derive temporary keys from it for protection of bulk data.

We describe now the Authentication and Key Agreement (AKA) mechanism at a general level. The authentication procedure can be started after the user has been identified in the serving network. The identification occurs when the identity of the user (i.e. permanent identity IMSI or temporary identity TMSI) has been transmitted to the VLR or the SGSN. Then the VLR or the SGSN sends an authentication data request to the Authentication Centre (AuC) in the home network.

The AuC contains the permanent keys of the users and, based on the knowledge of the IMSI, the AuC is able to generate authentication vectors for the user. The generation process consists of executions of several cryptographic algorithms, which are described later in more detail. The generated vectors are sent back to the VLR/SGSN in the authentication data response. These control messages are carried by the MAP protocol.

In the serving network, one authentication vector is needed for each authentication instance; that is, for each run of the authentication procedure. This means that the (potentially long-distance) signalling between the SN and the AuC is not needed for every authentication event, and it can in principle be done independently of the user actions after the initial registration. Indeed, the VLR/SGSN may fetch new authentication vectors from the AuC well before the number of stored vectors runs out.

The serving network (the VLR or the SGSN) sends a user authentication request to the terminal. This message contains two parameters from the authentication vector, called RAND and AUTN. These parameters are transferred to the USIM that exists inside a tamper-resistant environment, the Universal IC card (UICC). The USIM contains the permanent key K, and using it with the parameters RAND and AUTN as inputs, the USIM carries out a computation that resembles the generation of authentication vectors in the AuC. This process also consists of executions of several algorithms, as is the case in the corresponding AuC computation.

As a result of the computation, the USIM is able to verify whether the parameter AUTN was indeed generated in the AuC, and, in the positive case, the computed parameter RES is sent back to the VLR/SGSN in the user authentication response. Now the VLR/SGSN is able to compare the user response RES with the expected response XRES which is part of the authentication vector. When they match, the authentication has been successful.

The keys for radio access network encryption and integrity protection, namely CK and IK, are created as a by-product in the authentication process. These temporary keys are included in the authentication vector and, thus, are transferred from the AuC to the VLR/SGSN. These keys are later transferred further to the RNC in the radio access network when the encryption and integrity protection are started. On the other side, the USIM is able to compute CK and IK as well after it has obtained RAND (and verified it through AUTN). These temporary keys are subsequently transferred from the USIM to the mobile equipment where the encryption and integrity protection algorithms are implemented.

Discussion of UMTS AKA

We discuss here briefly some of the properties and prerequisites of UMTS AKA. This discussion is similar to one presented elsewhere [Horn and Howard 2000]. UMTS AKA relies on the following prerequisites.

- *Trust prerequisites.* The user has to trust his home network in all respects concerning this protocol. The serving network trusts the home network to send correct authentication vectors, and not disclose them to unauthorized entities. As the home network delegates authentication checking to the serving network, the former must place corresponding trust in the serving network.

- *Prerequisites on interface security.* It is assumed that the core network interfaces carrying authentication data between the serving network and the home network, and between two adjacent serving networks, are adequately secure. UMTS AKA can, however, be run securely without additional assumptions on the security of the interface between the UE and the serving network entities.
- *Prerequisites on cryptographic functions.* UMTS AKA makes use of symmetric key-based cryptographic functions f1 to f5, f1* and f5*. All seven functions[1] are implemented in the USIM and the AuC, respectively. Furthermore, there needs to be a random number generator in the AuC. None of these cryptographic functions need to be standardized; they are all up to agreements between operators and their AuC and UICC vendors. One such realization of these functions is the set of algorithms provided by the specification of MILENAGE [TS35.205].

UMTS AKA achieves the following protocol goals.

- *Entity authentication.* As for GSM AKA, the serving network obtains assurance that the user with the claimed identity was involved in the current protocol run. On the other hand, the user obtains the somewhat lesser assurance that the authentication challenge RAND, AUTN, and consequently the keys CK and IK derived from RAND, were generated by the user's home environment and not used in a previous successful UMTS AKA run. Serving network authentication is not achieved by UMTS AKA. The user only obtains assurance 'that he is connected to a serving network that is authorised by the user's HE to provide him services; this includes the guarantee that this authorisation is recent' [TS33.102]. Serving network authentication was not deemed necessary when 3G security was designed as there was an assumption of mutual trust among all UMTS operators. This assumption was, however, considered no longer valid for the entire lifetime of EPS; see the discussion on design decisions for EPS in section 6.3. Consequently, for EPS, UMTS AKA was enhanced to also provide serving network authentication – see section 7.2. A proposal to introduce serving network authentication already for 3G can be found in the publication [Zhang and Fang 2005].
- *Session key agreement.* As a result of UMTS AKA, the ciphering key CK and the integrity key IK are agreed between UE and VLR or SGSN. These keys are mutually implicitly authenticated in the sense that the keys can only be held by the legitimate entities, under the assumption that the entities and interfaces in the core network are secure, and the authentication algorithms are secure.
- *Session key freshness.* This property is obtained by the fact that the session keys are derived using RAND, and that RAND is guaranteed to be fresh by the use of the sequence number in the message authentication code protecting RAND. Both the UE and the VLR or SGSN have to trust the home environment for generating a new RAND for every protocol run. Due to key freshness, a compromise of the session keys agreed in a previous UMTS AKA run does not affect the session keys from the new UMTS AKA run.
- *User identity confidentiality.* The confidentiality of the user identity-related information on the interface between the user and the server is provided in the same way as in GSM, by

---

[1] The same seven functions are used in an identical way with EPS AKA. Their use with EPS AKA is described in detail in Chapter 7.

using a temporary identity TMSI. This implies that an eavesdropper on the interface between the user and the serving network cannot gain information on the user identity from reading UMTS AKA messages. However, active attacks using so-called 'IMSI catchers' are still possible. Prevention of active attacks was discussed at length in 3GPP, and it was concluded that symmetric key techniques bore the risk of shutting out legitimate users during a crash on the network side, and the cost of introducing a public key mechanism for this purpose would be too high.

- *Mitigation of pre-play attacks.* A number of RAND, AUTN pairs may be obtained from the network by an attacker at one point in time and used towards the user at some time later. However, such an attacker cannot know the agreed keys. The mandatory use of the integrity key IK rules out threats arising from this attack.
- *Mitigation of compromise of authentication vectors.* If authentication vectors stored in, or sent to, the VLR or SGSN became known to an attacker he could impersonate the user towards the compromised network, or he could impersonate any 3G network towards the user, using the integrity key IK in local authentications. Furthermore, the attacker could eavesdrop on sessions. But as soon as a new run of UMTS AKA has been performed successfully, the attacker can no longer benefit from the compromised authentication vectors.
- *Re-synchronization procedure.* UMTS AKA provides a mechanism to re-synchronize the sequence numbers used between USIM and Authentication Centre. While, in most cases, a rejection of an authentication request by a USIM due to a sequence number being out of range will be due to the VLR or SGSN using an outdated authentication vector it still has in storage, the re-synchronization procedure of UMTS AKA even allows securely resetting any sequence numbers held in the AuC to bring it into line with the sequence number held in the USIM.
- *Sequence number management.* The generation and verification of sequence numbers in the AuC and the USIM, respectively, need not be standardized in detail. However, the sequence number management scheme needs to be carefully considered. Therefore, [TS33.102] provides an informative annex for guidance. The schemes described in this annex are resilient against denial of service through accidental or malicious modification of authentication vectors in the network and against out-of-order use of authentication vectors by network entities. The selection of the scheme is important to ensure minimizing of re-synchronization events when AKA is used for multiple purposes, such as IMS or GBA.
- *Precomputation of authentication vectors in the AuC.* This is possible as a UMTS AKA run is not bound to any properties of the requesting entity (VLR or SGSN).

4.2.2   Ciphering Mechanism

Once the user and the network have authenticated each other they may begin secure communication. As described earlier, a cipher key CK is shared between the core network and the terminal after a successful authentication event. Before encryption can begin, the communicating parties have to agree on the encryption algorithm also. The encryption algorithms are discussed in section 4.3.

The encryption/decryption takes place in the terminal and in the RNC on the network side. This means that the cipher key CK has to be transferred from the core network to the radio access network. This is done in a specific Radio Access Network Application Protocol

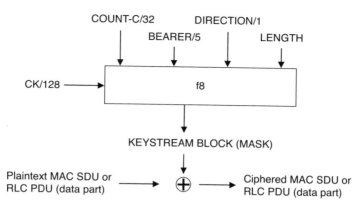

Figure 4.1 Third-generation encryption. Reproduced from Niemi and Nyberg, *UMTS Security*, John Wiley & Sons, Ltd. © 2003

(RANAP) message called Security Mode Command. After the RNC has obtained CK it can switch on the encryption by sending an RRC Security Mode Command to the terminal.

The 3G encryption mechanism is based on a stream cipher as described in Figure 4.1. See section 2.3 for details of the concept of a stream cipher.

The encryption occurs in either the Medium Access Control layer (MAC) or in the Radio Link Control layer (RLC). In both cases, there is a counter that changes for each Protocol Data Unit (PDU). In MAC this is Connection Frame Number (CFN) and in RLC a specific RLC Sequence Number (RLC-SN). If these counters were used as such as input for the mask generation, replay of messages could still occur since these counters wrap around very quickly. This is why a longer counter called a Hyperframe Number (HFN) is introduced. It is incremented whenever the short counter (CFN in MAC case and RLC-SN in RLC case) wraps around. The combination of HFN and the shorter counter is called COUNT-C and is used as an ever-changing input to the mask generation inside the encryption mechanism.

In principle, the longer counter HFN could also eventually wrap around. Fortunately, it is reset to zero whenever a new key is generated during the authentication and key agreement procedure. The authentication events are in practice frequent enough to rule out the possibility of HFN wrap-around.

The radio bearer identity BEARER is also needed as an input to the encryption algorithm since the counters for different radio bearers are maintained independently of each other. If the input BEARER was not in use then this could again lead to a situation where the same set of input parameters would be fed into the algorithm, and the same mask would be produced more than once. Consequently, replay of messages could occur, and the messages (this time in different radio bearers) encrypted with the same mask would be exposed to the attacker.

The parameter DIRECTION indicates whether uplink or downlink traffic is encrypted. The parameter LENGTH indicates the length of the data to be encrypted. Note that the value of LENGTH affects only the *number* of bits in the mask bit stream; it does not have an effect on the bits themselves in the generated stream.

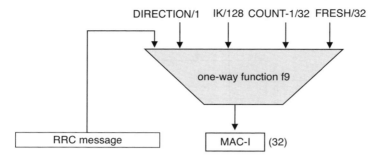

Figure 4.2 Third-generation integrity protection. Reproduced from Niemi and Nyberg, *UMTS Security*, John Wiley & Sons, Ltd. © 2003

4.2.3   Integrity Protection Mechanism

The purpose of integrity protection is to authenticate individual control messages. It is important to do this since the entity authentication procedure UMTS AKA gives assurance of the identities of the communicating parties *only* at the time of the authentication. This leaves a door open for the following attack: a *man-in-the-middle* acts as a simple relay and delivers all messages in their correct form until the authentication procedure is completely executed. After that, the man-in-the-middle may begin to manipulate messages freely. On the other hand, if messages are protected individually, deliberate manipulation of messages can be observed and false messages can be discarded.

The integrity protection is implemented at the RRC layer. Thus, it is used between the terminal and the RNC, just as for encryption. The integrity key IK is generated during the authentication and key agreement procedure, again similar to how the cipher key CK is generated. Also, IK is transferred to the RNC together with CK in the Security Mode Command.

The integrity protection mechanism is based on the concept of a Message Authentication Code: a one-way function that is controlled by the secret key IK. The function is denoted by f9, and its output is called MAC-I: a 32-bit random-looking bit string. On the sending side, the MAC-I is computed and appended to each RRC message. On the receiving side, the MAC-I is also computed and it is checked that the result of the computation equals the bit string that has been appended to the received message. Any change in any of the input parameters affects the MAC-I in an unpredictable way.

The function f9 is depicted in Figure 4.2. Its inputs are IK, the RRC message itself, a counter COUNT-I, direction bit (uplink/downlink) and a random number FRESH. The parameter COUNT-I resembles the corresponding counter for encryption. Its most significant part is a HFN that consists of 28 bits in this case where the four least significant bits contain the RRC sequence number. COUNT-I protects against replay of earlier control messages: it guarantees that the set of values for input parameters is different for each execution of the integrity protection function f9.

The algorithms for 3G integrity protection are discussed in section 4.3.

The parameter FRESH is chosen by the RNC and transmitted to the UE. It is needed to protect the network against a maliciously chosen start value for COUNT-I. Indeed, the most significant part of HFN is stored in the USIM between connections. An attacker could masquerade as the USIM and send a false value to the network forcing the start value of

HFN to be too small. If the authentication procedure is not run then the old IK is taken into use. This would create a chance for the attacker to replay RRC signalling messages from earlier connections with recorded MAC-I values if the parameter FRESH was not involved. By choosing FRESH randomly, the RNC is protected against this kind of replay attacks, which are based on recording earlier connections. As already explained, it is the ever-increasing counter COUNT-I that protects against replay attacks based on recording during the *same* connection as FRESH stays constant over a single connection. From the terminal point of view, it is still essential that the value COUNT-I never repeats itself even between different connections because a false network could send an old FRESH value to the UE in order to try a replay attack in the downlink direction.

Note that the radio bearer identity is not used as an input parameter to the integrity algorithm, although it is an input parameter to the encryption algorithm. Because there are several parallel radio bearers for the control plane also this seems to leave room for a possible replay of control messages that were recorded within the same RRC connection but on a different radio bearer. There is a historical reason for this state of affairs: at the time of freezing the requirements for the integrity protection algorithm design work, the specification for UTRAN contained only one signalling radio bearer.

Instead of changing the algorithm structure retrospectively, the following trick was introduced into the integrity protection mechanism in order to remove the security hole. The radio bearer identity is always appended to the message when the message authentication code is calculated, although it is not transmitted with the message. Therefore, the radio bearer identity has an effect to the MAC-I value, and we have protection also against replay attacks based on recordings on different radio bearers.

Clearly, there are a few RRC control messages whose integrity cannot be protected by the mechanism. Indeed, messages sent before the integrity key IK is in place cannot be protected. A typical example is the RRC Connection Request message sent from the UE. There is a list in [TS33.102] about messages that are not integrity protected.

4.2.4   Identity Confidentiality Mechanism

The permanent identity of the user in UMTS is the IMSI (as is the case also in GSM). However, the identification of the user in UTRAN is in almost all cases done by temporary identities: the TMSI in the circuit-switched domain or the P-TMSI in the packet-switched domain. This implies that confidentiality of the user identity is protected almost always against passive eavesdroppers. Initial registration is an exceptional case where a temporary identity cannot be used since the network does not yet know the permanent identity of the user. After that it is in principle possible to use temporary identities.

The mechanism works as follows. Assume the user has been identified in the serving network by the IMSI already. Then the serving network (VLR or SGSN) allocates a temporary identity (TMSI or P-TMSI) for the user and maintains the association between the permanent identity and the temporary identity. The latter is only significant locally, and each VLR/SGSN simply takes care that it does not allocate the same TMSI/P-TMSI to two different users simultaneously. The allocated temporary identity is transferred to the user once the encryption is turned on. This identity is then used in both uplink and downlink signalling until the network allocates a new TMSI (or P-TMSI). Paging, location update, attach and detach are examples of signalling procedures that use (P-)TMSIs.

The allocation of a new temporary identity is acknowledged by the terminal, and, after that, the old temporary identity is removed from the VLR (or SGSN). If the allocation acknowledgement is not received by the VLR/SGSN it will keep both the old and new TMSIs and accept either of them in uplink signalling. In downlink signalling, the IMSI must be used because the network does not know which temporary identity is currently stored in the terminal. In this case, the VLR/SGSN tells the terminal to delete any stored TMSI/P-TMSI and a new re-allocation follows.

Still one problem remains: how does the serving network obtain the IMSI in the first place? Since the temporary identity has a meaning only locally, the identity of the local area has to be appended to it in order to obtain an unambiguous identity for the user. This means that the Location Area Identity (LAI) is appended to the TMSI and the Routing Area Identity (RAI) is appended to the P-TMSI.

If the UE arrives in a new area then the association between IMSI and (P-)TMSI can be fetched from the old location area or routing area if the new area knows the address of the old area (based on LAI or RAI). At the same time, unused authentication vectors may also be transferred from the old VLR/SGSN to the new VLR/SGSN (if there are any). If the address of the old area is not known, or a connection to the old area cannot be established, then the IMSI must be requested from the UE.

There are some specific places, such as, airports, where lots of IMSIs may be transmitted over the radio interface as people are switching on their mobile phones after the flight. This means also that the identity of the people arriving can in principle be found when an eavesdropper knows their IMSIs. On the other hand, tracking of people is usually also otherwise easier in this kind of special place.

Altogether, the user identity confidentiality mechanism in UMTS does not give one hundred per cent protection but it offers a relatively good protection level. Note that the protection against an active attacker is not very good since the attacker may pretend to be a new serving network to which the user has to reveal its permanent identity. The mutual authentication mechanism does not help here since the user has to be identified before he/she can be authenticated.

The details of handling of temporary identities can be found from [TS43.020] and [TS23.060].

4.3   Third-generation Cryptographic Algorithms

The time when 3G security was designed coincided with an important watershed time for commercial cryptography. As explained in Chapter 3, export control restrictions had been greatly harmonized and also liberalized just a few years earlier. This gave the opportunity to opt for strong cryptographic algorithms with a state-of-the-art key length.

The 3GPP document [TR33.901] lists the design criteria for 3G cryptographic algorithms. Two important decisions had to be made in the beginning. First, it had to be decided whether to aim for publicly available algorithms or secret algorithms (as was the case in GSM). Second, it had to be decided for each algorithm whether it is obtained by

- selecting an already existing off-the-shelf algorithm (with adaptations to fit into the 3G security architecture); or

- inviting submissions from cryptography experts and/or the security community at large for a new algorithm; or
- commission a specific group of experts to carry out the design work in a task force project.

The first decision was easier to take: many aspects seemed to favour public algorithms over secret ones. Maybe the most important fact was that trust in algorithms is much higher if they are available for analysis by the whole cryptographic community. Another important point that made the decision quite easy was the fact that, in cases where the algorithm needs to be implemented in all 3G terminals, details of the algorithm would eventually be reverse-engineered by some people and leaked to the public anyhow.

The second decision was made somewhat more difficult by another important concurrent process in cryptography. The National Institute of Standards and Technology [NIST] had a big competition ongoing for finding a new standard general-purpose encryption algorithm that would, for example, replace the Data Encryption Standard (DES) algorithm in US governmental use. The difficulty lay in the fact that, although the NIST competition for finding a new Advanced Encryption Standard (AES) algorithm was in many ways very good for industrial purposes, it was scheduled to be closed significantly later than when specifications for first 3G cryptographic algorithm were needed for implementations. Here it should be noted that, after the political agreement was reached to establish 3GPP and aim for a truly global system, extremely ambitious schedules were set for when the first release of the specifications had to be available.

This issue with tight schedules had two effects. The most obvious candidate (i.e. AES) for the 'off-the-shelf selection' option would not have been available in time. On the other hand, it was evident that there was not enough time for any kind of 3GPP cryptography competition that would run in parallel with the (at that time) ongoing NIST competition. Hence, the second option of inviting submissions was not seen as viable at all. Also, the timing for going with the 'off-the-shelf' option was not good at all. Of course, possibilities like choosing one of the AES candidate algorithms or existing standard algorithms (e.g. from ETSI or FIPS) was considered but no satisfactory solution was found. For these reasons, 3GPP chose the third option and delegated the ETSI body Security Algorithms Group of Experts (SAGE), see [ETSI], to create a task force for the design and evaluation work for the first 3G cryptographic algorithms.

As explained earlier in this chapter, cryptographic algorithms for AKA were left out of scope of standardization. Therefore, the first two algorithms were needed for encryption and integrity protection between UE and RNC. There were tight performance requirements for both algorithms, and it was required that the chosen algorithms would perform well in both hardware and software implementations.

4.3.1   KASUMI

The time schedule was tight also for the commissioned task force organized by ETSI SAGE. Therefore, the task force did not start their design work from scratch but instead looked for a suitable existing algorithm that could serve as a starting point for their work. Suggestions were also welcome from experts in the industry. In this way, the design process actually contained some elements of all three basic options.

The task force then chose the algorithm MISTY [Matsui 1997], designed in Japan, as starting point for the design work. The main designer of MISTY, Mitsuru Matsui, also joined the task force. Several changes were made to MISTY (or, more specifically, MISTY1), mainly for the purpose of making the hardware implementations simpler or faster. The resulting algorithm was given the name KASUMI, which is the Japanese word for mist.

The task force had its own evaluation and testing teams but, in addition, three different academic expert groups were invited to carry out the evaluation of KASUMI. The report of the task force project can be found in [TR33.908].

The KASUMI algorithm is a block cipher that uses a block size of 64 bits and a key size of 128 bits. Details of the algorithm can be found in [TS35.202]. KASUMI has a Feistel structure, and it contains 8 rounds with very similar structures. The atomic nonlinear functions are called S-boxes (S_7 and S_9) and they can be implemented by a small amount of combinational logic. In the S-box S_7 (resp. S_9), the output of 7 bits (resp. 9 bits) is calculated by bit operations on 7 (resp. 9) input bits.

KASUMI was not designed to be used as a stand-alone block cipher but only as a building block for the UMTS encryption and integrity algorithms.

4.3.2   UEA1 and UIA1

The first 3G encryption algorithm UEA1 and the first 3G integrity protection algorithm UIA1 were designed by the task force of ETSI SAGE as special modes of operation of KASUMI. For UEA1, a mode was needed that made a block cipher usable as a stream cipher. Two popular modes for that purpose are 'Counter Mode' and 'Output Feedback Mode', and the task force ended up with a new mode that is a combination of both of these modes. For UIA1, a mode was needed that uses a block cipher for creating a message authentication code. Details of the specifications of UEA1 and UIA1 can be found in [TS35.201]. Test data is available in [TS35.203] and [TS35.204] for implementation and conformance testing purposes. Niemi and Nyberg have discussed KASUMI-based algorithms and the work of the task force [Niemi and Nyberg 2003].

Because the algorithms UEA1 and UIA1, together with their core building block KASUMI, were made publicly available, there have been cryptanalytic efforts on them. Several papers have also been published on the subject, targeting typically either KASUMI or MISTY1, but many results on MISTY1 are also applicable to KASUMI. A typical approach on doing cryptanalysis against algorithms with an iterative structure is reducing the number of rounds and trying to attack the resulting weakened version of the algorithm. At the time of writing, there exist attacks that are faster than exhaustive search for a 6-round version of KASUMI [Kühn 2001; Dunkelmann and Keller 2008]). Very strong attacks have been found in the related key attack and the chosen plaintext scenario [Dunkelmann et al. 2010]. See section 2.3.6 for more information about these attack models. Fortunately, related key attack scenarios do not apply to the 3G security architecture and chosen plaintext attacks are hard to realize in the UEA1 mode of operation.

4.3.3   SNOW3G, UEA2 and UIA2

Back in 2004, 3GPP decided that it was time to start specification work for another algorithm set, in addition to the KASUMI-based algorithms UEA1 and UIA1. New attacks had been introduced, called algebraic attacks, which were not taken into consideration when KASUMI was designed. Still, there were no signs that algorithms in the KASUMI family would be broken soon. Nevertheless, it was seen that the dependence on one single algorithm creates a kind of 'single point of failure'. Furthermore, sometimes it has happened that the time interval has been relatively short between first theoretical cryptanalytic results showing weaknesses in an algorithm and practical attacks exploitable against the algorithm.

The design and specification work takes time, and implementation work followed by wide-scale deployment adds to the delay. Especially when algorithms are implemented in hardware, their introduction to products is a slow process. This fact further emphasized the need for a proactive approach in introducing another algorithm set.

Sometimes breakthroughs in cryptanalysis lead to successfully breaking many algorithms almost simultaneously. To minimize the chances that both 3G algorithm sets would be broken in a relatively short time, it was required that the new algorithm set should be built on design principles as different as possible from those used for KASUMI-based algorithms.

This point also ruled out the introduction of AES as the basis for the new algorithms, since the atomic nonlinear functions inside both KASUMI and AES follow very similar design principles. The same design options – i.e. 'off-the-shelf', 'competition' and 'commission' – were again considered, now without restrictions imposed by a tight time schedule. It was felt that, after ruling out AES, the 'off-the-shelf' method would not work well. Arranging a 3GPP-specific design competition was seen as too heavy a process, even if more time was available than when the first algorithm set was designed. Good experiences with the special task forces also encouraged choice of the third option again for the design process.

ETSI SAGE was delegated to form a special task force for the design and evaluation work. It was also required that external experts would be used for independent evaluations. Furthermore, it was decided to leave some time at the end for voluntary evaluation and cryptanalysis work by experts in the industry and academia. The 3GPP document [TR35.919] contains a report of the task force project.

Guided by the differentiation requirement, the special task force looked for a suitable genuine stream cipher that could be used as a starting point for the work. Such an algorithm was found in SNOW 2.0 [Ekdahl and Johansson 2002]. The first version of SNOW had been submitted to the European NESSIE project (New European Schemes for Signatures, Integrity and Encryption) that had been ongoing for some time with the goal of identifying appropriate cryptographic algorithms for different purposes. Building on the feedback on the first version, the designers created a new version, which had been subject to cryptanalysis for a couple of years without showing any signs of weaknesses.

Similar to the case of KASUMI, some changes were made to SNOW 2.0 in order to adapt it to the requirements of the 3G environment and thwart the newly discovered algebraic attacks. SNOW 3G has indeed a classical stream cipher structure, producing a continuous key stream. It is built on a Linear Feedback Shift Register (LFSR) and a Finite State Machine (FSM); see Figure 4.3 for a graphical description of the SNOW 3G structure.

Each of the 16 cells in the LFSR are 32-bit words. The three registers R1, R2 and R3 of the FSM also contain 32 bits each. The additive operations are bit-wise exclusive or (xor) and addition modulo 2^{32}. The operations S_1 and S_2 are nonlinear substitutions (S-boxes) on the registers. The S-box S_1 is taken directly from SNOW 2.0 and is originally part of the AES round function. The S-box S_2 was designed to have a good resistance against algebraic attacks. In the feedback function of the LFSR, contents of cells s_{11} and s_0 are multiplied by appropriate constants, before applying the xor-function with each other and content of cell s_2. Details of the SNOW 3G algorithm can be found in [TS35.216].

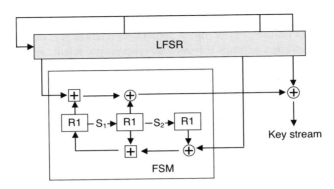

Figure 4.3   Structure of SNOW 3G

The SNOW 3G algorithm is used as the core component of both UEA2 and UIA2. The construction of UEA2 is fairly straightforward. Input parameters for the encryption function are used to fill in initial values for the LFSR (using simple formulas), then SNOW 3G is clocked, first as part of the initialization phase and, after that, producing as many key stream words as needed to mask all of the plaintext. See [TS35.215] for a detailed description of UEA2.

Application of SNOW 3G for integrity protection is more complicated. The structure of UIA2 is given in Figure 4.4.

Similar to UEA2, the 16 cells are populated by input parameters of the integrity protection mechanism, using simple formulas that can be found in [TS35.215]. Then SNOW 3G is used

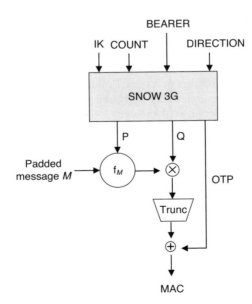

Figure 4.4   Structure of UIA2

to create the bit strings P and Q, both of length 64 bits, and one 32-bit value OTP. Also this phase is very similar to the generation of the (first 160 bits of the) key stream in UEA2. After a certain padding, the message is transformed into a polynomial over the finite field $GF(2^{64})$ of 2^{64} elements. The polynomial is evaluated in the secret point P, and the result is multiplied by the other secret 64-bit value Q. Then truncation is applied to obtain 32 bits, which are finally xor-ed with the third secret value OTP.

The design of UIA2 is based on the principle of universal hashing [Carter and Wegman 1979; Stinson 1992; Bierbrauer et al. 1993], and is similar to the Galois MAC construction [McGrew and Viega 2004] with the important difference that fresh P and Q are generated for each message. Further details of UIA2 and its design principles can be found in [TS35.215] and [TR35.919].

4.3.4 MILENAGE

Although the 3G security architecture does not require standardization of the cryptographic algorithms needed for authentication and key agreement protocol, a special task force very similar to the one described in previous subsections was formed to create an informative example algorithm set for that purpose. This set is called MILENAGE, and its specifications can be found in [TS35.205], [TS35.206], [TS35.207] and [TS35.208]. The report of the task force project is in [TR35.909].

The MILENAGE algorithms use a core function of a block cipher, in which both block size and key size are 128 bits. For instance, the (basic form of) the AES algorithm can be used as the core function. Niemi and Nyberg have discussed MILENAGE and its design process [Niemi and Nyberg 2003].

4.3.5 Hash Functions

The hash functions are also needed for 3G security purposes. The network domain security features, discussed in section 4.5, use hash functions for creating digital signatures, especially in certificates. Another use of hash functions is in creating a message authentication code, in particular for network domain security features. For the purposes of WLAN-3G interworking, discussed in Chapter 5, a hash function is also used for key derivation.

A popular choice of algorithm for all of the above purposes is SHA-1. As explained in section 2.3, SHA-1 cannot now be assumed to be collision-resistant. This implies other choices of hash functions, such as SHA-256, should be considered for those use cases of the hash function that depend on the property of collision-resistance. Use for a digital signature certainly depends on collision-resistance of the hash function, but, on the other hand, collision-resistance is not crucial for use cases of message authentication code and key derivation. Therefore, SHA-1 can still be considered adequate for the latter purposes.

4.4 Interworking between GSM and 3G security

The radio interfaces of GSM and 3G are completely different, but still most terminals that support 3G support also GSM. On the other hand, the 3G core network is a straight evolution from that of GSM. This makes it easier to roam from one system to another, and furthermore, even handovers between the two systems are possible. Since the security features of the two systems differ from each other, it is tricky to define how security is managed in the interoperation cases.

When 3G was introduced, it was seen that a smooth transition was needed from a pure GSM network into a mixed network with both wide coverage provided by GSM and hotspots provided by 3G. In order to ensure that the transition is indeed smooth, it was decided that access to UTRAN is possible with SIM cards, and the use of a USIM is not mandatory. Then a user may switch to a 3G terminal without a need to change his or her smart card.

The disadvantage of enabling smooth transition was of course the fact that the new security enhancements provided by USIM could not be assumed for every terminal accessing a 3G network. When the SIM card is used to access UTRAN, there is no authentication of the network and, furthermore, a SIM provides only 64 bits of key material (in the form of Kc) per authentication. But for the integrity protection and encryption on the UTRAN side, two 128-bit keys are needed. To solve this mismatch, the 64-bit key Kc is expanded into two 128-bit keys by using specific conversion functions. However, it should be noted that the resulting security level is comparable to that of GSM because the conversion functions make keys longer only nominally.

Another interworking case is when a 3G subscriber with a proper USIM moves out of 3G coverage. Then there is the opposite need to compress the longer 3G keys CK and IK into 64 bits in order to be able to use GSM encryption. Also in this process the security benefit of longer keys vanishes and the user obtains the GSM security level, at least as regards encryption.

4.4.1   Interworking Scenarios

All possible interworking scenarios in a mixed GSM/3G environment are systematically studied in [TR31.900]. There are five basic entities involved in these scenarios: the security module, the terminal, the radio network, the serving core network and the home network. Each of these entities could be classified into either 2G or 3G. Some of these entities may actually be mixed cases, but from the security point of view it is simpler to try to define a clear-cut division between 2G and 3G for each entity.

- The security module can be either a SIM card (= 2G case) or a UICC (= 3G case). It is important to note that a UICC may contain a SIM application in addition to a USIM application; when a SIM application is in use, we have a 2G case from a security point of view.
- The ME is classified as 2G if it only supports the GSM radio access network. Otherwise the ME is 3G, it supports either UTRAN only or both GSM radio access and UMTS radio access.
- The division for radio access networks is clear: GSM is the 2G case while UTRAN is the 3G case.
- The serving network VLR/SGSN is classified as 2G if it only supports GSM authentication or it can only be attached to a GSM Base Station Subsystem (BSS). Otherwise, the VLR/SGSN is 3G – i.e. it supports both the UMTS AKA and GSM AKA – and it can be attached to both UTRAN and GSM BSS. Furthermore, a 3G serving network is assumed to support the conversion functions.
- The HLR/AuC is 2G if it only supports authentication triplet generation for 2G subscriptions. A 3G HLR/AuC supports authentication vector generation for 3G subscriptions, and it

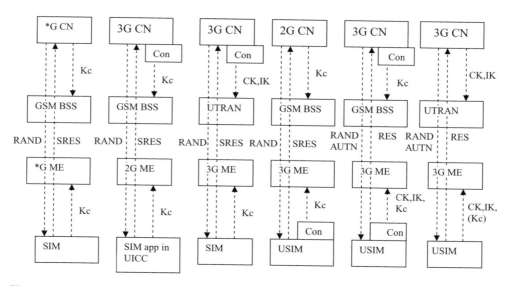

Figure 4.5 Main 2G–3G interworking cases. Reproduced from Niemi and Nyberg, *UMTS Security*, John Wiley & Sons, Ltd. © 2003

also supports conversion functions to support GSM authentication. A 3G HLR/AuC may additionally support pure triplet generation for 2G subscriptions.

Altogether, we have 32 different combinations of 2G/3G entities. If we additionally count the SIM application in the UICC as a third possible case for the security module, we have 48 combinations. All combinations are listed and analysed in [TR31.900]. In this chapter, we highlight only scenarios that differ essentially from each other. In Figure 4.5 we combine the core network entities: we say that the core network is 3G if both the serving network and the home network are 3G, otherwise we say the core network is 2G. Altogether, six essentially different cases are depicted in the figure.

4.4.2   Cases with SIM

We have three essentially different cases where a SIM is used as an access module.

SIM and GSM BSS

If a SIM is used for accessing a GSM BSS, then we have a pure GSM case from the security point of view. The security features are 2G authentication and 2G encryption.

SIM application and GSM BSS

A slight variant of the previous case is when a UICC is used in a 2G ME. The use of a 2G ME implies that the radio access network must be a GSM BSS. We have exactly the same security features executed as in the previous case: 2G authentication and 2G encryption. On the other hand, a SIM application is used in the UICC, which implies that conversion functions must be available in the core network in order to be able to produce authentication triplets.

SIM and UTRAN

In this case, both the core network and the ME must be 3G, as they both support UTRAN. The GSM encryption key Kc is expanded to CK and IK by the conversion functions, both in the core network and in the ME. The special case of a SIM application in the UICC is similar here. We have 2G authentication while encryption and integrity protection are provided by 3G mechanisms but used with a 2G key, thus resulting in 2G level security for encryption.

4.4.3   Cases with USIM

There are, again, three essentially different cases when a USIM is used as the security module. In all cases, both the ME and the home network must be 3G, since they must support a USIM.

USIM and GSM BSS, with 2G serving network

In this case the home network must produce authentication triplets with the help of conversion functions, because the serving network only supports triplets. On the terminal side, the USIM itself applies a conversion function to derive a GSM encryption key Kc. We have 2G authentication and 2G encryption. In order to address certain vulnerabilities [Meyer and Wetzel 2004a; Meyer and Wetzel 2004b], 3G authentication should be performed as soon as the terminal with the USIM attaches to a 3G serving network. An analysis of the two aforementioned papers can be found in [3GPP 2005].

USIM and GSM BSS, with 3G serving network

In this case the authentication vectors can be used. Even if the radio access network is only 2G, it is possible to run the UMTS AKA, as this protocol is transparent to the radio network. On the other hand, keys CK and IK cannot be used. Thus, a conversion function has to be applied both in the USIM and in the core network to generate a GSM encryption key Kc or Kc_{128}, depending on the selected GSM encryption algorithm. Note that CK and IK are, however, transferred to the ME in order to support potential future handovers to UTRAN. We have now 3G authentication but 2G encryption (in case of Kc_subscript_128 encryption key length is the same as for 3G encryption).

Pure 3G case

In this case all elements are 3G, and the full set of UMTS security features is in use. Note additionally that the converted GSM key Kc may be derived in order to support potential future handovers to GSM BSS.

There are no technical restrictions that would prevent running GSM authentication also in this case. Indeed, a USIM does not know whether the ME is connected to a UTRAN or to a GSM BSS. Therefore, in order to guarantee full 3G security, the ME has to abort GSM authentication attempts when it is connected to UTRAN and contains a USIM.

Only in this case do we have all 3G security features: 3G authentication, 3G encryption and 3G integrity protection (with 128-bit keys).

4.4.4 Handovers between GSM and 3G

The procedures for handovers are somewhat different, depending on whether they are done in the circuit-switched or the packet-switched domain. It is in principle easier to start sending packets via a new cell; for a circuit-switched bit stream the transition from one cell to another has to be planned more carefully. This difference is visible also in the case of inter-RAT (Radio Access Technology) handovers. For both cases, all MSCs or SGSNs that support handover from GSM to UMTS should support and use 3G authentication [3GPP 2005].

CS handovers from UTRAN to GSM BSS

The encryption algorithm must be changed during a handover from UTRAN to GSM BSS. The 3G algorithm UEA is replaced by a GSM A5 algorithm. Also, the UTRAN key CK is replaced by the converted Kc. The information about supported/allowed GSM algorithms together with the key is transferred inside the system infrastructure before the handover can take place. Naturally, integrity protection is stopped at handover to GSM BSS.

CS handovers from GSM BSS to UTRAN

If the handover is done from GSM BSS to UTRAN, then the encryption algorithm is changed from A5 to UEA. Before the actual handover may happen, the GSM BSS requests the UE to send information about its UTRAN security capabilities and security parameters, such as information about CK and IK. This information is transferred inside the system infrastructure to the target RNC before encryption and integrity protection can be started on the UTRAN side.

Intersystem change for PS services

There are a couple of notable differences between intersystem handovers for CS services and corresponding intersystem changes for PS services. First, the GPRS encryption terminates in the core network, so transfer of keys is somewhat simpler. Second, there is a difference in the case where also the core network changes in addition to change of the radio network. If the UE moves to the area of a new MSC/VLR, the old MSC/VLR still remains as the anchor point for the call. On the other hand, if the UE moves to the area of a new SGSN, then this new SGSN becomes also the anchor point for the connection.

4.5 Network Domain Security

4.5.1 Generic Security Domain Framework

With 3GPP Release 5, the need for confidentiality, integrity, authentication and anti-replay protection for control plane traffic on core network interfaces was identified. Therefore, a general framework for security of IP-based protocols used in 3GPP and fixed broadband networks was introduced. This framework is specified in [TS33.210], and is known under the acronym NDS/IP as abbreviation of 'Network Domain Security for IP-based protocols'. Besides describing the network architecture and security mechanisms to be used for protection of IP traffic between different security domains and between single network entities within one security domain, it also contains a basic framework for cryptographic key management that mandates support for pre-shared keys only. Later in Release 6 this specification was augmented with a separate specification on Public Key Infrastructure (PKI) support in [TS33.310]. This specification is often referred to as NDS/AF for 'Network Domain Security / Authentication Framework'.

In order to be compliant with the NDS/IP framework, a network needs to apply it for control plane traffic only, while user plane traffic is not covered by NDS/IP in general. Still the framework may also be applied to user plane traffic, either based on decision of the operator, or even mandatorily if required by other 3GPP specifications. One example is the security of user plane data for the eNB backhaul link as required by [TS33.401] – see Chapter 8.

NDS architecture

A basic concept for NDS is the notion of a security domain, which is defined as being a part of or a complete network administrated by a single authority, and typically providing the same level of security and the same type of security services to all elements and connections contained within this domain. A security domain is normally confined to a single operator, while any operator is free to subdivide their network into separate security domains. Different security domains are connected by secure channels, which terminate in Security Gateways (SEGs) at the borders of these security domains. The reference point for these secure channels between security domains is called Za.

The above requirement implicitly disallows any direct connection between network elements (NEs) located in different security domains. But the specification does not prevent an NE and the related SEG to be co-located. The only restriction here is that such a co-located SEG should only act on behalf of this one NE, and not as a general SEG at the border of a security domain.

Additionally, it is often assumed implicitly that the connections between the NEs within one security domain are secure, and thus do not need any explicit security measures in the NDS/IP framework. This assumption is not always fulfilled, so NDS/IP provides also security mechanisms for intra-domain connections using the reference point Zb. As SEGs are not required within security domains, the Zb reference point may be used between NEs and SEGs, but also directly between two different NEs of the same security domain. Anyway, NDS/IP is clear on the fact that the application of cryptographic security to intra-domain connections is up to operator policy.

Figure 4.6 shows the basic architecture of NDS/IP with the Za and Zb reference points. It can be clearly seen that the underlying security model of this architecture is 'hop-by-hop' security in a tunnel/hub-and-spoke model, where each pair of SEGs and their tunnel in between act as hub and the connections to other network elements constitute the spokes.

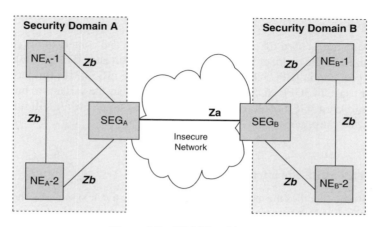

Figure 4.6 NDS/IP architecture

Security services provided by NDS/IP

The following security services are provided by NDS/IP:

- data integrity;
- data origin authentication;
- anti-replay protection;
- confidentiality;
- limited protection against traffic flow analysis (only when confidentiality is applied).

Note that confidentiality protection is optional. When confidentiality protection is used, there is also limited protection against traffic flow analysis when tunnel mode (see later) is used. Then only the IP addresses of the SEGs are visible, and the internal traffic sources and destinations within the security domains are hidden.

Security gateways (SEGs)

The security gateways have a twofold task, as they terminate the secure connections over the Za reference point, and act as policy enforcement point for the security policies bound to the security domain. The mechanisms for the secure connection are detailed in the NDS/IP and NDS/AF specifications and described further below, while the policies are out of scope of standardization. Such policies depend a lot on the type of Za interface, be it intra-operator or inter-operator, and, in the latter case, on the agreements between different operators. Mechanisms deployed for the enforcement of such policies may be simple packet filters or more complex firewall rules, but also sophisticated content-related inspections, relating to network signalling or even application-level content.

Certificate profiles

If a PKI infrastructure is used for authentication, the format and content of the certificates has to be standardized. In the realm of NDS/AF, only X.509 certificates [ITU X.509] are used with the particular profiling in [RFC5280]. Clause 6 of NDS/AF [TS33.310] further profiles these certificates for usage in 3GPP NDS/AF. This profiling reduces the variants to be implemented and thus helps to keep 3GPP-conformant implementations simple, while still fully functional, for 3GPP network purposes.

Support for TLS certificate infrastructure

While NDS/IP only specifies the usage of IKE and IPsec (see the next section), the specification on NDS/AF [TS33.310] extends the definition of certificate profiles and interconnection/cross-signing infrastructure to entities using Transport Layer Security (TLS). Connections secured by TLS are not handled in NDS/IP [TS33.210], therefore no SEGs are defined for TLS usage either. As there is no restriction on TLS client and server location, the roles of client and server are only defined by the fact of who initiates the TLS handshake. Thus this TLS certificate usage applies to both connections between entities within the same security domain and in different security domains.

TLS is not specifically used in EPS procedures; but, for example, the security specification for IMS [TS33.203] allows the usage of TLS between IMS CSCF elements and other SIP proxies. For details on IMS the reader is referred to Chapter 12.

4.5.2   Security Mechanisms for NDS

The first releases of NDS/IP and NDS/AF were built on the set of IETF RFCs on the Security Architecture for IP (IPsec) ([RFC2401] and following RFCs) from November 1998. The first release of NDS/IP mandated only support for authentication and key agreement based on pre-shared keys. For a detailed description of the key management used with IKEv1 and the features related to the update to the new RFCs mentioned below the reader is pointed to the NDS/IP specification [TS33.210]. As a new feature in Release 6, a PKI-based certificate infrastructure was added with creation of the NDS/AF specification [TS33.310]. This was mainly done to allow a flexible, yet simple, architecture for connection of two different security domains.

In December 2005, the IETF published a whole new series of RFCs ([RFC4301] and following RFCs) on IPsec, which superseded the old set of RFCs. Accordingly, 3GPP now references the new RFCs from Release 8 onwards, and implementations according to the new RFCs are recommended. This also included introducing the new authentication and key agreement protocol IKEv2 [RFC4306] into [TS33.210] and [TS33.310].

Usage of IPsec

In particular, references to [RFC2406] were updated to [RFC4303]. This new RFC enhanced the Encapsulating Security Payload (ESP) packet format, and put the algorithm selection to a separate RFC [RFC4305]. The latter RFC takes account of updated security requirements on algorithms, such as deprecating the single DES algorithm, no longer requiring MD5 as algorithm mandatory to support, and introducing and recommending AES based algorithms.

In addition, Annex E of NDS/IP [TS33.210] gives an overview of influences on the ESP packet format induced by the switch from [RFC2406] to [RFC4303].

Usage of IKE

The NDS/IP specification requires SEGs to implement both versions of the Internet Key Exchange (IKE) protocol, to allow interworking with SEGs and NEs from older releases supporting only IKEv1. But it encourages the usage of IKEv2 as it has certain security and performance advantages over IKEv1. In addition, it allows that other 3GPP specifications may impose more stringent requirements on IKE usage. This was done for example in [TS33.401], where implementation of IKEv2 is mandated (see section 8.4.2). For a detailed description of the message flows of IKEv2 including the information elements transferred, see [RFC4306].

Mechanisms on the Za reference point

The secure connection between two SEGs is implemented as two unidirectional IPsec Security Associations (SAs). The specification does not require multiple SAs per direction as a single SA per direction is sufficient to protect all traffic between SEGs.[2] As traffic over Za is not originated and terminated in the SEGs, but only forwarded by the SEGs, the usage of ESP in tunnel mode is mandatory. Application of a non-NULL ESP authentication transform is mandatory, while for the ESP encryption transform also ESP_NULL is allowed. For details on mandatory and optional algorithms the reader is referred directly to the NDS/IP specification [TS33.210] as this specification is updated from time to time in line with the progress of cryptanalytical work, which may render certain algorithms insecure in the future.

The establishment of a secure connection between SEGs is performed in two phases:

- The authenticity of the peers is mutually assured.
- The establishment of a secure transport channel is based on session key(s) established during the authentication phase.

It should be noted that authorization to access the operator network based on the above-mentioned authentication only provides a very simple access control method, meaning that every element authenticated is also provided service. A more fine-grained access control is not in scope of the NDS/IP specification. Thus the additional application of packet filter or firewall functionality is to be considered if explicit authorization for access to the network or single network elements is to be enforced.

---

[2] It should be noted that NDS/IP is specified for IP-based control plane traffic. Thus one SA per direction is seen as sufficient here. But if the same mechanisms are referenced for other kinds of traffic, multiple SAs per direction may be appropriate, for example allowing different quality of service (QoS) levels. This applies to [TS33.401], for instance, which refers to NDS/IP for user and management plane traffic also – see section 8.4.2.

Depending on the usage of the Za reference point – i.e. if it is used between different operators and authentication is based on PKI – an SEG has to validate the certificate of another SEG signed by a root Certification Authority (CA) of another operator. The NDS/AF specification [TS33.310] provides guidance on the establishment of a cross-signing infrastructure including certificate issuing, renewal and revocation.

Mechanisms on the Zb reference point

Basically the same mechanisms as for the Za reference point are deployed. Differences stem from the optional character of the reference point:

- The implementation of Zb in NEs is optional so as not to burden the implementation of all NEs with additional functionality.
- The implementation and usage of ESP in transport mode (in addition to tunnel mode) is allowed.
- Cross-signing of certificates is not necessary as, by definition, all NEs and SEGs belong to the same security domain.

4.5.3   Application of NDS

During the first phase of introduction of NDS to 3GPP specifications, particular use cases and scenarios were added to the NDS/IP specification [TS33.210] in normative annexes. The following gives the application to GSM, UMTS and IMS security.

For future standardization work it is preferred to add such normative text to the 3GPP specification where the use case for NDS is specified, as done for example for EPS in [TS33.401] – see Chapter 8. The definition of the mentioned reference points can be found in [TS23.002].

Control plane traffic in the GPRS and 3G core networks

For all control plane traffic carried over the Gn and Gp reference points (so-called GTP-C messages, see [TS29.060]), Annex B of [TS33.210] gives the rules how NDS/IP has to be applied. All GTP-C messages carried over Gp between different networks or over Gn between different security domains of the same network have to be protected using the mechanisms specified for the Za reference point.

User plane traffic carried in GTP-U messages is not subject to mandatory protection by NDS/IP. Still the security policy of an operator may decide that NDS/IP is to be applied also to GTP-U traffic. Easy discrimination between GTP-C and GTP-U traffic is possible from 3GPP Release 4 onwards, as GTP-C uses a dedicated UDP port. Anyway, it is completely optional to apply NDS/IP to a system that is compliant only to 3GPP Release 99 (but not to later releases).

Control plane traffic between core network and access network

Annex D of [TS33.210] mandates the usage of the NDS/IP Za reference point mechanisms for all Iu control plane traffic using IP transport and traversing borders of security domains. In contrast to general NDS/IP deployment, for this traffic confidentiality protection is mandatory on Za, thus disallowing the ESP encryption transform ESP_NULL. The reason is that Iu may carry subscriber specific security keys, which are vital for end-user security.

Control plane traffic in IMS

Annex C of [TS33.210] defines for IMS that confidentiality and integrity protection for SIP-signalling has to be provided in a hop-by-hop fashion. In the realm of NDS/IP this applies to all signalling between IMS core network elements, both within the same security domain (Zb interface) and between different security domains and IMS operator domains (Za interface). When sensitive information is carried over Za interfaces (e.g. IMS session keys), both encryption and integrity protection must be used. For more information on IMS, see Chapter 12.

5

3G–WLAN Interworking

This chapter has a close relationship with section in 11.2 on interworking with non-3GPP networks. It is also helpful when reading the authentication procedures for home base stations in section 13.4, but is not required for an understanding of other parts of the book. Therefore, readers not interested in the topics dealt with in Chapters 5 and sections 11.2 and 13.4 may safely skip this chapter.

This chapter is about security procedures for the case of a user accessing a 3GPP core network via a WLAN radio access network. While the two preceding chapters on the security of second- and third-generation mobile systems were meant to lay the foundations for understanding the security mechanisms applied when a user accesses the Evolved Packet Core via an LTE access network, this chapter is intended to provide the foundations for the case when a user accesses the Evolved Packet Core via an access network not defined in 3GPP specifications, for example via a Wireless Local Area Network (WLAN) [IEEE 802.11], or a WiMAX network [WiMAX], or a cdma2000® HRPD network defined in 3GPP2 specifications [3GPP2].

5.1 Principles of 3G–WLAN Interworking

5.1.1 The General Idea

3GPP performed its work on interworking between WLANs and 3GPP networks, in short 3G–WLAN interworking, as part of its Release 6, around the year 2004. LTE was not even on the horizon at that time, but it turned out that the same framework that is used for 3G–WLAN interworking could also be applied to the interworking of other types of access networks with a 3GPP-defined core network, either a 3G core, or an Evolved Packet Core.

Around 2004, wireless LANs had already found widespread adoption, in particular on laptop computers, but also on mobile phones. IP traffic from the laptop or the mobile phone could be carried over the WLAN air interface to the WLAN access point, and from there over fixed lines (e.g. DSL lines) further on into the Internet, or into external IP networks, such as IP networks provided by 3G operators. Users could, in this way, use packet-based services such

LTE Security   Dan Forsberg, Günther Horn, Wolf-Dietrich Moeller, and Valtteri Niemi
© 2010 John Wiley & Sons, Ltd

as the IMS (see Chapter 12) without using cellular radio technology. This seemed attractive to users as the available bandwidth was much higher than for 2G or 3G cellular radio interfaces. The support for seamless mobility, an advantage of cellular networks over WLAN, was not always needed by users.

3G–WLAN interworking defines two mechanisms:

- *direct IP access*, which provides access to the WLAN and a locally connected IP network (e.g. the Internet); and
- *3GPP IP access*, which allows users to establish, via access to the WLAN, connectivity with external IP networks, such as 3G operator networks, corporate intranets or the Internet via the 3GPP system.

The reader may wonder what the role of 3GPP, largely responsible for cellular access and the corresponding core networks, would be in this set-up as WLANs and IP networks are available independently of 3GPP technology. It turned out that there was a missing link in this set-up, which 3GPP technology could fill: operators offering services always need to know who their users are so that they can appropriately restrict service access to authorized users only and charge them for the use of the service. In other words, a AAA infrastructure was required, where AAA stands for 'Authentication, Authorization and Accounting'. While the protocols for AAA were available from IEEE specifications [IEEE 802.1X], and IETF specifications (references given below), and corresponding products were in the market, the infrastructure for distributing credentials for user authentication and user profiles for user authorization was not available on a global scale. But SIMs were already ubiquitous in mobile phones, and USIMs were also gaining traction fast. So, a technology was needed such that SIMs and USIMs, and the user profiles available in GSM and 3G cellular networks, could fill the gap. This technology is 3G–WLAN interworking.

The fact that an attractive use case appeared to be the use of laptops was accommodated in two ways:

- by designing dedicated WLAN cards for laptops with integrated SIMs or USIMs;
- by defining so-called 'split user equipments'.

While the former seems rather straightforward, the latter needs a bit of explanation. The idea is that a user has both a laptop and a mobile phone. The laptop would terminate the WLAN radio interface, and run the applications over IP, while the mobile phone would hold the SIM or USIM. A local radio connection between the laptop and the mobile phone, possibly by means of Bluetooth [Bluetooth], would allow a software module on the laptop to access the SIM or USIM on the mobile phone. In doing so, the laptop would use a protocol originally designed for the communication between a phone without (U)SIM built into a car and a mobile phone, with a (U)SIM, for example in the driver's pocket. The technical details of split user equipments are not essential for the main purpose of this chapter. We therefore refer the reader interested in these details to [TS33.234].

3G–WLAN interworking builds on a key technology called EAP, which we therefore describe next.

5.1.2   The EAP Framework

The Extensible Authentication Protocol (EAP) allows bringing together security mechanisms from three different areas:

- credential infrastructures and domain-specific authentication protocols, such as (U)SIMs, Authentication Centres, and AKA protocols defined by 3GPP;
- AAA protocols defined by the IETF;
- link layer-specific security protocols, such as those defined for WLAN by the IEEE in [IEEE 802.11i].

EAP is an authentication framework that supports multiple authentication methods, called EAP methods. We only present the very basics of EAP required for our purposes here. EAP is specified in [RFC3748], while the EAP key management framework is specified in [RFC5247]. EAP methods are defined in separate RFCs. The EAP methods relevant in the context of this book are:

- *EAP-SIM* as specified in [RFC4186] – this method allows using SIMs and GSM authentication vectors and cryptographic functions within the framework of EAP;
- *EAP-AKA* as specified in [RFC4187] – this method allows using USIMs and UMTS authentication vectors and cryptographic functions within the framework of EAP;
- *EAP-AKA'* as specified in [RFC5448] – this method also allows using USIMs and UMTS authentication vectors and cryptographic functions within the framework of EAP; in addition, EAP-AKA' provides a binding of derived keys to the access network identity. For more details see section 11.2.

The EAP architectural model

The basic entities are, according to [RFC3748]:

- *Authenticator* – the end of the link initiating EAP authentication;
- *Peer* – the end of the link that responds to the authenticator;
- *EAP server* – the entity that terminates the EAP authentication method with the peer.

In all scenarios relevant in the context of this book, the authenticator operates in the so-called pass-through mode and, hence, does not perform any EAP method-specific operations.

The EAP layering and forwarding model

Figure 5.1, which is based on [RFC3748], shows the forwarding of an EAP response packet across the protocol layers at the peer, pass-through authenticator, and EAP server respectively.

One can see from the figure that the transport of EAP messages between the peer and the authenticator differs from that between the authenticator and the EAP server. EAP messages between the peer and the authenticator are typically carried directly on the link layer, in a way specific to the particular link layer. This is the case for 3G–WLAN interworking with direct

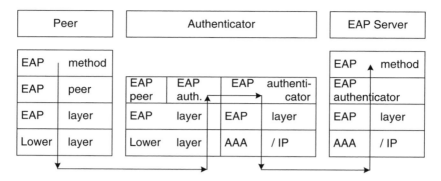

Figure 5.1 Forwarding of EAP response packet across protocol layers

IP access, as described in this chapter, where EAP messages are carried over WLAN using a mechanism known as EAPOL (or 'EAPoverLAN' [IEEE 802.11i]). It is also the case for trusted access to the Evolved Packet Core, as described in section 11.2, where EAP messages are carried using a mechanism specific to, for example, cdma2000® HRPD or WiMAX. But EAP messages between the peer and the authenticator may also be encapsulated within a protected channel created by IKEv2 [RFC4306], the key management protocol for IPsec security associations. This is the case for 3G–WLAN interworking with 3GPP IP access, as described in this chapter, or for untrusted access to the Evolved Packet Core, as described in section 11.2.

EAP messages between the authenticator and the EAP server are carried using AAA protocols such as RADIUS/EAP [RFC3579] and Diameter EAP Application [RFC4072]. This implies that the authenticator includes the functionality of a AAA client.

According to [RFC3748], the EAP layer receives and transmits EAP messages via the lower layer, implements duplicate detection and retransmission, and delivers and receives EAP messages to and from the EAP peer and authenticator layers. EAP methods implement the authentication algorithms and receive and transmit EAP messages via the EAP peer and authenticator layers.

Mapping of EAP architectural entities to a 3GPP setting

In the scenarios covered in Chapter 5 and section11.2, the EAP architectural model is applied as follows.

- *Authenticator*. There are two cases for the allocation of the authenticator: for 3G–WLAN interworking with direct IP access, and for trusted access to the Evolved Packet Core the authenticator resides in the non-3GPP access network. In a WLAN, the authenticator often coincides with the WLAN Access Point terminating the WLAN radio interface. For 3G–WLAN interworking with 3GPP IP access, and for untrusted access to the Evolved Packet Core, the authenticator resides on the Packet Data Gateway (PDG) and the evolved Packet Data Gateway (ePDG), respectively, which act as the responder in IKEv2.

- *Peer.* The peer resides on the user equipment. It requires the use of SIM functionality for the purposes of EAP-SIM, and USIM functionality for the purposes of EAP-AKA and EAP-AKA'.
- *EAP server.* In all cases described in this book, the EAP server resides on the 3GPP AAA server [TS23.234; TS23.402]. The 3GPP AAA server communicates with the HSS or HLR as part of the EAP method execution to retrieve authentication vectors.

The role of link layer security

In general, authentication is not sufficient for ensuring secure communication. When the access link is prone to eavesdropping or tampering, encryption and/or integrity protection is required. The encryption and integrity keys need to be generated as part of the authentication and key agreement protocol. All three EAP methods relevant in our context, EAP-SIM, EAP-AKA and EAP-AKA', provide mutual authentication and key agreement. The agreed keys are called Master Session Key (MSK) and Extended Master Session Key (EMSK) in EAP terminology [RFC5247]. Both the peer and the EAP server derive both the MSK and the EMSK. On the network side, while the EMSK remains in the EAP server, and is used in, for example, [TS33.402] to derive mobile-IP-specific keys (see section 11.2), the MSK is sent by the EAP server to the authenticator. For direct WLAN access, the WLAN-specific keys defined in [IEEE 802.11i] are derived from MSK in the peer and the authenticator; [RFC5247] explains the relationship of these keys with MSK.

When EAP is used in conjunction with IKEv2, the key MSK is used to compute the parameter AUTH, which is part of the IKEv2 exchange (of which more later in this chapter), but MSK plays no role in the derivation of the keys used to protect IP packets in the IPsec tunnel set up by IKEv2 [RFC4306].

Use of EAP with IKEv2

IKE [RFC2409] and IKEv2 [RFC4306] are key exchange protocols for generating security associations to be used, for example, with IPsec ESP (IP Encapsulating Security Payload, [RFC4303]). The IKE mutual authentication is based on either shared keys or on certificates on both sides. There are deployments, however, where neither of these two possibilities is well suited: the use of shared keys does not scale, and the distribution of certificates to a large number of users of a public service may also be difficult from an administrative point of view. Therefore, version 2 of IKE offers another possibility for authentication, namely the use of an EAP method for authenticating the client (the 'initiator' in IKEv2 terminology). This possibility adds more flexibility by allowing the re-use of existing authentication infrastructures. It further allows the separation of the IKEv2 termination point on the network side (the responder in IKEv2 terminology, e.g. a VPN gateway) from the backend authentication server. But IKEv2 still requires certificates for the authentication of the responder.

While the distribution of certificates to the responders in a network may be considered practically feasible, as they are not very numerous, the requirement to have a Public Key Infrastructure may still be a burden in some usage scenarios. One may therefore wonder whether responder authentication by certificates is required at all when the EAP method provides mutual authentication and key agreement, as many EAP methods do. And, indeed, at the time of writing this book, efforts are under way at the IETF to remove the requirement of responder authentication by certificates under certain conditions – see [draft-ietf-ipsecme-eap-mutual-00].

When doing so one of the problems to be taken into account is the 'lying authenticator' problem: even if the EAP method provides mutual authentication and key agreement, and the MSK is used to establish a secure link between the EAP peer and the authenticator, the EAP peer will, in general, know after a successful EAP protocol run only that the authenticator is some entity entrusted by the EAP server to receive the key MSK. The peer does not know from the EAP protocol run to which authenticator it is connected. Consequently, the authenticator could lie about its identity. There are scenarios where this may matter. For example, a user in a 3G–WLAN interworking scenario may want to know whether he is connected to a WLAN access point meant to provide direct IP access to the Internet, or an operator-controlled PDG meant to provide 3GPP IP access to the operator's network.

When EAP is used with IKEv2, and the responder, assuming the role of the authenticator, is authenticated by a certificate binding the public key to the authenticator's identity, the authenticator cannot lie about its identity. So while, in principle, responder authentication by certificates could be replaced by EAP mutual authentication, this would be admissible only in scenarios where

- the lying authenticator problem did not matter; or
- the EAP server coincided with the authenticator/responder; or
- the EAP method also provided authentication of the authenticator (or a group of authenticators) to the peer.

The third condition is fulfilled by EAP-AKA′ [RFC5448] in the sense that it authenticates the group of authenticators identified by having the same 'access network identifier' to the peer, but this condition is not fulfilled by EAP-AKA.

5.1.3   Overview of EAP-AKA

The most important example of an EAP method in the context of this book is EAP-AKA [RFC4187], which is based on the use of the USIM. We therefore give an overview of EAP-AKA here. As the focus of this book is LTE security, and SIMs are no longer allowed for accessing LTE and the Evolved Packet Core, we only briefly address the case of EAP-SIM further below.

EAP-AKA full authentication

EAP-AKA allows using USIMs, UMTS authentication vectors and UMTS cryptographic functions within the framework of EAP. The USIM and the Authentication Centre perform the same functions as in UMTS AKA (see section 4.2). The message flow and the message formats differ, however, between UMTS AKA and EAP-AKA. Figure 5.2 shows the basic message flow of a successful EAP-AKA full authentication.

1. The procedure starts with the authenticator requesting the peer's identity. From this point onwards the authenticator only passes EAP messages through.

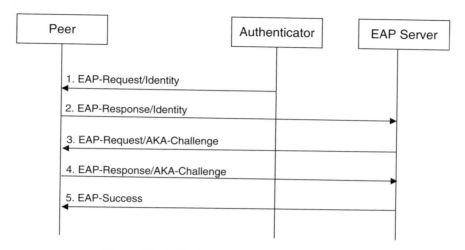

Figure 5.2   EAP-AKA full authentication procedure

2. The peer replies with sending its identity. This identity may be a permanent identity or a temporary identity (pseudonym).
3. The EAP server fetches UMTS authentication vectors and sends RAND and AUTN encoded in the EAP-Request/AKA-Challenge message to the peer. The EAP server first derives a master key MK from the user identity, and the keys CK and IK. From MK, the EAP server then derives the keys MSK and EMSK (see section 5.1), as well as Transient EAP Keys (TEKs), the encryption key K_encr and the integrity key K_aut, for protecting the EAP-AKA messages. For the details of EAP-AKA key derivation, see [RFC4187]. If the EAP server supports user identity privacy it may also include a pseudonym protected with this encryption key. The peer may use this pseudonym in the next full authentication.
4. The peer first hands the received RAND and AUTN to the USIM, which processes them as for UMTS AKA. The peer obtains CK and IK from the USIM and performs the same key derivations as the EAP server. The peer decrypts the encrypted parts of the EAP-Request/AKA-Challenge message and checks the integrity of the message. The peer then sends the AT_RES attribute, which includes RES, in the EAP-Response/AKA-Challenge message to the EAP server.
5. The EAP server checks RES against XRES in the UMTS authentication vector it received from the HSS/HLR, and then checks the integrity of the message. If the checks are successful the EAP server sends an EAP-Success message to the peer. The AAA message, in which the EAP-Success message is carried between the EAP server and the authenticator, may also carry the key MSK. MSK is not forwarded by the authenticator to the peer as the peer derived MSK in step 3.

The EAP-AKA full authentication procedure ends with step 5. It may be followed by a procedure, such as a four-way handshake according to [IEEE 802.11i], to establish link layer security using the key MSK shared between the peer and the authenticator.

EAP-AKA fast re-authentication

The fast re-authentication procedure differs from a full authentication procedure in that it does not consume a new UMTS authentication vector, nor does it involve the USIM. Authentication and key derivation are based on the keys derived during the preceding full authentication procedure. The same TEKs are used while fresh keys MSK and EMSK are derived from MK. Fast re-authentication identities, different from the permanent user identity, are used with this procedure. The EAP-AKA fast re-authentication procedure has no equivalent in UMTS AKA; but one could argue that EPS AKA has a somewhat similar feature in that EPS-AKA produces the local master key K_{ASME}, from which new session keys can be generated without involving the USIM or consuming new authentication vectors (see Chapter 70).

EAP-AKA identities

The identities of the peer have the form of a Network Access Identifier (NAI), as defined in [RFC4282]. A NAI is composed of a username and, optionally, a realm, and has the form 'username@realm'. The username uniquely identifies a user within a realm (when there is a realm). [TS23.234] specifies that the NAI representing the permanent user identity in EAP-AKA for 3G–WLAN interworking shall be derived from the user's IMSI as defined in [TS23.003]. A permanent user identity then takes the form

'0<IMSI>@wlan.mnc<MNC>.mcc<MCC>.3gppnetwork.org'

Permanent user identities are only used in EAP-AKA full authentications. Temporary identities, also called 'pseudonyms', are used for the purpose of user identity privacy. They are only used in EAP-AKA full authentications. Pseudonyms may be generated by the EAP server in an implementation-dependent manner, as only the EAP server needs to be able to map the pseudonym username to the permanent identity. 3GPP, however, specified a particular way of generating pseudonyms from permanent user identities in [TS33.234] in order to allow for the case where a user may be served by different servers within an operator's network at different points in time (e.g. for load-balancing purposes). The 3GPP-defined mechanism, which is described in section 5.3, then allows all these servers to understand the pseudonyms. As we have seen above, pseudonyms are sent by the EAP server to the peer in an encrypted part of an EAP message. If they were not encrypted user identity privacy would be endangered.

Fast re-authentication identities are generated by the EAP server in a way similar to pseudonyms.

In the following section, the principles laid out above are applied to securing the 3G–WLAN interworking procedures for direct IP access and 3GPP IP access.

5.2 Security Mechanisms of 3G–WLAN Interworking

5.2.1 Reference Model for 3G–WLAN Interworking

We show a simplified reference model in Figure 5.3, which is taken from [TS23.234]. The link from the box 'WLAN Access Network' directly to 'Intranet/Internet' refers to WLAN direct IP access functionality. The shaded area refers to WLAN 3GPP IP access functionality.

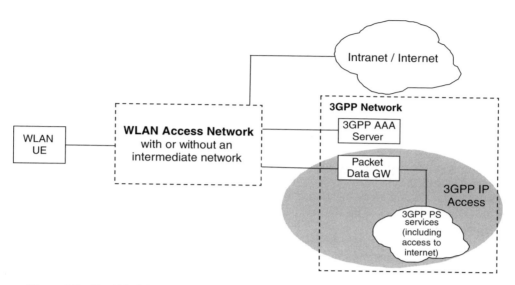

Figure 5.3   Simplified WLAN network model. Adapted with permission from © 2009, 3GPP™

The role of the 3GPP AAA server, which is used with both functionalities, has already been partly explained in section 5.1, and will be detailed further in sections 5.2.2 and 5.2.3. 3GPP packet-switched services are accessed via a Packet Data Gateway (PDG).

Regarding security, the PDG plays the role of IKEv2 responder in authentication and key agreement, and it terminates the IPsec ESP tunnel from the user equipment. It may also filter out unauthorized or unsolicited traffic with packet filtering functions. It further performs various non-security related functions, such as routing, IP address handling, and quality-of-service related functions; for details see clause 6 of [TS23.234].

Independence of direct IP access and 3GPP IP access

Direct IP access and 3GPP IP access are independent procedures. It seems obvious that direct IP access can be used without 3GPP IP access if only access to the Internet, and not to an operator-controlled IP network, is desired. But the converse is also true: 3GPP access can be used without direct IP access. 3GPP IP access presupposes IP connectivity because IKEv2 used with 3GPP IP access is an IP-based protocol. But any method to establish IP connectivity over the particular access network may be used; it is not necessary that direct IP access, as defined in [TS33.234], be used in conjunction with 3GPP IP access.

5.2.2   *Security Mechanisms of WLAN Direct IP Access*

Both users with a SIM and users with a USIM may use direct IP access. We treat only the case of successful USIM-based access authentication, which uses EAP-AKA, as it helps to understand LTE security, the focus of this book. Indeed, SIM-based authentication is not allowed for LTE access. For WLAN direct IP access using EAP-SIM, we refer the reader to [TS33.234]. We also limit ourselves to presenting the full authentication procedure as the fast re-authentication procedure is very similar.

USIM-based access authentication

The authentication and key agreement procedure using an EAP-AKA full authentication can be seen in Figure 5.4.

The role of the EAP server/backend authentication server from the general EAP model is assumed here by the 3GPP AAA server in conjunction with the HLR or HSS. The authenticator resides in the WLAN access network (WLAN AN). As mentioned before, the authenticator is often realized as part of the WLAN access point. The peer is realized in the WLAN User Equipment (WLAN UE).

The numbering of the steps in Figure 5.4 is the same as that in Figure 4 of [TS33.234] so as to make it easier for the reader to compare the text explaining the figure here with the somewhat more detailed text in the 3GPP specification. We do this even if it makes the figure a little more complicated than necessary as, in many cases, the WLAN AN simply passes the EAP messages through transparently without any further action on the EAP messages themselves. The textual description in this book is shortened compared to [TS33.234] as not all details presented there are essential for the understanding of the principles of 3G-WLAN interworking.

1. A connection is established between the WLAN UE and the WLAN AN, using a wireless LAN technology specific procedure (out of scope for 3GPP).
2. The authenticator in the WLAN AN sends an EAP-Request/Identity message to the WLAN UE.
3. The WLAN UE sends an EAP-Response/Identity message containing an identity in NAI format, either the permanent identity or a pseudonym.
4. The message is routed towards the proper 3GPP AAA server based on the realm part of the NAI.
5. The 3GPP AAA server receives the EAP-Response/Identity message, encapsulated in a AAA message.
6. The 3GPP AAA server determines the IMSI from the identity received in step 5 and the EAP method to be used. According to [TS24.234], if the 3GPP AAA server is not able to map the user identity received in EAP-Response/Identity to a subscriber identity (e.g. because of an obsolete pseudonym), but it recognizes the EAP method, the 3GPP AAA server requests a new identity using the EAP method indicated by the WLAN UE. If this EAP method is EAP-AKA, the 3GPP AAA server proceeds to step 7. If the 3GPP AAA server is able to map the user identity received in EAP-Response/Identity to a subscriber identity (IMSI), it checks whether it has an authentication vector for this IMSI. If not it fetches one, or more, from the HSS/HLR.
7. The 3GPP AAA server requests the user identity again, using the EAP-Request/AKA Identity message.
8. The WLAN AN forwards the EAP-Request/AKA Identity message to the WLAN UE.

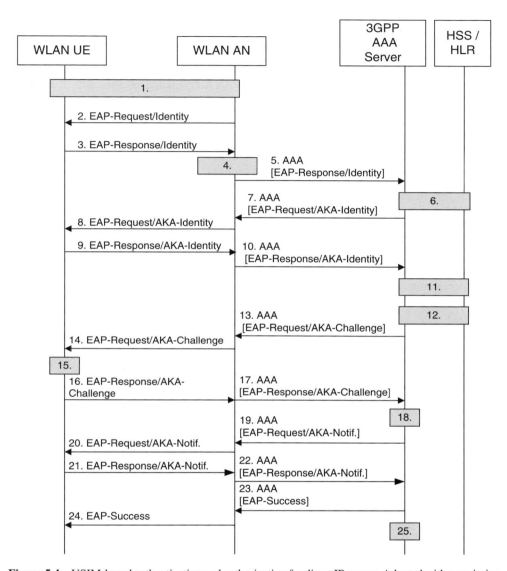

Figure 5.4 USIM-based authentication and authorization for direct IP access. Adapted with permission from © 2010, 3GPP™

9. The WLAN UE responds with an identity that depends on the parameters contained in the message received in step 8.

10. The WLAN AN forwards the EAP-Response/AKA Identity message to the 3GPP AAA server. The identity received in this message will be used by the 3GPP AAA server in the rest of the authentication process. If an inconsistency is found between the identities received in steps 5 and 10, new authentication vectors need to be retrieved from the HSS/HLR.

11. The 3GPP AAA server checks that it has the WLAN access profile of the subscriber available. If not, the profile is retrieved from HSS/HLR. The 3GPP AAA server verifies that the subscriber is authorized to use the WLAN service.
12. The 3GPP AAA server derives keys as described in section 5.1.
13. The 3GPP AAA server sends an EAP-Request/AKA-Challenge message. The 3GPP AAA server may also indicate that it wishes to protect the success result message at the end of the process (if the outcome is successful), depending on the home operator's policies.
14. The WLAN AN forwards the EAP-Request/AKA-Challenge message to the WLAN UE.
15. The WLAN UE processes the EAP-Request/AKA-Challenge message as described in section 5.1.
16. The WLAN UE sends the EAP-Response/AKA-Challenge message to the WLAN AN.
17. The WLAN AN forwards the EAP-Response/AKA-Challenge packet to the 3GPP AAA server.
18. The 3GPP AAA server processes the EAP-Response/AKA-Challenge message as described in section 5.1.
19. If all checks in step 18 are successful, the 3GPP AAA server sends the EAP-Request/AKA-Notification message if the 3GPP AAA server requested in step 13 to use protected successful result indications.
20. The WLAN AN forwards the message to the WLAN UE.
21. The WLAN UE sends the EAP-Response/AKA-Notification.
22. The WLAN AN forwards the EAP-Response/AKA-Notification message to the 3GPP AAA server.
23. The 3GPP AAA server sends the EAP-Success message to the WLAN AN optionally including the key MSK (see section 5.1).
24. The WLAN AN forwards the EAP-Success message to the WLAN UE. Now the EAP-AKA exchange has been successfully completed, and, if an MSK was sent in step 23, the WLAN UE and the WLAN AN share keying material for protecting the WLAN access network.
25. The 3GPP AAA Server registers the WLAN UE to the HSS/HLR if not done before. The 3GPP AAA Server also performs a session update, if needed; for details see [TS33.234].

If the EAP-AKA run terminates owing to a failure, the 3GPP AAA server informs the HSS/HLR of this event.

5.2.3   Security Mechanisms of WLAN 3GPP IP Access

Both users with a SIM and users with a USIM may use WLAN 3GPP IP access. For the same reasons as already mentioned for direct IP access, we treat only the case of a successful USIM-based authentication for 3GPP IP access, which uses EAP-AKA. We also limit ourselves to presenting the full authentication procedure as the fast re-authentication procedure is very similar.

USIM-based authentication for WLAN 3GPP IP access

For WLAN 3GPP IP access, an IPsec tunnel is established between the UE and the PDG, spanning across the WLAN access network. Once established, the tunnel protects from any vulnerabilities in the access network, so that the strengths and weaknesses of WLAN security

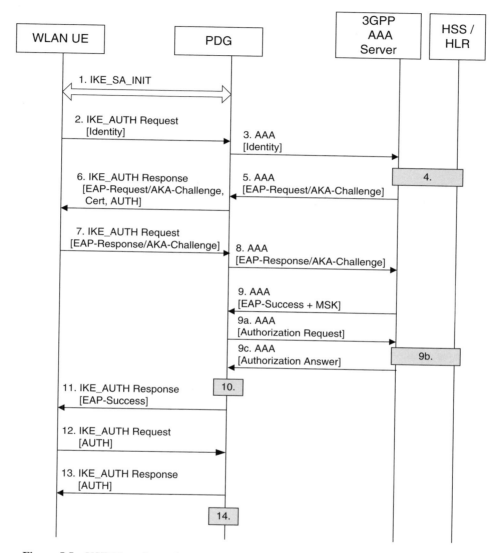

Figure 5.5   USIM-based tunnel set-up, authentication and authorization for 3GPP IP access

become irrelevant for WLAN 3GPP IP access. For protecting IP packets between the UE and the PDG, IPsec ESP [RFC4303] in tunnel mode is used. For setting up the corresponding security association, IKEv2 combined with EAP-AKA is used. The combined IKEv2-EAP-AKA procedure using an EAP-AKA full authentication for 3GPP IP access can be seen in Figure 5.5.

The role of the EAP server/backend authentication server is assumed by the 3GPP AAA server in conjunction with the HLR or HSS. The authenticator resides on the PDG. The peer is again realized in the WLAN user equipment.

The numbering of the steps in Figure 5.5 is the same as that in Figure 7A of [TS33.234] so as to make it easier for the reader to compare the text explaining the figure here with the somewhat more detailed text in [TS33.234].

1. The WLAN UE and the PDG exchange the first pair of messages, known as IKE_SA_INIT, in which the PDG and the WLAN UE negotiate cryptographic algorithms, exchange nonces and perform a Diffie–Hellman exchange.
2. The WLAN UE sends the user identity in the form required for EAP-AKA in this first message of an IKE_AUTH exchange. In accordance with [RFC4306], the WLAN UE omits the AUTH parameter in order to indicate to the PDG that it wants to use EAP over IKEv2.
3. The PDG sends an appropriate AAA message to the 3GPP AAA server, containing the user identity. The PDG includes a parameter indicating that the authentication is being performed for tunnel establishment. This will help the 3GPP AAA server to distinguish between authentications for direct IP access and authentications for 3GPP IP access. When Diameter is used, the messages between the PDG and the 3GPP AAA server are specified in [TS29.234], which in turn relies on the Diameter EAP application specified in [RFC4072].
4. The 3GPP AAA server fetches the user profile and authentication vectors from the HSS/HLR (if these parameters are not yet available in the 3GPP AAA server) and determines the EAP method (EAP-SIM or EAP-AKA) to be used.
5. The 3GPP AAA server initiates the authentication challenge by sending an EAP-Request/AKA-Challenge message, encapsulated in a AAA message, towards the PDG. The user identity is not requested again as the user identity received in step 3 could not have been modified or replaced by any intermediate node.
6. The PDG sends its identity, a certificate, and an AUTH parameter to the WLAN UE. The PDG generates this AUTH parameter by computing a digital signature over parameters in the first message it sent to the WLAN UE (in step 1). The PDG also includes the EAP-Request/AKA-Challenge message received in step 5.
7. The WLAN UE verifies AUTH using the public key in the certificate received in step 6 and sends the EAP-Response/AKA-Challenge message towards the PDG.
8. The PDG forwards the EAP-Response/AKA-Challenge message to the 3GPP AAA server, encapsulated in a AAA message.
9. When all checks are successful, the 3GPP AAA server sends the EAP-Success, encapsulated in a AAA message, which also contains the key MSK.
9a. The PDG sends an authorization request to the 3GPP AAA server.
9b. The 3GPP AAA server checks, based on the user's subscription, if the user is authorized to establish the tunnel.
9c. The 3GPP AAA server sends the authorization answer to the PDG. The 3GPP AAA server includes the IMSI if it received only a pseudonym in step 9a. This provides the PDG with the means to identify the user by a permanent identity across all runs of this procedure with varying pseudonyms.
10. The PDG generates another two AUTH parameters by computing message authentication codes over parameters in the two messages exchanged in step 1 using the shared key MSK. Note that the PDG could defer the generation of these two AUTH parameters until receiving the message in step 12.

11. The PDG forwards an EAP-Success message to the WLAN UE over IKEv2.
12. The WLAN UE generates two AUTH parameters, using the locally derived MSK, in the same way as the PDG in step 10, and then sends the AUTH parameter protecting the first message from the UE to the PDG (sent in step 1).
13. The PDG verifies the AUTH parameter received in step 12 by comparing it with the corresponding value computed in step 10. The PDG then sends the other AUTH parameter computed in step 10 to the WLAN UE. The WLAN UE verifies the received AUTH parameter by comparing it with the corresponding value computed in step 12.
14. The PDG deletes any old IKE SAs relating to the same WLAN access point name (see [TS23.234]) and informs the WLAN UE of this deletion in an INFORMATIONAL exchange (not shown in the figure).

5.3   Cryptographic Algorithms for 3G–WLAN Interworking

As explained earlier in this chapter, several security mechanisms in 3G–WLAN interworking are based on cryptographic algorithms.

Authentication of the subscriber is based on either USIM or SIM, and in each case the corresponding algorithms, described in sections 4.3 and 3.4 respectively, apply also for 3G–WLAN interworking. Additionally, both EAP-AKA and EAP-SIM bring extensions to the mechanisms provided by USIM/SIM, and some of these extensions contain cryptographic components; see [RFC4187] and [RFC4186] for details.

Protection of the user data and signalling data between the UE and the WLAN access point also involves cryptographic algorithms. These protection mechanisms are out of scope of 3GPP specifications but, for the sake of understanding the whole system, they have been reviewed in an informative Annex A of [TS33.234].

For 3GPP IP access, the security mechanisms between the UE and the PDG are also based on cryptography. The security tunnel between the UE and the PDG is based on IPsec ESP and IKEv2, and the cryptographic algorithms that are available for these security protocols are defined in the relevant IETF RFCs; see section 5.2 for references. In addition, 3GPP has narrowed down the number of options by selecting specific profiles that must be supported in 3GPP–WLAN interworking; see clauses 6.5 and 6.6 of [TS33.234].

In the GSM and 3G systems (and similarly in EPS), the generation of the temporary identity TMSI (or P-TMSI) has been left out of the 3GPP specifications. The reason for this is that there are no interoperability issues around the generation of a TMSI. It is generated inside a single network entity, and once it has been generated there is no need for any entity to know any more how the generation actually happened. The only basic requirement is that temporary identities should be unpredictable, and therefore chosen essentially at random or using a pseudorandom generator.

Temporary identities are also in use for 3G–WLAN interworking but there the issue of generating these identities is slightly more complicated. In this context, it is possible that the temporary identity (e.g. pseudonym) is processed by a network entity that does not know the relationship of the temporary identity to any permanent identity. For this reason, generation of temporary identities is standardized, and it is also possible to reverse the process; that is, to find the permanent identity IMSI based on the temporary identity and some auxiliary information that is provided only to authorized network entities.

A simple encryption mechanism is specified in clause 6.4 of [TS33.234] for this purpose. To create the temporary identity, the IMSI is first coded as a bit string, and then padded to obtain a 128-bit value. Then one single run of the AES algorithm [FIPS 197] is applied to this value, using a specific key that is shared between WLAN AAA servers in the same operator network. The encrypted 128-bit bit string is used as the temporary identity.

6

EPS Security Architecture

6.1 Overview and Relevant Specifications

The Evolved Packet System (EPS) brings two new major ingredients into the 3GPP environment: the radio network E-UTRAN with a new radio interface, and the flat IP-based core network Evolved Packet Core (EPC). The security functions and mechanisms that are part of GSM and 3G security architectures are mostly based on designs and principles that are generic enough and usable in many other environments. But still both GSM and 3G security architectures have a tight coupling with other functions and mechanisms in these systems; security functions have been embedded into the overall architecture in an optimal and efficient manner.

The design of the EPS security architecture follows the same principle of maximizing, from a system point of view, the synergies between security functions and other functions. In particular, this implies that:

- GSM and 3G security mechanisms offer a good basis for the EPS security architecture; but
- to a certain extent, each GSM or 3G mechanism, if reused, needs to be adapted from the original context and embedded to the EPS architecture.

The EPS must also be able to interwork with legacy systems, so these adaptations have to be done in a backward-compatible manner. In addition to adaptations from security functionalities already existing in legacy systems, many new extensions and enhancements have been introduced in the EPS security architecture.

In the following, we show how major security features (further discussed in section 6.2) fit into the EPS architecture. This is illustrated by Figure 6.1.

After the User Equipment (UE) has been identified, the Mobility Management Entity (MME – described in Chapter 2) in the serving network fetches authentication data from the home network. Next the MME triggers the authentication and key agreement protocol with the UE. After this protocol has been successfully completed, the MME and the UE share a secret key, K_{ASME}, where the acronym ASME refers to Access Security Management Entity. In the EPS, it is the MME that takes the role of the ASME.

LTE Security   Dan Forsberg, Günther Horn, Wolf-Dietrich Moeller, and Valtteri Niemi
© 2010 John Wiley & Sons, Ltd

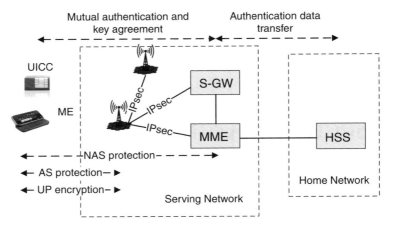

Figure 6.1 EPS security architecture

Now the MME and the UE are able to derive further keys from the K_{ASME}. Two derived keys are used for confidentiality and integrity protection of the signalling data between the MME and the UE. This is represented in the figure by the arrow with 'NAS protection' (Non-Access Stratum).

Another derived key is transported to the base station (eNB). Three more keys are subsequently derived both in the base station and in the UE. Two of these keys are used for confidentiality and integrity protection of the signalling data between the eNB and the UE – see the arrow with 'AS protection' (Access Stratum). The third key is used for confidentiality protection of the user plane data between the eNB and the UE – see the arrow with 'UP encryption'.

In addition to the protection of signalling and user plane data originated or terminated by the UE, there is also confidentiality and integrity protection for the signalling and user data carried over the interface between the base station and the core network (Evolved Packet Core, EPC). The signalling data is transferred between the UE and the MME over the S1-MME interface while the user data is transferred between the UE and the Serving Gateway (S-GW) over the S1-U interface. If cryptographic protection is applied to the S1-interfaces the protection mechanism used is IPsec. (More on the conditions for applying IPsec can be found in section 8.4.) The needed keys are not specific to the UE.

The X2 interface between two base stations is similarly protected by IPsec with keys that are not specific to the UE in case cryptographic protection is applied.

Let us next take a look at how confidentiality and integrity protection mechanisms are embedded in the protocol stack of EPS. In Figure 6.2, the relevant signalling plane protocols are depicted.

The integrity protection and ciphering for Non-Access Stratum signalling is further explained in section 8.2. The integrity protection and ciphering for Access Stratum signalling protects the messages of the Radio Resource Control protocol (RRC) see section 8.3. The IPsec protection on the interface S1 (and similarly for the interface X2) is profiled as defined in 3GPP specifications for Network Domain Security/IP layer (NDS/IP) see section 8.4.

Figure 6.3 provides an illustration of how user plane protection is provided.

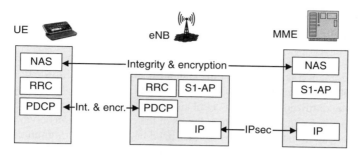

Figure 6.2   EPS signalling plane protection

For both signalling and user data, the (optional) confidentiality protection between the UE and the base station is embedded into the Packet Data Convergence Protocol (PDCP). Integrity protection is not applied on user data between the UE and the base station. For X2 and S1 interfaces, cryptographic protection for user data is provided in a way similar to that for the corresponding control plane interfaces, by means of the IPsec protocol.

6.1.1   Need for Security Standardization

The main purpose of this book is to explain how the standardized part of EPS security works. To some extent we also discuss ingredients of the security that do not need to be standardized. The fact that something need not be standardized does not make it less important from a security point of view. For example, internal protection mechanisms inside network elements or terminals are of utmost importance for guaranteeing the integrity of functions in these elements, and therefore also in guaranteeing the correct and secure functioning of the overall system. But, from the system interoperability point of view, it does not matter whether elements in the system use similar or different internal protection mechanisms. What matters is that the protection is there and each element should be protected in a way that is optimal for that particular element.

For mechanisms that involve several elements, e.g. the terminal and the base station, interoperability is a key issue. The mechanism does not work unless both communicating parties

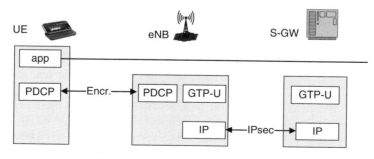

Figure 6.3   EPS user plane protection

are able to figure out what the other party is up to. For this purpose, standardizing the behaviour of the parties is indispensable. That said, there are also caveats regarding the need for standardization in communications, which apply also more widely than only to security protocols and mechanisms. If both communicating parties are always controlled by the same administrative domain it is enough that the standardization happens inside that domain.

As an example, the communication between a client application and an application server need not to be standardized when both client and server program are exclusively developed by one single company. As another example, details of the communication between the UICC and the operator backend do not need to be standardized because both communicating parties are owned by the same operator. In particular, for security, those operations in the authentication and key agreement protocol that are carried out exclusively at either end of the chain (i.e. in the UICC or in the home operator backend) do not need to be standardized. This applies to the choice of cryptographic algorithms and also to the management of sequence numbers.

Sometimes it is good to provide standards also for situations where interoperability does not strictly necessitate standardized behaviour. The example of authentication and key agreement also applies here: 3GPP has provided a standard choice for the cryptographic algorithms in the form of the MILENAGE algorithms (see Chapter 4), and appropriate mechanisms for sequence number management are provided in an informative annex of [TS33.102]. But in these cases, the standard solution is provided only as a guideline or as a recommendation. The purpose of this type of standard is to help companies in developing and deploying secure solutions in a cost-efficient manner.

It is a useful general principle to leave room for introducing better solutions without the delay caused by standardizing these better solutions first. For security, having some heterogeneity in security mechanisms has two effects. On the positive side, when only part of the system is protected by a particular mechanism, the value for an attacker of breaking or circumventing this mechanism becomes smaller. On the negative side, there are many more targets for an attacker to work with and, inevitably, there are going to be some weak ones among the many mechanisms, especially because some organisations may not have the skills or the resources to develop and evaluate appropriately secure mechanisms.

There is another point worth mentioning in the context of how and when to standardize a communication protocol. Sometimes it is enough that the behaviour of one side of the communication is standardized. Especially in the case where there is some asymmetry between the positions of the communicating parties, one of the parties could adjust its own actions based on the predictability of actions from the side whose behaviour follows a standardized pattern. An example could be, again, the communication between the terminal and the network: because the network is in control of the proper operation of the whole system, it is enough that the network knows:

- how the terminal would react to requests, inquiries and other messages from the network;
- which circumstances would trigger the terminal to initiate the communication from its side.

Indeed, some specifications of radio layer protocols in 3GPP follow these principles. Only the UE actions are specified in detail while much more freedom is left on the network side regarding when to invoke procedures and how to react on messages received from the UE side.

From a security point of view this kind of situation is problematic. A security protocol is effective only when used in the right manner. It is not sufficient that the protocol is sound and provides the required security goal once it is run; this is not much of help if the protocol is not run at all. Leaving too much freedom to one side of the communication does not help in guaranteeing to the other side that all security measures are applied in situations where they need to be applied. More specifically, the network side must not take certain actions, such as initiating certain procedures, unless certain security procedures, such as start of signalling integrity protection, have been successfully run first. Thus, sometimes security requires that the behaviour of communicating parties is standardized to a greater extent than would be needed for enabling the communication as such.

6.1.2   Relevant Non-security Specifications

When considering security for a hugely complex system such as the EPS, it is difficult to state that some specifications would not be relevant from a security point of view. Indeed, any functionality in the system could potentially be misused, so any functionality is also a potential subject for security considerations. However, the three-stage model of 3GPP specifications (see section 2.4) helps with this issue. Addressing security in an inclusive manner in specifications for stages 1 and 2 is a much more attainable goal than adding security considerations into every single stage 3 level specification. Because the stage 3 specifications are built on stage 2, addressing security exhaustively on the latter level ensures that security is also adequately covered in stage 3 specifications.

The service requirements for EPS are captured in [TS22.278]. Clause 9 contains high-level requirements for security and privacy. The service principles applied to the whole 3GPP environment are covered in [TS22.101].

The system architecture of EPS is defined in [TS23.401]. This is a stage 2 level document that gives the overall picture of the whole EPS: what entities are there, their functions, what types of interface exist between different entities, what types of procedure are run over these interfaces, and so on. Because security functions and security procedures are important special cases, there are many references to security in [TS23.401].

The most important component of the EPS architecture is the LTE radio network, E-UTRAN. The stage 2 description of E-UTRAN is given in [TS36.300]. This specification also describes how home base stations fit into the overall architecture. Clause 14 of [TS36.300] is devoted to security.

There are a big number (almost one hundred) of stage 3 specifications for EPS, and for E-UTRAN in particular. Perhaps the most relevant of these are [TS24.301], which defines the non-access stratum procedures, and [TS36.323] and [TS36.331], which define the most relevant access-stratum protocols. The Packet Data Convergence Protocol (PDCP) is specified in [TS36.323] while the Radio Resource Control protocol is specified in [TS36.331]. Refer back to Figures 6.1 and 6.2 for the role of these procedures and protocols in the EPS security architecture.

In order to get some feeling about the complexity of these specifications, it may be interesting to note that, at the time of writing, the most recent versions of these specifications had the following numbers of pages:

- stage 1 EPS requirements [TS22.278]: 25 pages
- stage 2 EPS architecture [TS23.401]: 254 pages

- stage 2 E-UTRAN architecture [TS36.300]: 178 pages
- stage 3 NAS procedures [TS24.301]: 286 pages
- stage 3 PDCP procedures [TS36.323]: 24 pages
- stage 3 RRC procedures [TS36.331]: 233 pages.

In addition to E-UTRAN, there are many other access technologies that may be used by the EPS. Some of these have their specifications maintained in 3GPP, like the GSM and 3G radio technologies. But it is also possible to attach non-3GPP access technologies to the EPS. The stage 2 architecture description of non-3GPP interworking aspects is given in [TS23.402].

6.1.3   Security Specifications for EPS

The main specification for EPS security is [TS33.401]. It contains the stage 2 description of the EPS security architecture, including all EPS security features. It is also the most important single reference for this book. Care has been taken to ensure that the security architecture of [TS33.401] is fully aligned with the system architecture of [TS23.401]. The document [TS33.401] contains also many security requirements. Similarly, care has been taken to ensure that these requirements are aligned with service-related requirements of [TS22.278]. At the time of writing, the most recent version of [TS33.401] contains 100 pages.

The specification [TS33.401] describes the security functions for access to the EPC via E-UTRAN, and it also covers the security architecture for the cases where other 3GPP access technologies (GERAN and UTRAN) have been attached to the EPC. The security aspects of the cases where a non-3GPP access technology (e.g. CDMA) is attached to the EPC are described in another stage 2 security specification, namely [TS33.402]. Again, care has been taken to ensure that [TS33.402] is aligned with the corresponding system level description [TS23.402].

The security architecture for home base stations is specified in [TS33.320]. This specification addresses two cases, access via UTRAN (Home NodeB) and E-UTRAN (Home eNodeB). The security architecture of [TS33.320] is aligned with the overall architecture for home base stations. The latter is described in [TS25.467] for the case of UTRAN and in [TS36.300] for the case of E-UTRAN.

The EPS security architecture obtains many important ingredients from earlier 3GPP security specifications. The 3G authentication and key agreement protocol UMTS AKA, which is the central building block in EPS AKA, is described in the 3G security architecture [TS33.102]. Similarly, the EPS user/subscriber identity confidentiality mechanism is the same as the one described for 3G in [TS33.102]. The USIM application from 3G is usable as such for the EPS. It is described in [TS31.102]. There are also specifications for cryptographic algorithms, originally created for 3G but applicable also for EPS; these are covered in Chapter 10. Security functions for interfaces between two EPS network elements are also very similar to corresponding functions in the case of 3G network elements. Therefore, specifications for network domain security – [TS33.210] and [TS33.310] – are applicable to EPS as well. The key derivation function for EPS purposes is based on the one defined originally for the Generic Bootstrapping Architecture in [TS33.220]. This brief list of earlier specifications that have been, at least to some extent, reused for the specifications of EPS security is not exhaustive; other references are handled in specific sections in the remainder of this book.

In the design process of EPS security, feasibility study reports were started for both LTE-based security architecture and for non-3GPP interworking aspects. The first one, [TR33.821], paved the way for [TS33.401], while [TS33.822] paved the way for [TS33.402]. Some time later, aiming for Release 9, a similar approach was taken for home base station security: [TR33.820] paved the way for [TS33.320].

In an optimal situation, a separate set of reports would have been created for analysis purposes. However, creating the specifications themselves had an obvious priority and it was decided that the technical report [TR33.821] would also serve the purposes of analysis and guideline for EPS and E-UTRAN, documenting why each chosen mechanism addresses certain threats and why some other mechanisms under consideration have been left out of the specifications. A word of warning is needed here: because of the time pressure in finalizing Release 8, and the relatively short time that was allowed for creating Release 9 specifications, it was impossible to bring the technical reports fully in line with the contents of the corresponding technical specifications. This kind of caveat was also included explicitly in both [TR33.821] and [TS33.822].

6.2   Requirements and Features of EPS Security

As explained in the previous section, there are two sources of requirements for EPS security: [TS22.278] and [TS33.401]. The former provides high-level and service-related requirements, including security requirements, while the latter provides implementation and security requirements derived from analysing the threats.

The high-level security requirements of [TS22.278] can be summarized as follows.

(**H-1**) EPS shall provide a high level of security.

(**H-2**) Any security lapse in one access technology must not compromise other accesses.

(**H-3**) EPS should provide protection against threats and attacks.

(**H-4**) EPS shall support authenticity of information between the terminal and the network.

(**H-5**) Appropriate traffic protection measures should be provided.

(**H-6**) EPS shall ensure that unauthorized users cannot establish communications through the system.

The more service-related security requirements of [TS22.278] can be summarized as follows.

(**S-1**) EPS shall allow a network to hide its internal structure from the terminal.

(**S-2**) Security policies shall be under home operator control.

(**S-3**) Security solutions should not interfere with service delivery or handovers in a way noticeable by end-users.

(**S-4**) EPS shall provide support for lawful interception.

(S-5) Rel-99 (or newer) USIM is required for authentication of the user towards EPS.

(S-6) USIM shall not be required for re-authentication in handovers (or other changes) between EPS and other 3GPP systems, unless requested by the operator.

(S-7) EPS shall support IMS emergency calls.

The privacy-related requirements can be summarized as follows.

(P-1) EPS shall provide several appropriate levels of user privacy for communication, location, and identity.

(P-2) Communication contents, origin and destination shall be protected against disclosure to unauthorized parties.

(P-3) EPS shall be able to hide user identities from unauthorized parties.

(P-4) EPS shall be able to hide user location from unauthorized parties, including another party with which the user is communicating.

In the following subsections we go through all standardized security features that are included in the EPS security architecture in order to meet the above requirements. For each feature, there are also more detailed requirements associated with it.

6.2.1 Threats against EPS

There are many security threats associated with communication in general. Most of these are also of concern for EPS. In addition, there are EPS-specific threats that stem from the particularities of the EPS architecture, trust model, characteristics of radio interface, and so on. Security threats for EPS are included in [TR33.821]. We do not go through all of these threats in this book. Instead, we just list here the broader categories of threats seen as relevant to EPS, and give examples of threats in each category.

- *Threats against user identity*. These are already explicitly addressed by requirements P-1 and P-3 above.
- *Other threats against privacy*. These are explicitly addressed by the privacy requirements above.
- *Threats of UE tracking*. Examples are tracking a user based on an IP address that could potentially be linked to an IMSI or another identity, or tracking a user based on handover signalling messages.
- *Threats related to handovers*. An example is forcing a handover to a compromised base station by a powerful signal.
- *Threats related to base stations and last-mile transport links*. Examples are the threat of injecting packets directly into the last-mile transport link, or the threat of physical compromise of base stations in vulnerable locations.
- *Threats related to multicast or broadcast signalling*. An example is broadcasting false system information that would prevent proper functioning of the network.

- *Threats related to denial of service*. Examples are radio jamming or launching a distributed attack from many UEs towards certain parts of the network or DoS attacks against other UEs.
- *Threats of misusing network services*. Examples are flooding the network from inside the network by compromised elements, or from the outside from the Internet.
- *Threats against the radio protocols*. An example is faking or modifying the first radio connection establishment messages from the UE side.
- *Threats related to mobility management*. An example is the threat of disclosure of sensitive data about users' locations.
- *Threats against manipulation of control plane data*. These are addressed by requirements H-4, H-5 and H-6 above.
- *Threats of unauthorized access to the network*. These threats are already explicitly addressed by the requirement H-6 above.

As can be seen from the list, several of the threats are already addressed by the high-level requirements or privacy requirements listed above. Most of the other threats are addressed by more specific requirements. However, there is one type of threat that is difficult to address completely. This is the denial-of-service type of threat (DoS) against the network. Indeed, it is extremely difficult to find logical countermeasures against radio jamming, for example. Of course, it is difficult to launch a radio jamming attack without exposing oneself to the risk of getting caught; the source of disturbing radio traffic can usually be located by physical means. These types of physical means are out of scope of the EPS security architecture, but the idea of radio jamming attack is useful in determining what type of logical DoS attacks are worth protecting against.

The line of thinking is roughly as follows. If there is a logical DoS attack whose impact is smaller than that of a radio jamming attack, then there is no point in adding specific countermeasures against such an attack, even when the cost of such countermeasures would be relatively small. An example of this type of attack could be flooding the radio waves by fake requests for radio connection establishment. What is common in this attack and the radio jamming attack is that the network recovers the normal functionality as soon as the attacker either stops jamming or stops flooding the network with fake requests.

The guiding principle in finding protection against DoS attacks in the EPS context (and more widely in other 3GPP contexts) has been to focus on DoS attacks that have a persistent nature: there is a longer-standing negative impact on the network functions even after the attacker has gone away after making its malicious actions.

6.2.2 EPS Security Features

This subsection lists the security features provided by EPS security architecture. Some of the crucial security features came along with the security design for the LTE architecture. These design decisions are explained in more detail in section 6.3. For most of these features, a more detailed description of the feature is given in the remaining chapters of the book.

Confidentiality of the user and device identities

This feature addresses privacy requirements P-1 and P-3. The purpose of the feature is to prevent eavesdroppers from getting information to identify the communicating parties. There are two different identities involved. The subscriber identity IMSI is stored in the UICC. The device identity, which comes in two variants – International Mobile Equipment Identity (IMEI) and the IMEI and Software Version number (IMEISV) – is stored in the ME, as explained in Chapter 7. There are no straightforward ways of generally linking any of these identities to the identity of the actual person. On the other hand, as a phone and a UICC are used for a long time, a person may be identified by any of these identities during this time once the link to the person has been established.

This feature is copied from 3G and GSM security. The details of the mechanism are defined in [TS33.102]. It is also discussed in sections 3.3 and 4.2 of this book. For the device confidentiality there are some enhancements created for EPS: the device identity is not sent to the network before security measures for traffic protection have been activated.

Authentication between the UE and the network

This feature addresses the high-level requirements H-2, H-4 and H-6. The purpose of the feature is to verify the identities of the communicating parties. This is a cornerstone of the correct functioning of the whole system because, without authentication, it would be impossible to securely connect users to each other. The feature provides also the possibility for the UE to verify the identity of the network that it is connected to.

This feature is also mainly copied from the 3G security architecture – see [TS33.102] and section 4.2. The authentication of subscribers is already present for GSM – see [TS43.020] and section 3.3. There is an enhancement property in EPS authentication that provides means for the UE to directly verify the serving network identity. 3G authentication only provides assurance that the serving network is authorized by the home network to serve the user. This enhancement partially addresses the requirement H-2.

There is another important security function tightly integrated with authentication: in addition to verifying identities of each other the terminal and the network also agree shared secret keys that can be subsequently used for the features of confidentiality and integrity protection of data. Chapter 7 is devoted to the feature of EPS Authentication and Key Agreement (EPS AKA).

Confidentiality of user and signalling data

This feature addresses the high-level requirement H-5 and privacy requirements P-1 and P-2. The purpose of the feature is to encrypt (another word is cipher) the digital communication in order to make it incomprehensible to eavesdroppers, especially on the radio interface. A similar feature exists in the 3G security architecture. However, the different system architecture of EPS, compared to that of 3G, imposes differences also for this feature. Most notably, the endpoint of the encryption on the network side (for user data and radio network signalling data) is in the base station for EPS, while it is the radio network controller (RNC) in 3G. The reason for this change is explained in section 6.3. Another big change is that an additional confidentiality protection mechanism is introduced for signalling between the UE and the core network. Similar to the situation with GSM and 3G, providing data confidentiality is optional for the network operator. This feature is described in detail in Chapters 8 and 10.

Integrity of signalling data

This feature addresses the high-level requirements H-4, H-5 and H-6. The purpose of this feature is to verify the authenticity of each signalling message separately; that is, to ensure that the signalling message is not modified in transit but instead received in exactly the same form in which it has been sent. As in 3G, no integrity protection is provided for user data, for the same reasons. In both cases, it was felt that the risk of successfully exploiting any modification of encrypted user data sent over the air was relatively small, and the overhead added by integrity protection would have been significant, especially for services with short packet sizes, such as voice. Furthermore, the security gain provided by the 'proof-of-origin' part of integrity protection (see Chapter 2) would have been relatively small unless integrity protection was provided in true end-to-end fashion between the endpoints of the user data communication (e.g. between two terminals). Supporting this would have required major extensions in the key management.

Similar to the confidentiality feature above, certain changes have been necessary when compared to the corresponding feature in 3G. This feature is also covered in Chapter 8 and 10.

Visibility and configurability of security

This feature is present already in both 3G and GSM. The purpose is to give the user some options to benefit from information about the security features. For the visibility purpose, there is a *ciphering indicator* in the UE that shows whether the feature of data confidentiality is applied by the network or not. For the configurability purpose, the user has the option of applying password (PIN) based access control to the UICC.

Platform security of the eNodeB

The importance of platform security for base stations (i.e. evolved NodeBs) is emphasized in EPS for two reasons:

- eNodeB is a termination point for major EPS security mechanisms;
- eNodeBs are expected to be installed in more vulnerable locations than 3G base stations when EPS is deployed.

Similar trends are also present in the most recent evolution of 3G technology. The High Speed Packet Access (HSPA) architecture contains an option where RNC and node B functionalities are in the same node. Also, the concept of home base station applies to both UTRAN and E-UTRAN base stations. It is clear that base stations in people's homes (for example) are in a more vulnerable location than macro cell base stations controlled by the operator. In order to address these issues, requirements on the secure implementation of eNodeBs are included in [TS33.401]. They are described in more detail in section 6.4. For the case of home base stations, there is a complete security specification in [TS33.320]. Home base station specific security features are described in detail in Chapter 13.

Lawful interception

This feature addresses the service requirement S-4. The purpose of the feature is to provide access for law enforcement to the content of communications and related information, such as identities of the communicating parties and times of the communications. Lawful interception (LI) has a special role among the security features because it constrains the choice of the other security mechanisms in the system. There is a certain contradiction between the service requirement S-4 of providing lawful interception and the privacy requirements. In this sense, the interception goes against the other security features and should rather be seen as a controlled exception to the other security features.

The conditions under which the lawful interception can be activated by the law enforcement side are out of scope of the 3GPP specifications. They are a matter of the legislation of the country where the interception is to be done. A typical way is to require a court order before the lawful interception can be started.

Lawful interception is one of the EPS security features that are present also for 3G and GSM. The 3GPP specifications for lawful interception have been arranged in such a manner that, for every new feature, the existing lawful interception specifications are extended to cover the arrangements needed for providing lawful interception aspects for the new feature. This is a handy practice from a referencing point of view. The stage 1 specification [TS33.106] contains lawful interception requirements for all 3GPP features, the stage 2 specification [TS33.107] contains the lawful interception architecture, and the stage 3 specification [TS33.108] contains the bit-level description of the interface by which the needed information could be handed over to the law enforcement side.

The LTE radio technology as such does not bring many new issues from the lawful interception point of view. The information that falls into the LI scope is still roughly the same as for GSM and 3G.

Emergency calls

This is another feature that, in a certain sense, interferes with other security features. In some countries, the legislation requires that emergency calls should be possible even in cases where security measures mandatory for normal calls are not present. An example case is when there is no UICC inserted in the terminal. The feature addresses the service requirement S-7. Special arrangements done for emergency calls, and emergency sessions in general, are described in sections 8.6 and 13.6.

Interworking security

This feature is rather an enabler for the other security features but that does not make it less important than the other features. The purpose is to ensure that security holes do not appear in situations where there is a change from one system to another, such as when moving from EPS to 3G or vice versa. Equally important are situations inside EPS where coordination between several network entities is needed, possibly being under different administrative domains, such as handovers between two different operator networks. The features of data confidentiality and data integrity are based on the existence of shared secret keys. In the interworking situations a big part of this feature is in key management, ensuring that the correct keys are in the correct places at the correct time. Security for transitions and mobility inside EPS is described in Chapter 9. The interworking security with other systems, including both other 3GPP systems and non-3GPP systems, is described in Chapter 11.

Network domain security (NDS)

This feature is inherited from 3G. Its purpose is to protect the traffic between network elements. Mutual authentication between the communicating parties, data confidentiality and integrity are all ingredients of this feature. The details of the feature are described in [TS33.210] and [TS33.310]; see also sections 4.5, 8.4 and 8.5 of this book.

IMS security for voice over LTE

The EPS is an IP packet-based system. This implies that the voice calls have to be provided by some means other than what has been customary for GSM and 3G; that is, other than by a circuit-switched solution. There is a ready-made solution for this issue already in Release 5 of 3GPP, namely the IP Multimedia Subsystem (IMS) which is based on the SIP protocol [RFC3261]. The IMS is an overlay system that works for any access technology, including LTE.

The fact that IMS is independent of the access technology has implications for security: there have to be security features that guarantee correct functioning of IMS regardless of the security functions that the access technology (potentially) provides. The 3GPP security specification for IMS is [TS33.203]. Chapter 12 addresses the IMS-based security features for voice over LTE.

6.2.3   *How the Features Meet the Requirements*

An important part of the design of any system is the comparison of the included features against the requirements that guide the design. This is especially true for the design of a security architecture because leaving any requirement unaddressed may potentially undermine the whole purpose of the design, which is providing a secure system. It is also possible to leave some requirements unaddressed but this should, of course, rather be a conscious decision than an oversight. In the case of a security requirement, it may be that the requirement has been added because of a remote threat that has minor consequences from the system point of view. If only countermeasures can be found that would add significant complexity and cost to the system, then it may be decided that the cost-optimal solution is to leave the requirement unaddressed.

Table 6.1 summarizes how the security features listed in this chapter address the requirements of [TS22.278]. The features are listed in the same order as above, 'LI' is lawful interception, 'EC' refers to emergency calls, 'I/W' to interworking security, and so on.

In the table a cell has been marked whenever the particular feature is relevant for meeting the particular requirement. No distinction has been made on whether the feature addresses the requirement completely or only partly. Some of these connections are also rather indirect. For

Table 6.1 Requirements versus features

	ID conf.	Auth.	Data conf.	Data int.	Vis.	eNB	LI	EC	I/W	NDS	IMS
H-1	×	×	×	×	×	×			×	×	×
H-2		×							×	×	
H-3	×	×	×	×	×	×			×	×	×
H-4		×		×							×
H-5			×	×						×	×
H-6		×		×							×
S-1										×	×
S-2		×								×	×
S-3	×	×	×	×	×	×	×	×	×	×	×
S-4							×				×
S-5		×									
S-6									×		
S-7								×			×
P-1	×		×							×	
P-2			×							×	
P-3	×									×	
P-4	×		×							×	

example, it is marked that network domain security (NDS) addresses the service requirement S-1 of hiding the network internal structure from the terminal. The connection is quite indirect. On the one hand, network domain security puts a security gateway on the border of one network and therefore enables hiding of the network structure behind it from other networks. On the other hand, network domain security protects also network internal interfaces and therefore an eavesdropper on such an interface does not learn much from studying the traffic.

There are a couple of requirements that do not seem to be very well addressed based on Table 6.1. The first such requirement is S-1, which we have just discussed. The main protection against finding out the network internal structure is provided by the architecture of the system. Using suitable practices for selecting addresses and identities for the network elements could mitigate attempts at learning the network structure. The flat structure of the EPS network makes this kind of hiding more difficult.

Another requirement that is addressed only indirectly is the privacy requirement P-4. The network knows the location of terminals attached to it with rather good accuracy. Therefore, there is an obvious danger of leaking this information somehow out to other users of the system or to outsiders. Also in this case, the main protection comes from an appropriate design of protocols and procedures in such way that information about one user's whereabouts does not affect content or context of procedures relating to other users, even in cases where two users are communicating with each other. In addition, specific care is taken on protecting the location-related information on its way from the base station further back to the network.

It is important to note here that there are a number of location-based services that are based on the idea of providing the location of one user to another. Examples are finding friends in the locality, monitoring children and fleet management. But these services are provided on the application layer and are, presumably, dependent on the consent of the users involved.

6.3 Design Decisions for EPS Security

The preceding section presented the requirements placed on EPS security and their reasons. This section highlights a few of the major design decisions that 3GPP took when deciding how to satisfy the requirements. These decisions led to the EPS security architecture being quite different from the 3G security architecture.

The allocation of security functions to functional entities and protocol layers is a fundamental task to be performed when designing a security architecture. Let us briefly recapitulate the major elements of the 3G security architecture, as described in Chapter 4, and then explain why the EPS security architecture had to be extended compared to the 3G security architecture. As stated earlier, in view of the success of the 3G security architecture, 3GPP endeavoured to deviate from it only where it was made necessary by differences in the overall EPS architecture compared with the overall 3G architecture, and by differences in the security requirements (due to, for example, changing business requirements or deployment scenarios).

Permanent security association

The 3G security architecture is anchored in a permanent security association between a USIM application on a UICC in the UE and the Authentication Centre (AuC) in the Home Location Register (HLR). The corresponding permanent key is never visible outside the security module and the AuC. This permanent key is used in the Authentication and Key Agreement (AKA) protocol. This principle of a permanent security association is kept in EPS.

Interfaces in UE and HSS/HLR

The interface between the ME on the one side and the UICC and the USIM on the other is fully standardized to allow interoperability between MEs, produced by handset vendors, producing MEs, and UICCs with USIMs, produced by smart card vendors. The standardization of this interface also ensures that the lifetimes of handsets and smart cards are completely decoupled, which is an important business consideration. The picture is different on the HLR side: here it was not felt necessary to standardize the interface between AuC and the (rest of) the HLR, rather the AuC is considered part of the HLR. These principles are kept in EPS, with the obvious modification that an HSS is used instead of an HLR.

Reuse of 3G USIMs

As we will see below, the authentication and key agreement protocol in EPS, called EPS AKA, has evolved from UMTS AKA, which is used in 3G. Although the differences are not very big, they exist and raise the valid question, discussed in 3GPP, whether special support from an evolved USIM is needed, or desirable, for EPS AKA. The decision in 3GPP was that EPS AKA must be designed in such a way that the reuse of USIMs as used in 3G handsets (i.e. USIMs according to Release 99 specifications) is possible. There is an overwhelming business case that can be made to support this decision: a very large number of 3G USIMs have already been shipped to subscribers, and it would incur significant cost to operators if they had to exchange all these 3G USIMs for EPS-enabled ones before subscribers could enjoy EPS services. Furthermore, when a 3G USIM can be reused for EPS then all a subscriber needs to do for being able to use EPS is buy a new handset and insert his old 3G USIM (provided the conditions of his subscription are compatible with it).

Nevertheless, security advantages of allocating certain security functions and keys to an EPS-enabled USIM, and not the ME, were cited in the discussion in 3GPP; and indeed such advantages exist. The main advantage is that certain cryptographic keys are not available in the ME, but only in the more secure environment of the UICC, when the UE is in the deregistered state. However, while in registered state, these keys must be available in the ME anyway, so the advantage of storing them on the USIM is quite limited.

So, 3GPP had to trade off a clear business advantage against a moderate gain in security. The 3GPP decision was that, while the reuse of 3G USIMs had to be possible, EPS-enabled USIMs were also specified. In this way, operators are given the possibility to perform the trade-off between business requirements and security according to their particular requirements. We also mention here that there are enhancements to the USIM for EPS that are not related to security.

This approach is quite similar to the one taken in the introduction of 3G security. Although the differences between GSM authentication and UMTS AKA are much more substantial than the ones between UMTS AKA and EPS AKA, at the time of 3G standardization it was decided to allow access to 3G radio access networks using 2G security modules (GSM SIMs).

No reuse of 2G SIMs in EPS

We have seen now that both 3G and EPS allow the reuse of the security modules of the respective previous system generation. However, 3GPP decided that it was not allowed for EPS to go back even two generations, so 3GPP forbade the reuse of GSM SIMs for access to LTE radio networks.

Obviously, with GSM SIMs, only the GSM authentication and key agreement protocol is possible; and the security disadvantages of GSM AKA over EPS AKA are quite significant. On the other hand, the business case for reusing GSM SIMs for LTE radio access networks is much weaker now than the business case for reusing GSM SIMs for 3G was ten years ago (when 3G was introduced), because now significant numbers of USIMs are in the field.

Delegated authentication

In both GSM and 3G, it is the VLR (for the circuit-switched domain) and the SGSN (for the packet-switched domain), respectively, not the HLR, that run the actual authentication procedure with the UE. The VLR or the SGSN fetches authentication vectors from the HLR, and, at some later time chosen at the discretion of the VLR or the SGSN, the VLR or the SGSN sends an authentication request to the UE and checks the correctness of the response. The VLR or the SGSN is also responsible for the distribution of the session keys to the endpoints of protection. In this sense, the HLR delegates the control of authentication checking and session key distribution to the VLR or the SGSN. This implies that, in the roaming case, the home network delegates these tasks even to the visited network.

3GPP decided to keep this principle also for EPS. This means that the MME requests authentication vectors from the HSS, checks the authentication response and distributes session keys to the endpoints of cryptographic protection. An advantage of this decision is that the same model of interaction with the HSS as in 3G can be maintained and that the HSS need not keep state during the run of an authentication protocol with the user. It also implies that the EAP authentication framework (see section 5.1) does not apply.

This delegation of an important security task from the home network to the visited network also implies a certain amount of trust of the home network in the visited network. Any risks arising from the (unlikely) case that there should be a breach of this trust are mitigated in EPS by a new feature enhancing the AKA protocol, namely cryptographic network separation (discussed below).

Reuse of the fundamental elements of UMTS AKA

3GPP decided to build on UMTS AKA, which has served 3G security well and has stood up to analysis for ten years now, and enhance it with additional functions only as far as needed. It turned out that only one enhancement was considered necessary, namely cryptographic network separation.

Cryptographic network separation and serving network authentication

This feature limits the effects of any security breach in a network to that network and prevents a spillover of the effects of the breach to other networks. It therefore addresses requirement H-2 from section 6.2. This is achieved by binding any EPS-related cryptographic keys, which leave the HSS, to the identity of the serving network, to which the keys are delivered. It also enables the UE to authenticate the serving network. In 3G, a UE cannot authenticate the serving network but only ascertain that it communicates with a serving network authorized to do so by the UE's home network (see Chapter 4).

It should be mentioned that the principle of cryptographic network separation is strictly adhered to only in authentication procedures. 3GPP decided that keys obtained by one serving network may be forwarded to another serving network in mobility events (handover or idle state mobility) and used there until the next authentication, which then requires new keys bound to the new serving network. This decision is again a trade-off between security and efficiency, in this case the efficiency that results from minimizing the impact on the AuC and reducing delays in mobility events. A more detailed description of this feature can be found in section 7.2.

Termination point for encryption and integrity protection extending from the UE

It is clear for every radio system that the air interface, as the most vulnerable part of the system, needs to be protected by providing confidentiality and, depending on the type of data, also integrity protection. So, as the UE is one endpoint of the air interface, it is clear that the range of this protection extends from the UE. It is less obvious what the network endpoint of this protection should be. This question was answered differently even for the different

3GPP-defined mobile systems, and it turned out to be one of the most crucial security decisions that 3GPP had to take.

In the circuit-switched service of GSM, encryption terminates right at the network termination of the air interface, at the Base Transceiver Station (BTS). The designers of 3G security saw this as one of the weak points of GSM security because the BTS is often placed at an exposed location, and the link to the Base Station Controller (BSC), the next node further up in the network, is an often unprotected microwave link. Therefore, 3GPP decided in 1999 that encryption (and integrity protection, which is not provided in GSM) should extend further back into the network and terminate at the Radio Network Controller (RNC), which was considered to be at a physically secure location and connected to the core network via a secure link.

In GPRS, the 2G packet-switched service, encryption extends even further into the network, namely up to the SGSN. However, this was not done for security reasons, but rather for reasons that had to do with particular characteristics of GPRS [Hillebrand 2001].

The difficulty the designers of EPS security were now facing stemmed from the fact that one of the major overall design goals of EPS was to achieve a flat network hierarchy and dispense with intermediate nodes like an RNC. This means that the Radio Resource Control (RRC) protocol, which terminates in the RNC in 3G systems, now terminates in the eNB in EPS; that is, again right at the edge of the air interface and at an exposed location. But then the protection of RRC messages also has to terminate at the eNB. This is in seeming contradiction to the decision by 3G security designers that such a termination point would constitute a security weakness. The seeming contradiction was resolved in EPS by accepting the priority of having a flat overall architecture, but at the same time acknowledging the particular vulnerability of the eNB and putting (for the first time for a 3GPP-defined network node) stringent platform security requirements on the eNB. These requirements are described in more detail in section 6.4. Once it was established that the eNB would be physically secured there was no fundamental objection any more to terminate also user plane security at the eNB. This decision made protocol design significantly simpler.

On the other hand, Non-Access Stratum (NAS) signalling extends between the UE and the MME, a controller in the core network. While it would have been possible to provide protection for NAS signalling in a hop-by-hop fashion, with one hop extending between the UE and the eNB, and a second hop extending between the eNB and the MME, it was decided to provide protection for NAS signalling end-to-end between the UE and the MME. As NAS signalling is required whenever a user registers to a network, or periodically re-registers, this decision also helps to mitigate any potential remaining security risks of terminating protection for RRC and user plane in the eNB. Furthermore, the NAS security context remains stored in the UE and the MME while the UE is in idle state. This allows NAS signalling to be secured even before the Access Stratum (AS) security extending between the UE and the eNB is set up after the transition from idle state to connected state. However, the decision also comes at a cost: as opposed to GSM and 3G, in EPS we now have different endpoints for protection extending from the UE in the network, namely the eNB and the MME. This is one of the reasons for the more elaborate key hierarchy in EPS compared to 3G.

New key hierarchy in EPS

In GSM and 3G, the key hierarchy is quite simple: there is a permanent key shared between (U)SIM and AuC, and there are session keys Kc and (CK, IK) respectively, which are directly used with the encryption and integrity algorithms. As we will see in section 7.3 in more detail, the key hierarchy in EPS is considerably more elaborate, which can be easily seen already from a mere glance at the key hierarchy diagram in section 7.3. We only mention the main reasons for this new key hierarchy here.

There is a local master key K_{ASME} at the core network level, which is distributed from the HSS to the MME, and between MMEs, and is also generated in the ME. The introduction of this key became necessary through the decision to reuse 3G USIMs, and hence obtain (CK, IK) from the USIM, and the new requirement of cryptographic network separation, which implies a binding of keys to the serving network identity, a property that is not fulfilled by (CK, IK). The introduction of this local master key K_{ASME} has another very desirable effect, namely that it reduces the frequency with which authentication vectors need to be fetched from the HSS. K_{ASME} is not directly used in encryption and integrity algorithms, so it does not need to be renewed as often as (CK, IK) in 3G. K_{ASME} is less exposed also because it is never transferred to the radio access network – it remains in the core network.

There is another intermediate key at the radio access network level, called K_{eNB}, which is distributed to the serving eNB from the MME. Its introduction was primarily motivated by the fact that keys used for RRC control plane and user plane protection in the eNB are bound to certain parameters specific to an individual eNB and that handovers between eNBs should not necessarily involve the MME before the completion of the handover procedure (the so-called X2 handover described in Chapter 9). Therefore, a new level of key hierarchy was required for an intermediate key, which was for use at the eNB level, but was not yet bound to the parameters specific to an individual eNB and hence could be used in handovers without MME involvement. The details of how this is exactly done are tricky. A part of the complication arises from another security requirement introduced to limit the consequences of a security breach in an eNB, namely key separation in handovers, discussed below.

At the bottom of the key hierarchy, there are the keys directly used with the encryption or integrity protection algorithms to protect the NAS, RRC, or user plane protocols.

Key separation in handovers

For efficiency reasons, there are handover preparations that do not involve the core network. For these X2 handovers, the source eNB provides a key of type K_{eNB} to the target eNB for use after the handover. If the K_{eNB} was handed over unchanged then the target eNB would know which K_{eNB} was used by the source eNB. In order to prevent this, not the K_{eNB} used at the source eNB itself, but rather the image of a one-way function applied to K_{eNB}, is forwarded to the next eNB. This ensures so-called backward key separation in handover.

But backward key separation solves only one part of the problem: for a fast-moving user, there may be a whole chain of handovers, and, if the image of a one-way function applied to K_{eNB} was forwarded to the next eNB in this chain of handovers, then all eNBs in that chain would know the K_{eNB} used further downstream in that chain, and one compromised eNB in that chain would put all other downstream eNBs in the chain at risk (although, by the property of backward key separation, not the eNBs upstream from it in the chain, the eNBs the UE visited before the compromised eNB). In order to prevent this, the requirement of forward key separation in handovers (also called 'forward security' in [TS33.401]) was introduced to ensure that the MME provides a fresh key for the next hop immediately after handover if it was not possible during the handover. Details can be found in Chapter 9.

It should be noted here that the terms 'forward key separation', 'backward key separation' and 'forward security' used in this book and in 3GPP specifications are somewhat at odds with similar terms used in other parts of the security literature. In particular, the term 'perfect forward secrecy' [Menezes *et al.* 1996] denotes a property more akin to 'backward key separation' as defined here.

Homogeneous security concept for heterogeneous access networks

EPS provides a framework for connecting heterogeneous access networks to a single core network, the Evolved Packet Core (EPC). These include not only access networks defined by 3GPP, namely GERAN, UTRAN and LTE, but also access networks defined by other standardization bodies, such as cdma2000®HRPD defined by [3GPP2] and WiMAX defined by [WiMAX], and possibly many more to be defined in the future. Also, there is no requirement to restrict access to the EPC only to wireless access networks.

As it would be technically difficult and inefficient to design different procedures for all these different access networks, a framework had to be found that could accommodate the various access technologies. For authentication, this framework is provided by EAP, the Extensible Authentication Protocol [RFC3748]. EAP allows carrying authentication messages over a variety of transports and, thus, makes authentication independent of the particular nature of the access networks. For access networks that are deemed untrusted by the EPC, EAP is combined with the use of IKEv2 [RFC4306] and IPsec ESP [RFC4303] to provide protection against any potential weaknesses in the access network security. Details can be found in Chapter 11.

6.4   Platform Security for Base Stations

6.4.1   *General Security Considerations*

[TS33.401] does not consider common security principles for all network element platforms, but only handles features specific to the evolved NodeBs (eNBs). But it should be mentioned here that common 'good engineering practices' for security design are necessary for all network elements. This includes hardening of the elements (e.g. disabling of unused services and network ports) and secure software (SW) design to avoid as much as possible vulnerabilities caused by design or implementation flaws. If third-party software is used, such as open-source SW and/or libraries provided with compilers, such SW must comply with the standards of secure SW design.

6.4.2   *Specification of Platform Security*

As described in the preceding subsection, it was a design decision for EPS that the RRC control and user plane security should terminate in the base station. Additionally, the EPS architecture allows locating the eNB outside the security domain(s) of the mobile network operator, in physically insecure locations. These two facts together create the situation, as opposed to former solutions, that sensitive communication and configuration data is available at locations outside conventional security domains. Thus, for the first time in 3GPP standardization, specific requirements for platform security are addressed in a related specification. Still, these new requirements will not eliminate the need for, nor detract from the importance of, the good engineering practices mentioned above.

As in EPS only the base stations can be placed in an exposed location, the remainder of this section only handles platform security issues applicable to eNBs. All other network elements used in EPS are still located within the security domain of the operator, and thus are not subject to standardized platform security requirements.

6.4.3 Exposed Position and Threats

Attacks against base stations may happen locally or remotely. By performing a local attack an attacker may, for example, get physical access to the base station and interfere with the internal elements or use a direct connection to the base station antenna and network interfaces to intercept or inject data. Remotely an attacker may manipulate a possibly insecure backhaul link connection between a base station and the Security Gateway (SEG) of the operator network or the direct connections between different base stations. Attacks from inside the operator network via the backhaul link are not considered a threat for platform security, as for EPS the assumption still holds that, from a standardization point of view, the security inside the security domain of the operator is left to operator policies.

There may also be attacks on the physical implementation, such as by direct wire-tapping of internal lines used for eavesdropping on data, or for injection of malicious SW or configuration parameters. Such physical attacks require the physical presence of the intruder, at least during the preparation of the attack.

An entirely different category of attacks relies on pure SW methods to alter the functionality of the platform itself, either by causing the base station to malfunction resulting in denial-of-service, or by targeting the attacks so that the attacker gains partial or full control of the base station. These attacks may be performed locally or remotely. Mostly such attacks are targeted to vulnerabilities within the platform SW (e.g. within the operating system), or the communication protocol stacks, or the application layer SW. They may comprise addition or modification of SW, or modification of operational parameters.

A third attack category focuses on the intentions of the attacker. In some cases the attacker may want to get information without further interfering with the platform operations. Such 'passive attacks' may be hard to detect, as the platform functionality is not altered. Attacks of this kind may be targeted at eavesdropping on long-term keys (e.g. keys used in authenticating the base station), on medium-term user-specific keys (e.g. the intermediate key K_{eNB} and the Next Hop parameter NH), or on short-term session keys used for protecting the backhaul and air interfaces. Also, the user plane traffic is available in cleartext within the base station and thus the confidentiality may be at stake. On the other hand, if the attacker wants to change the behaviour of the platform (by, for example, pushing it to transmit with higher power than configured by the management system), or to deny services to certain users, then such attacks may also be detected based on the functional behaviour of the base station.

6.4.4 Security Requirements

The above-mentioned threats led to security requirements for the base station platforms as specified in clause 5.3 of [TS33.401]. This clause states requirements on the platform and

communications security of the base stations. While communications-related security requirements are handled in section 8.4, here we deal with the platform-related security requirements.
They are categorized as follows.

Base station setup and configuration

These requirements deal with the SW and the configuration data used within the base station.
All SW loaded into the base station, either locally in the factory or on-site, or remotely
via an Operations and Management (O&M) system, must be authorized for use in the base
station. The text in the specification does not explicitly state who the responsible authority
is, but both the manufacturer of the base station and the mobile network operator should be
considered here. Only the manufacturer can ensure the correct operation of the base station in
terms of its software, and on the other hand only the operator can determine the settings of
many operational configuration parameters like transmission frequencies and power levels. A
prerequisite for authorized SW installation is the integrity-protected transfer of the SW to the
base station, as otherwise any authorization does not make sense. Also confidentiality of the
SW transfer has to be ensured, so as to not disclose the SW to unauthorized third parties having
access to the network used for the backhaul link between the base station and the operator
network. To ensure that only authorized SW is loaded and executed in the base station, a
secure environment within the base station is required for enforcement. This is described
later.

Key management inside the base station

All keys used for providing confidentiality and integrity protection within the base station
shall be secured. Most keys are used only inside the base station; they shall never leave the
secure environment within the platform. This applies to long-term keys, such as the secrets
used for authenticating the base station to the operator network. Also, the session keys used for
securing subscriber-specific sessions must remain within the secure environment. This applies
to keys used for RRC signalling security and for the encryption and decryption of the user
plane data. Only when the specified operation of EPS requires the transfer of keys, such as the
K_{eNB}^* transferred in X2-handovers, are such keys allowed to leave the secure environment.
For securing such keys during transfer into and out of the secure environment, see below.

Handling of user and control plane data

All ciphering and deciphering of user and control plane data shall take place inside the
secure environment of the base station where also the related keys are stored. For the control
plane, also the control procedures, the message integrity checks, and the replay protection
procedures have to be executed within the secure environment. NAS signalling is not affected
by this requirement, as the base station forwards only protected NAS messages, without any
interpretation of them. The transfer of unencrypted user plane data within the base station
between the Uu and S1/X2 reference points is not explicitly mentioned in the specification,
but it is obvious that also this transport has to take place within the secure environment or has

to be secured by other (e.g. cryptographical) means, otherwise the protection of user plane traffic would be incomplete.

Secure environment

The text in clause 5.3 of [TS33.401] mentions the term 'secure environment' and describes some of its features. Nevertheless it does not give a detailed description of a secure environment and does not enforce certain mechanisms related to it, like secure boot. Instead it relies on a state-of-the-art interpretation of the term and leaves the details to the implementer. Only some properties are explicitly mentioned – the support of the secure environment given to the boot process of the base station, the storage of sensitive data, and the functionality required for cryptographic security functions.

Based on these properties, it is obvious that the secure environment must contain a root of trust, which is either unalterable, or can only be changed by applying mechanisms with a high security level. This root of trust is then used for SW integrity checking during SW download and/or boot processes. The specification does not require that all of the base station SW must be running within the secure environment. Using mechanisms as described by [TCG Mobile Phone Work Group 2008] as an example, the secure environment may be used to measure all SW loaded during the boot process, and enforce that only authorized SW be executed.

An additional difficulty with specifying a secure environment in a generic way is that an attacker always looks for the weakest point in the implementation. As the specification did not want to require a certain implementation of the base station platform, any manufacturer is free to, for instance, do a partitioning of the functionality according to their own design, and to use any SW as long as this SW fulfils the security requirements. Thus a general risk analysis for the base station platform is not possible, so each manufacturer has to provide their own security analysis of their respective security design.

Physically securing the base station is not explicitly mentioned in the specification. But the requirements clearly do not only apply to SW-based attacks on the base station, but also to any physical attack. This means that physical tampering with the base station platform has to be prevented, be it for probing of circuits within the platform for eavesdropping, or for unauthorized modification of SW and data. On the other hand, it is commercially not viable to raise the physical security of the base stations above a certain level, as otherwise both the capital expenditure for manufacturing as well as the operational expenditure for maintenance would exceed acceptable levels for a network element deployed in huge numbers. This leaves the manufacturer of the base station with the task to define suitable platform security architectures and to assure their customers, the operators, of the security level of their implementations and of conformance to the specifications. Requiring evaluation of such architecture according to existing standards (e.g. Security Requirements for Cryptographic Modules [ISO/IEC 19790], the internationalised version of FIPS Publication 140-2 [FIPS 140-2]) was discussed during standardization but was dismissed. Such standards are mainly intended for specialized security subsystems, including crypto co-processors or modules, but not for complete functional systems like base stations that comprise a secure environment as a subsystem only. In addition, each new hardware or SW version would require a new evaluation, which would also increase recurring costs and time delays beyond acceptable ranges.

Extensions for special types of base station

Clause 5.3 of [TS33.401] states that the security requirements are valid for all types of base station. Specifications for specific base stations may not weaken these requirements, but only have more stringent requirements. The first type of such a specific base station is the Home eNodeB (HeNB). The security aspects of HeNBs are described in [TS33.320]. Within this book, HeNBs are handled in Chapter 13.

7

EPS Authentication and Key Agreement

This chapter describes how users are identified and authenticated for network access in EPS. Section 7.1 introduces the means to identify subscribers and terminals, and the mechanisms to protect the related identities. Section 7.2 then provides a detailed presentation of EPS AKA, the protocol used in EPS to authenticate subscribers and agree a local master key. Further keys are then derived from this local master key to protect signalling and user traffic over various interfaces between the UE and the network. The complete EPS key hierarchy resulting from this derivation process is described in section 7.3. In addition to keys, other security-related parameters need to be shared between two entities running a security protocol between them. These parameters, together with the keys, form a security context, and the various security contexts used in EPS are described in section 7.4.

7.1 Identification

We first describe the means to identify subscribers and terminals in EPS and explain the uses of the corresponding identities. We then proceed to describe the identity confidentiality features, which help to protect the user's privacy. These identities are specified in [TS23.003].

User identification

GSM, 3G and EPS all use the same type of permanent subscriber identity, the International Mobile Subscriber Identity (IMSI), to uniquely identify a subscriber. The IMSI is composed of three parts:

- The Mobile Country Code (MCC) identifies the country of domicile of the mobile subscriber.
- The Mobile Network Code (MNC) identifies the home network of the mobile subscriber in that country.
- The Mobile Subscriber Identification Number (MSIN) identifies the mobile subscriber within a home network.

LTE Security   Dan Forsberg, Günther Horn, Wolf-Dietrich Moeller, and Valtteri Niemi
© 2010 John Wiley & Sons, Ltd

The IMSI is crucial for EPS security, as it is for GSM and 3G security, because the permanent authentication key K used in EPS AKA, the Authentication and Key Agreement protocol used in EPS, is identified by the IMSI. The permanent authentication key K is stored in the Authentication Centre and in the USIM, but nowhere else.

There are a number of temporary identities associated with an IMSI in EPS, notably the GUTI and the C-RNTI. The GUTI is allocated for the purposes of user identity confidentiality. The C-RNTI [TS36.331] is used to identify a user equipment having a Radio Resource Control (RRC) connection within a cell. The only use of the C-RNTI in security procedures is with handover preparation (see section 9.4.4).

Terminal identification

GSM, 3G and EPS all use the same type of permanent terminal identity, the International Mobile Equipment Identity (IMEI). In all systems, IMEI is sometimes accompanied by a Software Version Number (SV) in which case the identity is called IMEISV. Because of possible software upgrades in the terminal, the SV may change during its lifetime, while IMEI remains the same.

7.1.1   User Identity Confidentiality

The EPS protects the confidentiality of the user identity against passive attacks in pretty much the same way as do GSM and 3G. In each of these systems, the network assigns the user a temporary identity sent in a message protected from eavesdropping. It is the purpose of this temporary identity to provide an unambiguous identification of the UE that does not reveal the user's permanent identity – the IMSI. The temporary identity can be used by the network and the UE to establish the permanent user identity during signalling between them.

The temporary user identity used in EPS is called Globally Unique Temporary UE Identity (GUTI). It is a bit different in structure from the TMSI used as a temporary user identity in the circuit-switching domain of GSM and 3G, and the P-TMSI used as a temporary user identity in the packet-switching domain of GSM and 3G.

The GUTI has two main components:

* the GUMMEI, which globally uniquely identifies the MME that allocated the GUTI;
* the M-TMSI, which uniquely identifies the UE within the MME that allocated the GUTI.

The GUMMEI (Globally Unique MME Identifier) is constructed from the MCC, the MNC and the MME Identifier (MMEI).

For certain procedures, such as paging and service requests, a shortened version of the GUTI is used, namely the S-TMSI. The S-TMSI consists of the M-TMSI and a part of the MMEI. The S-TMSI is to enable more efficient radio signalling procedures.

The MME may assign a GUTI to the UE in an Attach Accept message or in a Tracking Area Update Accept message. The MME may also assign a GUTI in a separate GUTI Reallocation procedure [TS23.401]. In each case, the MME sends the GUTI only after the protection for non-access stratum signalling has been enabled (see Chapter 8). If the network supports signalling confidentiality then an attacker listening on the link between the MME and the UE cannot read the GUTI, and so cannot associate the GUTI with the IMSI or an earlier GUTI sent in a message by the UE. This mechanism protects the confidentiality of the user identity against passive attacks (eavesdropping). It also prevents tracking a user by observing temporary identities consecutively assigned to the same user. If the network does not support signalling confidentiality then the user identity confidentiality protection is weakened as well because an eavesdropper can observe the relation between an IMSI sent over the air and a GUTI allocated by the network, or between two consecutive GUTIs.

As for GSM and 3G, there is no user identity confidentiality protection against active attacks; and the reason is again the same. In an active attack, an attacker would use a device known as an 'IMSI catcher', which incorporates a false base station, for sending an Identity Request message to the UE. The UE would then invariably respond with the IMSI. This Identity Request procedure is needed to recover from cases where the network lost the association between the temporary user identity and the IMSI, for example through a crash of the MME. Without such a recovery mechanism, the user could be permanently locked out of the system. 3GPP again discussed means to allow for recovery from such a situation while providing better protection against active attacks, but the only effective means seemed to be the use by the UE of public key certificates. For roaming cases, where the MME resides in another operator's network, this would assume the existence of a public key infrastructure spanning across all operators with mutual roaming agreements. While this would be possible in theory, 3GPP felt that mandating such an infrastructure would be too high a price to pay.

7.1.2   Terminal Identity Confidentiality

While the mechanism for protecting the user identity confidentiality in EPS is still pretty much the same as it was in GSM and 3G, there is an improvement in EPS with respect to GSM and 3G regarding the terminal identity confidentiality. In GSM and 3G it is possible that the network requests the terminal identity at any time, even before the signalling protection has been set up. Without signalling protection already set up, the UE would respond by sending the terminal identity in the clear. As a user tends to use the same terminal for an extended period of time, the terminal identity would also give strong hints regarding the user identity. This is no longer possible in EPS. In EPS, the UE shall not send IMEI or IMEISV to the network upon a network request before NAS security has been activated. (This does not apply to unauthenticated emergency calls.)

In particular, the MME may request the terminal identity in the NAS Security Mode Command message, and the UE then includes the terminal identity IMEISV in the NAS Security Mode Complete message, which is already ciphered (if the network supports confidentiality) – see Chapter 8.

7.2   The EPS Authentication and Key Agreement Procedure

The EPS AKA procedure is a combination of the following procedures:

- a procedure to generate EPS authentication vectors (AVs) in the HSS upon request from the MME, and to distribute them to the MME;[1]
- a procedure to mutually authenticate and establish a new shared key between the serving network and the UE;
- a procedure to distribute authentication data inside and between serving networks.

---

[1] Before the MME can request authentication vectors from the HSS, it needs to identify the UE – see section 7.1.

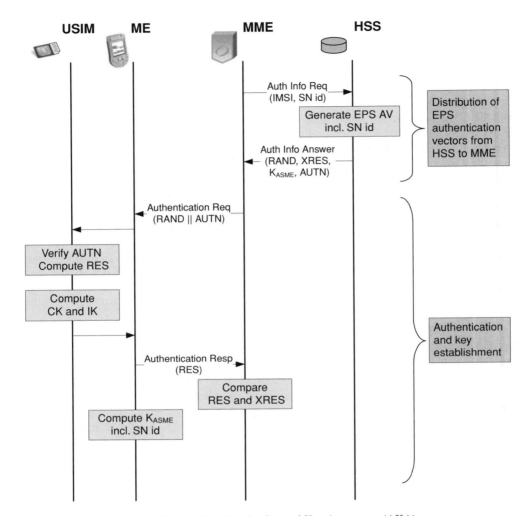

Figure 7.1   EPS Authentication and Key Agreement (AKA)

These procedures are described in the following, where also the terms 'EPS authentication vector' and 'authentication data' are explained. An overview of the EPS AKA procedure is shown in Figure 7.1.

There is no self-contained description of the EPS AKA procedure in the main reference for EPS security [TS33.401]. Rather, only the deltas to UMTS AKA are provided there. We describe the EPS AKA procedure here in full, adapting and explaining relevant text from [TS33.401] and [TS33.102].

An EPS AKA procedure needs to be run (apart from the case of emergency calls – see section 8.6) whenever the UE and the network want to communicate and do not share a security context. Security contexts are described in section 7.4. EPS AKA may be run, according to the network operator's policy, to renew a security context.

For understanding the differences between EPS AKA and UMTS AKA it is useful to compare the roles played by the involved entities. The MME plays a role in EPS AKA comparable to that of the VLR in UMTS AKA for circuit-switched 3G services and the SGSN in UMTS AKA for packet-switched 3G services. The MME performs, however, additional functions in EPS AKA, which have no equivalent in UMTS AKA. The ME and the HSS play similar, but not identical, roles in both protocols. The USIM from UMTS AKA may be reused for EPS AKA in an identical way, but optional EPS AKA-specific enhancements of the USIM have also been defined.

7.2.1   Goals and Prerequisites of EPS AKA

The design criteria for EPS AKA are presented in section 6.3 of this book. The prerequisites for EPS AKA, and the protocol goals achieved by EPS AKA, are quite similar to those for UMTS AKA listed in section 4.2. The one, seemingly small, enhancement of EPS AKA compared to UMTS AKA is that EPS AKA provides implicit serving network authentication, which UMTS AKA does not. Implicit serving network authentication is achieved by binding an appropriate key, K_{ASME}, to the serving network identity and successfully using the key with the messages following the authentication exchange. This seems straightforward, and indeed it does not require any changes to the USIM. In the ME and in the HSS, a few changes are required, however. As we will see below, one of these changes is due to the requirement that it must be possible to use UMTS AKA and EPS AKA simultaneously in a single operator's network, and even in a single HSS/HLR and with the same Authentication Centre (AuC). This simultaneous operation of UMTS AKA and EPS AKA is enabled by marking an authentication vector as 'for EPS use' or 'for legacy uses'. This marking is achieved by setting a specific bit in the Authentication Management Field (AMF), which is part of every authentication vector. More information on the AMF can be found below.

There is an additional trust prerequisite on EPS AKA compared to UMTS AKA that relates to the enhancement described in the previous paragraph, namely that the serving network trusts the home network to verify the identity of a serving network requesting authentication vectors and ensure that the serving network identity, to which the key K_{ASME} in an AV is bound, matches the verified identity of the serving network, to which the AV is sent. If this prerequisite was not fulfilled a serving network could obtain AVs with keys bound to the identity of another serving network, which would render serving network authentication by means of EPS AKA impossible.

On the other hand, the HSS need not trust the serving network in providing its correct identity as the HSS can and does verify this identity; if the verification fails the request for authentication vectors is denied.

There is also one additional cryptographic prerequisite: EPS AKA requires a key derivation function, residing in the ME and the HSS, outside the USIM and the AuC respectively, for deriving the local master key K_{ASME} – see below. As the key derivation resides in the ME, it needs to be standardized; see section 7.3 for further details on K_{ASME} and the derivation of the key hierarchy in general.

Both EPS AKA and UMTS AKA are based on the use of the same permanent secret key K which is shared between the USIM and the Authentication Centre in the user's Home Subscriber Server (HSS). K never leaves the USIM and the AuC. In addition, for both protocols, the USIM and the HSS keep track of counters SQN_{MS} and SQN_{HE} respectively to support network authentication (the subscript MS stands for Mobile Station while the subscript HE stands for Home Environment).[2] The sequence number SQN_{HE} is an individual counter for each user, which is used in the AuC for the generation of authentication vectors; the sequence number SQN_{MS} denotes the highest sequence number the USIM has accepted.

7.2.2 *Distribution of EPS Authentication Vectors from HSS to MME*

The MME invokes the procedure by requesting EPS authentication vectors from the HSS. The Authentication Information Request shall include the IMSI, the serving network identity 'SN id' of the requesting MME, and an indication that the authentication information is requested for EPS. The SN id is required for the computation of K_{ASME} in the HSS.

Upon the receipt of the Authentication Information Request from the MME, the HSS may have pre-computed authentication vectors available and retrieve them from the HSS database, or it may compute them on demand. The HSS sends an Authentication Information Answer back to the MME that contains an ordered array of n EPS authentication vectors AV(1 ... n). If $n > 1$ the EPS authentication vectors are ordered based on sequence number.

[TS33.401] recommends $n = 1$, so that only one authentication vector is sent at a time, because the need for frequently contacting the HSS for fresh AVs has been reduced in EPS through the availability of the local master key K_{ASME}, which is not exposed in a way similar to CK and IK in UMTS and, hence, does not need to be renewed very often. Based on the local master key, and keys derived from it, an MME can offer secure services even when links to the home environment are unavailable. Furthermore, pre-computed AVs are no longer usable when the user moves to a different serving network owing to the binding of the local master key K_{ASME} to the serving network identity. However, pre-computation may still be useful when the next request for authentication vectors is likely to be issued by an MME in the same serving network, which may be the case, for example, for a user in his home network.

Each EPS authentication vector is good for one run of the authentication and key agreement procedure between the MME and the USIM.

Generation of authentication vectors in the HSS

A UMTS authentication vector consists of a random number RAND, an expected response XRES, a cipher key CK, an integrity key IK and an authentication token AUTN (see section 4.2), while an EPS authentication vector consists of a random number RAND, an expected response XRES, a local master key K_{ASME} and an authentication token AUTN. Figure 7.2 shows the generation of a UMTS authentication vector by the AuC, and the generation of an EPS authentication vector from this UMTS authentication vector by the HSS.

Both UMTS authentication vectors and EPS authentication vectors play a role in EPS AKA. The AuC generates UMTS authentication vectors for EPS AKA in exactly the same format as for UMTS AKA. The HSS part outside the AuC derives K_{ASME} from the cipher and integrity keys CK and IK.

---

[2] It should be noted that the term 'Mobile Station' is no longer used in EPS. Nevertheless, we keep this notation here so as to make it easier for the reader to compare the presentation in this book with the description of sequence number handling in [TS33.102], which also applies to EPS AKA.

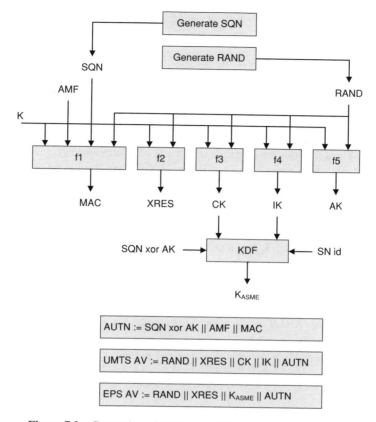

Figure 7.2   Generation of UMTS and EPS authentication vectors

The AuC starts with generating a fresh sequence number SQN and an unpredictable challenge RAND. For each user the HSS keeps track of the counter SQN_{HE}.

The HSS has some flexibility in generating sequence numbers, but some requirements need to be fulfilled by the mechanism used. According to [TS33.102], these requirements are:[3]

- *The generation mechanism shall allow the re-synchronization procedure in the HE (i.e. in the HSS) described in clause 6.3.5.*
- *When the SQN exposes the identity and location of the user, the AK may be used as an anonymity key to conceal it.* This needs some explanation. When the SQN of a particular user was predictable, and sufficiently different from the SQNs of other users in the area, it could be used to identify the user when eavesdropping on authentication messages. Whether there is a risk for the SQN to hint at a user's identity depends on the SQN generation scheme and needs to be decided by the operator. As it is always possible to use AK (there is a detailed description below), this second requirement is actually not a requirement on the SQN generation process itself, but a caveat on the use of SQNs in authentication vectors.

---

[3] Italicized text is reproduced with permission from © 2010, 3GPP™.

- *The generation mechanism shall allow protection against wrap-around [of] the counter in the USIM.*

It depends on the method of generating sequence numbers how exactly SQN_{HE} is used. Example methods for generating fresh sequence numbers are given in informative Annex C.1 of [TS33.102]. One method is based on using SQN_{HE} as a counter that is increased step by step, while another method is time-based. Combinations of these two methods are also possible. Furthermore, the SQN generation method can be chosen such that separate SQN spaces are used for different domains in which AKA authentication vectors are used – EPS, or 3G CS, or 3G PS, or IMS. The use of this latter feature is one way of minimizing synchronization failures; the handling of such failures is explained below. A full description of the SQN generation methods here would, unfortunately, require an amount of space and detail beyond the scope of this book. The interested reader is therefore referred to the referenced part of the specification.

An Authentication Management Field AMF is included in the authentication token of each authentication vector. The role of the AMF is discussed a bit further below.

Upon request from the HSS, the AuC computes the following values, as described in [TS33.102]:

- a message authentication code $MAC = f1_K(SQN \parallel RAND \parallel AMF)$, where f1 is a message authentication function;
- an expected response $XRES = f2_K (RAND)$, where f2 is a (possibly truncated) message authentication function;
- a cipher key $CK = f3_K (RAND)$, where f3 is a key generating function;
- an integrity key $IK = f4_K (RAND)$, where f4 is a key generating function;
- an anonymity key $AK = f5_K (RAND)$, where f5 is a key generating function or $f5 \equiv 0$.

Finally the authentication token $AUTN = (SQN\ xor\ AK) \parallel AMF \parallel MAC$ is constructed.

If the operator decides that no concealment of SQN is needed, then they set $f5 \equiv 0\ (AK = 0)$.

The following step is new for EPS compared to 3G. When the HSS receives the UMTS authentication vector from the AuC, the HSS applies the key derivation function KDF to CK, IK, SN id, and, for technical cryptographic reasons, (SQN xor AK). The result of the application of KDF is the key K_{ASME}. CK and IK can then be deleted in the HSS. The keys CK and IK used for computing EPS authentication vectors must never leave the HSS.

AMF usage for EPS AKA authentication vector identification

In earlier releases of the UMTS AKA specification, the use of the AMF was completely proprietary. Annex F of [TS33.102] lists example uses of the AMF. They are:

- indicating the algorithm and key used to generate a particular authentication vector when multiple algorithms and permanent keys are used;
- change of parameters relating to SQN verification in the USIM;
- setting threshold values for key lifetimes.

However, not much use was made of the AMF in practice; and it turned out, on the other hand, that the AMF was well suited to distinguish between authentication vectors for EPS use and those for legacy uses. 3GPP decided to use the most significant bit of the AMF for this distinction and call it the 'AMF separation bit', and to reserve the seven next most significant bits of the AMF for future standardization use, while leaving the remaining eight bits for proprietary use. Annex H of [TS33.102] defines this usage of the bits in the AMF. Annex H was introduced in Release 8, the 3GPP release where EPS was first specified.

The AuC shall set the AMF separation bit to '1' in authentication vectors for EPS use, and to '0' otherwise.

Readers may ask why the SQNs could not be used, instead of the AMF, to distinguish AVs for EPS use from those for legacy uses. After all, as explained above, SQNs can be generated such that separate SQN spaces are used for different usage domains of AVs. The answer is twofold:

- First and foremost, as we will see in section 7.2.3, it is the ME that must check whether the type of radio access network it is connected to (E-UTRAN) corresponds to the type of AV ('for EPS use') received from the network. But the ME cannot read the SQN if it is concealed by AK. The USIM cannot perform this check as it has no idea about the network connection. Furthermore, the USIM may be according to the Release 99 specifications and have no EPS-specific functionality.
- Second, the SQN management scheme is proprietary, and 3GPP prefers to keep it this way.

Only the AuC can set bits in the AMF. Therefore, in order for the AuC to be able to correctly set the AMF separation bit, the HSS must tell the AuC that the request for authentication vectors is for EPS use. The example uses of the AMF listed above can still be realized using the proprietary part of the AMF.

Lengths of authentication parameters

The lengths of authentication parameters are the same for EPS AKA as the ones specified for UMTS AKA in clause 6.3.7 of [TS33.102]. In particular, the permanent key K, RAND, CK and IK are all 128 bits long. It is true that K_{ASME} is 256 bits long, but it also has a key entropy of only 128 bits as it is derived from K. It should be noted, however, that EPS is specified in such a way that all keys can be extended to 256 bits of length if a need is seen in the future.

7.2.3   Mutual Authentication and Establishment of a Shared Key Between the Serving Network and the UE

The purpose of this procedure is the authentication of the user and the establishment of a new local master key K_{ASME} between the MME and the UE, and, furthermore, the verification of the freshness of the authentication vector and authentication of its origin (the user's home network) by the USIM. K_{ASME} is used in subsequent procedures for deriving further keys for the protection of the user plane, Radio Resource Control signalling, and Non-Access Stratum signalling (see section 7.3).

Authentication requests

The MME invokes the procedure by selecting the next unused EPS authentication vector from the ordered array of EPS authentication vectors in the MME database (if there is more than one). If the MME has no EPS AV it requests one from the HSS. The MME then sends the random challenge RAND and the authentication token for network authentication AUTN from the selected EPS authentication vector to the mobile equipment, which forwards it to the USIM. The MME also generates a key set identifier eKSI and includes it in the Authentication Request (see section 7.4).

Verification in the USIM

Upon receipt of RAND and AUTN, the USIM proceeds as shown in Figure 7.3, which is taken from Figure 9 in [TS33.102].

According to [TS33.102], the USIM first computes the anonymity key $AK = f5_K$ (RAND) and retrieves the sequence number $SQN = (SQN$ xor $AK)$ xor AK, where K is, as explained before, the permanent secret key shared between USIM and AuC. Remember that if no concealment is needed then $f5_K \equiv 0$ ($AK = 0$).

Next the USIM computes $XMAC = f1_K$ ($SQN \parallel RAND \parallel AMF$) and verifies that it equals the MAC included in AUTN.

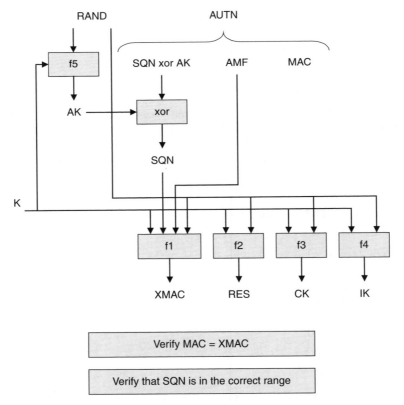

Figure 7.3   User authentication function in the USIM. Adapted with permission from © 2010, 3GPP™

Then the USIM verifies that the received sequence number SQN is in the correct range. The mechanism for the SQN verification in the USIM has not been standardized, for the same reason that the SQN generation in the HSS has not been standardized: both the USIM and the HSS are under the control of the same stakeholder, the operator. But, for those who do not want to specify their own mechanism, the informative Annex C.2 of [TS33.102] provides an example mechanism.

The SQN verification mechanism does have to satisfy certain requirements.

- The fundamental requirement is that no SQN shall be used twice. Once the USIM has successfully verified an AUTN it shall not accept another AUTN with the same SQN.
- It is additionally required according to [TS33.102] that the SQN verification mechanism shall, to some extent, allow the out-of-order use of sequence numbers. Out-of-order use of SQNs may occur, for example, when two different entities such as an MSC/VLR and an SGSN each request a batch of authentication vectors from the HSS (e.g. with SQNs 1 to 5 in the first batch and 6 to 10 in the second batch) and then use the AVs from the batches in AKA runs with the UE in an interleaved fashion. If the USIM simply kept track of the highest SQN received in a successfully verified AUTN and rejected all lower SQNs received later, then this would lead to so-called synchronization failures (a particular form of authentication failures explained below) when SQNs were presented to the USIM out of order. Therefore, the above-mentioned additional requirement was introduced to ensure that the authentication failure rate due to synchronization failures remained sufficiently low. For this purpose, the USIM must be able to store information relating to sequence numbers received in past successful authentication events. 3GPP even specified a threshold value for the out-of-order use of sequence numbers: if the received SQN is among the last 32 sequence numbers generated then it shall be accepted if it was not used in a previous successful authentication. This could be achieved, for example, by using a suitable window mechanism, but more sophisticated methods are available – see Annex C.2 of [TS33.102].
- However, a USIM may reject a time-based sequence number if it was generated too long ago. This check, if applied, takes precedence over the requirement in the previous paragraph.
- The SQN verification mechanism may additionally check that an SQN is not accepted if the jump from the last successfully received SQN is too big.

These various conditions on the SQN verification mechanism explain the formulation that the USIM checks whether the SQN is 'in the correct range'.

Authentication responses

If the sequence number is considered to be in the correct range, the USIM computes $RES = f2_K (RAND)$ and sends it to the ME, which includes RES in an Authentication Response message to the MME. The USIM also computes the cipher key $CK = f3_K (RAND)$ and the integrity key $IK = f4_K (RAND)$. The USIM sends CK and IK to the ME. A USIM may also support a key conversion function that converts the pair (CK, IK) into a GSM cipher key Kc. If the USIM does support this function, it also sends the so-derived Kc to the ME. If the ME supports 128-bit GSM encryption it also computes the GSM key Kc_{128} from CK and IK. For the use of these keys Kc and Kc_{128}, see section 4.4.

Upon receipt of the Authentication Response message the MME checks whether the received RES matches the expected response XRES from the selected authentication vector. If it does then the authentication of the user has been successful.

Up to this point, there are no functional differences between UMTS AKA and EPS AKA in the handling of authentication requests, verification in the USIM, and authentication responses.

EPS AKA additionally requires, however, that an ME accessing E-UTRAN shall check during authentication that the AMF separation bit is set to '1'. The ME ensures by performing this check that the authentication vector used in the current authentication run was marked by the AuC as 'for EPS use' indeed. This check is in turn a prerequisite for successful implicit serving network authentication. When the ME receives (CK, IK) from the USIM, the ME computes K_{ASME} using the same key derivation function and the same input parameters as the HSS. After this, CK and IK can be deleted in the ME.

Key storage on the USIM

In contrast to UMTS AKA, the ME does not request the USIM to store CK and IK resulting from an EPS AKA run. The reason for this is that EPS AKA shall work with USIMs from earlier 3GPP releases, and such a USIM may have already stored a pair (CK, IK) from a previous UMTS AKA run. If this pair was overwritten with the pair (CK, IK) generated during the recent EPS AKA run, this would lead to problems when EPS security context and UMTS security context had to be held simultaneously (for the purposes of interworking between E-UTRAN and 3G – see Chapter 11). It is only for EPS-enhanced USIMs that the ME requests the USIM to store an appropriate subset of the EPS security context upon certain events (see section 7.4).

Authentication failures

A detailed description of the behaviour of UE and MME upon an authentication failure, together with the cause values, is given in clause 5 of [TS24.301]. We give an overview here.

- *MAC code failure*. If the USIM determines that MAC differs from XMAC, it indicates this to the ME, which sends an Authentication Failure message back to the MME with an indication of the cause.
- *Synchronization failure*. This occurs when the USIM determines the sequence number to be not in the correct range. The behaviour of the USIM and the AuC in this case is the same for UMTS AKA and EPS AKA and is described in clause 6.3 of [TS33.102]. The USIM computes a parameter AUTS as shown in Figure 7.4, which is taken from Figure 10 in [TS33.102]. AUTS is included in an Authentication Failure message from the UE to the MME. The MME forwards AUTS to the HSS requesting new authentication vectors. The AMF used to calculate MAC-S is set to all zeros so that it does not need to be transmitted back to the HSS. The HSS uses AUTS to synchronize SQN_{HE} stored in the HSS with SQN_{MS} contained in AUTS. The details of how the HSS handles AUTS can be found in clause 6.3.5 of [TS33.102]. The only caveat is that the HSS needs to tell the AuC again that the request is related to EPS. The AuC cannot see this from the AMF as the AMF in AUTS is set to all

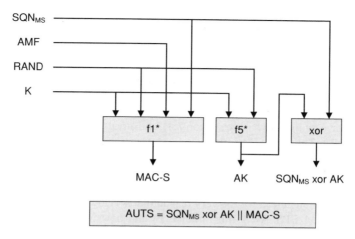

Figure 7.4   Construction of the parameter AUTS. Adapted with permission from © 2010, 3GPP™

zeros. After a possible synchronization of SQN_{HE} the HSS generates new AVs and sends them to the MME.

- *Incorrect type of authentication vector.* If the check of the AMF separation bit in the ME fails, then the ME sends an Authentication Failure message back to the MME with an indication of the cause.
- *Invalid authentication response.* If the MME determines that XRES differs from RES then, depending on the type of identity used, the MME may decide to initiate a new identification and authentication procedure towards the UE, or it may send an Authentication Reject message to the UE and abandon the procedure.
- *Authentication failure reporting.* For UMTS AKA, the VLR/SGSN shall report failed authentications to the HLR – see clause 6.3.6 of [TS33.102]. This is no longer required for EPS AKA as the usefulness of this reporting proved quite limited.
- *Reuse and retransmission of (RAND, AUTN).* The verification of the SQN by the USIM will cause the USIM to reject an attempt by the MME to re-use an authentication vector for establishing a particular key K_{ASME} more than once. In general, the MME is therefore allowed to use an authentication vector only once. There is one exception, however – see clause 5.4.2.3 of [TS24.301]. In the event that the MME has sent out an Authentication Request using a particular authentication vector and does not receive a response message (Authentication Response or Authentication Failure) from the UE, it may re-transmit the Authentication Request using the same authentication vector. However, as soon as a response message arrives no further re-transmissions are allowed.

7.2.4   Distribution of Authentication Data Inside and Between Serving Networks

When a user moves around, the MME serving the UE may change. When the UE then sends an Attach Request, or a Tracking Area Update Request [TS23.401], the UE will, in general, use its temporary identity, the GUTI, in order to protect the confidentiality of its permanent identity, the IMSI – see section 7.1. But the new MME is not able to make sense of the GUTI, so it has only two choices: request the permanent identity from the UE and break identity confidentiality in this way, or ask the old MME, which issued the GUTI, to translate the GUTI to the user's IMSI. The old MME will also send back authentication data to the new MME. Exactly what kind of authentication data is allowed to be exchanged between old and new MME depends on whether the two MMEs reside in the same or in different serving networks.

When the two MMEs reside in the same serving network then any EPS security context the old MME may have (at most two – see section 7.4), and any unused EPS authentication vectors, may be transferred. The new MME may use these transferred EPS authentication vectors as they are bound to the correct serving network identity, so serving network authentication in a new EPS AKA run using these authentication vectors will work fine.

When the two MMEs reside in different serving networks, then unused EPS authentication vectors must not be transferred for the very reason that serving network authentication would not succeed in a new EPS AKA run initiated by the new MME. But the transfer of the current EPS security context, and hence its use between the UE and the new MME, is allowed, depending on the security policy of the serving networks. From a procedural point of view, this will not create any protocol failures with serving network authentication. However, readers may wonder why this is allowed as the EPS security context was generated from EPS authentication vectors bound to a particular serving network identity and is now used in a serving network with a different identity. This decision by 3GPP is a result of the trade-off between complexity and risk, which has been performed also in many other instances during the design of EPS security. The reduction in complexity stems from the fact that the HSS need not be contacted, and no new round of EPS AKA is needed. This is important especially in situations where, due to the network topology, user movements may result in frequent changes between MMEs. The risk, on the other hand, is mitigated by the following facts.

- An EPS security context is forwarded only to a new MME trusted for this purpose by the old MME.
- As soon as EPS AKA is run another time, the serving network will be authenticated.
- EPS security would not be affected by any breach of security in a non-EPS network because EPS security context is not transferred to non-EPS nodes, such as an SGSN.

A serving network operator who deems the remaining risk still too high may adopt a policy of not forwarding EPS security context.

7.3   Key Hierarchy

As already explained in section 7.2, EPS AKA is an enhancement of UMTS AKA. This means that the key agreement is similar in EPS and in UMTS. But this is only part of the picture: in section 6.3 we discussed the reasons why the key agreement part of EPS AKA produces only a single intermediate key K_{ASME} instead of a set of keys that would be subsequently used in security mechanisms. The latter is the case for UMTS: ciphering key CK and integrity key IK are generated during execution of the UMTS AKA procedure.

All cryptographic keys that are needed for various security mechanisms are derived from the intermediate key K_{ASME} which can be viewed as a 'local master key' for the subscriber, in contrast to the permanent master key K (for this subscriber). On the network side, the intermediate local master key K_{ASME} is stored in the MME while the permanent master key K is stored in the AuC. The advantages of using an intermediate key are twofold.

- It enables cryptographic key separation, which implies that each key is usable in one specific situation (or context) only. Furthermore, knowing a key that is used in one context does not help in trying to find out, or guess, what kind of key could possibly be usable in another context.
- It also improves the system in terms of providing key freshness. That is, it is possible to more often renew the keys used in security mechanisms, for example in ciphering. As already explained in section 6.3, we do not have to run EPS AKA every time we want to renew keys used for protecting the radio interface, so we do not have to involve the home network to have fresh keys in place.

The obvious disadvantage of using intermediate keys is the added complexity: there are more types of key in the system, all of which need to be computed, stored, protected, kept in sync, and so on. Altogether, we have a typical security versus complexity trade-off situation. For EPS, the security benefits of using an intermediate key outweigh the added complexity, whereas at the design phase of 3G security, there was not enough justification for intermediate keys.

After the idea of using the intermediate key K_{ASME} was introduced in the design of EPS security, it was quite natural to take a further step: another intermediate key K_{eNB} was added that is stored in the base station eNB. Addition of K_{eNB} makes it possible to renew keys for protection of radio access without involving the MME. Furthermore, an appropriately modified K_{eNB} can be handed over between base stations in a so-called X2-handover (see section 9.4) without involving the MME. The keys used directly for protecting the RRC signalling and the user data on the radio interface would not be suited for this purpose as they are bound to particular cryptographic algorithms, which is not the case for K_{eNB}, and base stations may apply different cryptographic algorithms.

Figure 7.5 shows the whole key hierarchy of EPS. The UMTS key hierarchy is a small subset of this and consists of the two topmost layers only.

7.3.1   Key Derivations

In Figure 7.5, an arrow between two keys means that one key (the one to which the arrow points) is derived from the other. In all cases, there are also additional input parameters that are needed in the derivation. None of the additional parameters is assumed to be secret information. In practice, a potential attacker may not know the correct values for these additional parameters, but, to be on the safe side, it has to be assumed that the attacker is in a good position to make educated guesses about these values.

There is one special arrow in the figure, namely the loop arrow pointing from the box representing keys K_{eNB}/NH to itself; the next subsection has an explanation of these keys. For all other cases, each key is always derived from another key at a higher layer in the hierarchy. The special case refers to a situation where an intermediate key K_{eNB} for one base station has to be derived based on a key K_{eNB} or NH from another base station, without having access to keys higher up in the hierarchy. This restriction occurs in certain handover situations between eNodeBs where the MME is not involved. For details on the key handling in this situation, see section 9.4.

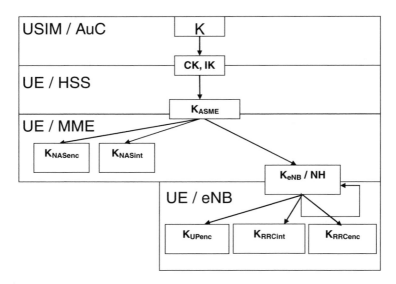

Figure 7.5   EPS key hierarchy. Adapted with permission from © 2010, 3GPP™

The most important property of the key derivation is that it meets the requirement of being one-way as described in section 2.3: starting from keys in lower layers of the key hierarchy it is impossible in practice to compute keys in the higher layers.

The topmost key derivation from K to CK and IK is different from the rest in the sense that details of it are not standardized. It is also the only key derivation that is also present in the 3G system. This first key derivation step happens, on the user side, inside the USIM and, on the network side, inside the AuC. Both USIM and AuC are controlled by the same operator and this is the reason why it is not necessary to standardize this step. However, 3GPP specified a set of algorithms called MILENAGE that may be used by operators – see section 4.3. For the rest of the derivations the situation is different: on the user side, key derivations happen in the ME, while on the network side, key derivation happens in the MME or the eNB, so it is necessary to standardize these functions.

From an implementation point of view, it makes a lot of sense that all the key derivations carried out in the UE share the same core cryptographic function. In fact, 3GPP has taken the approach that all specified key derivation functions make use of the generic key derivation function that is specified in [TS33.220]. In this generic key derivation function the core cryptographic primitive is the HMAC-SHA-256 algorithm – see sections 2.3 and 4.3.

Figure 7.6 shows how key derivations are done on the network nodes. Respective derivations need to be done also on the user side where all of them are carried out in the ME.

In the figure, 'KDF' denotes the generic key derivation function based on HMAC-SHA-256 and 'Trunc' denotes a simple truncation function that uses only the 128 least significant bits of a 256-bit value and throws away the most significant half. Note here that there is an inbuilt possibility in the EPS key hierarchy to take into use 256-bit keys for various security functions, such as ciphering. However, at least for Releases 8 and 9 of EPS, security provided by 128-bit keys is seen as adequate and the truncation is in use.

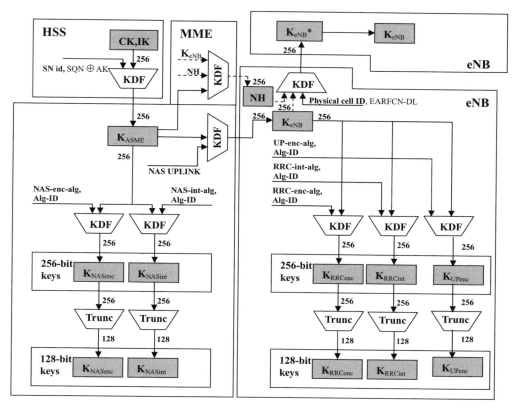

Figure 7.6 EPS key derivations on network side. Adapted with permission from © 2010, 3GPP™

7.3.2   Purpose of the Keys in the Hierarchy

As already explained, the key hierarchy contains one root key (K), several intermediate keys (CK, IK, K_{ASME}, K_{eNB} and NH) and several leaf keys (K_{NASenc}, K_{NASint}, K_{RRCenc}, K_{RRCint} and K_{UPenc}). Below we explain the purpose of all these keys and also briefly explain the input parameters that are needed in derivation of each key.

- K is the subscriber-specific master key, stored in the USIM and the AuC. It is not derived from any other key, but instead is a random 128-bit string.
- CK and IK are 128-bit keys derived from K, using additional input parameters, as described in the previous section.
- K_{ASME} is derived from CK and IK using two additional inputs. First, the serving network identity SN id, consisting of the Mobile Country Code MCC and the Mobile Network Code MNC, is used to tie the key to the network where it is supposed to be stored and used. Second, the bit-wise sum of two additional parameters, SQN and AK from the EPS AKA procedure, is used in order to more thoroughly use the variation of information available. Note that, although AK is another key derived during the EPS AKA, the value (SQN xor

AK) is part of the parameter AUTN that is sent in cleartext during EPS AKA, so it has to be assumed to be known by a potential attacker. The purpose of K_{ASME} is to be a local master key in the MME.

- K_{eNB} is derived from K_{ASME} and the additional input NAS uplink COUNT that is a counter parameter. This additional parameter is needed to ensure that each new K_{eNB} derived from K_{ASME} differs from the ones derived earlier. The purpose of this key is to be a local master key in an eNB.
- Next Hop (NH) is another intermediate key that is needed in handover situations (see section 9.4). NH is derived from $K_{ASME,}$ using either the newly derived K_{eNB} as an additional input for the initial NH derivation, or a previous NH as an additional input in case such an NH already exists.
- There is still another intermediate key needed in the process of deriving one K_{eNB} from another. This is called $K_{eNB}{}^{*}$, and it is derived from either K_{eNB} or a freshly generated NH if such a parameter exists. Additional parameters of physical cell id and downlink frequency are used to tie the key to the local context. In handovers, $K_{eNB}{}^{*}$ becomes the new K_{eNB} in the target base station. The reason for introducing a separate key $K_{eNB}{}^{*}$ is to bring clarity for the presentation of the key hierarchy in the specifications. An alternative would have been to refer to 'future K_{eNB}', 'potential new K_{eNB}', 'K_{eNB} for target base station' and so on. but all these formulations would have easily led to confusions.
- K_{NASenc} is a key that is used to encrypt NAS signalling traffic. It is derived from K_{ASME} and two additional parameters. The first one is called algorithm type distinguisher and, in the case of K_{NASenc}, it has a value indicating that this key is used for NAS encryption. The second one is the identifier of the encryption algorithm.
- Similarly as above, K_{NASint} is a key that is used to protect the integrity of NAS signalling traffic. It is derived from K_{ASME} and two additional parameters: the first one (algorithm type distinguisher) indicates that the key is used for NAS integrity, and the second one is the integrity algorithm identifier.
- K_{RRCenc} is a key that is used to encrypt RRC signalling traffic. It is derived from K_{eNB} and two additional parameters: the first one (algorithm type distinguisher) indicates that this key is used for RRC encryption, and the second one is the identifier of the encryption algorithm.
- Similarly, K_{RRCint} is used to protect the integrity of RRC signalling traffic. It is derived from K_{eNB} and two parameters: the first one indicates that this key is used for RRC integrity, and the second one is the integrity algorithm identifier.
- Finally, K_{UPenc} is used to encrypt user plane traffic. This key is derived from K_{eNB} and two parameters: the first one indicates that this key is used for UP encryption, and the second one is the encryption algorithm identifier.

There are even more keys used in EPS for the purpose of interworking with other systems, such as with 3G systems. For example, CK' is a key derived from K_{ASME} and it is needed after handover from EPS to 3G for encryption. These mapped keys are discussed in detail in Chapter 11.

7.3.3 Cryptographic Key Separation

One purpose of the complex key hierarchy is to provide key separation. It means that all keys are used in a single unique context for cryptographic protection of either user traffic or signalling traffic. Moreover, because all keys used for such protection are leafs in the hierarchy, it is infeasible to derive a key used in one protection context from another key (or set of keys) used in other contexts.

The intention is that attackers cannot find out any keys used in one context from keys used in any other context. But if it happens anyway that some protection keys are leaked in whatsoever manner to unauthorized parties, key separation prevents the leakage from expanding. Of course, cryptographic key separation does not help if there is a leakage of higher layer keys, perhaps because somebody has been able to get access to keys stored in MME. But, looking from another point of view, if some unauthorized party has been able to get access to the very core information stored in MME then, for example, protection of NAS signalling is less relevant for that kind of attacker: all NAS signalling terminate in MME and it is therefore visible in cleartext.

There are also purely cryptographic reasons why key separation is a useful goal in any circumstances; see section 2.3 for 'related key attack'.

Tying of a key to a particular context requires that this particular context is somehow affecting key derivation. Therefore, each of the additional parameters used in derivation of the keys in the hierarchy is motivated from this angle. Let us look, for example, at the keys K_{NASint} and K_{RRCint}. It is natural that the derivation of these two keys is similar because they are used for similar purposes. Both keys have two additional parameters, one of which is the same in both cases: the integrity algorithm. The other input parameter fixes explicitly that K_{NASint} is used for NAS integrity protection while K_{RRCint} is used for RRC integrity protection; hence there is a difference in this second additional parameter. For these two keys there is of course another difference: K_{NASint} is derived from K_{ASME} while K_{RRCint} is derived from K_{eNB}. This difference would already guarantee that these two derived keys are different in normal operations, but it is difficult to claim that no active attack scenario would exist where there would be a possibility to get either UE or network elements to derive K_{NASint} and K_{RRCint} from the same key. On the other hand, there is no particular need to minimize the number of input parameters to the ones that are absolutely necessary. Using the purpose of the key as an explicit parameter in key derivation is therefore a handy countermeasure in preventing the same key being used for two different purposes, either by accident, design flaw or as a result of an active attack.

7.3.4   Key Renewal

As mentioned above, another benefit of the complex key hierarchy is that keys can be renewed without affecting all other keys. When one key is changed, only the keys that are dependent on it have to be changed; the others may remain the same. For example, K_{eNB} can be re-derived without changing K_{ASME} in the process. As a consequence of changing K_{eNB}, all keys derived from K_{eNB} (e.g. K_{RRCenc} and K_{RRCint}) are changed as well.

There are several reasons why renewing keys is seen useful although, at a first glance, it seems to add unnecessary complexity: one key is replaced by another one that is used for exactly the same duties as the old one was. One reason is a cryptographic one: when a key is changed, the task of the attacker to find or guess the key is 'returned back to square one'. Another reason follows from a generic security principle: we should minimize the need to distribute the same secret to many elements. In the case of K_{eNB}, it is renewed whenever it is derived for a new eNB, thus preventing two base stations from using the same key.

However, not all keys that are leafs in the key hierarchy can typically be renewed without renewing the whole key hierarchy. Indeed, the security architecture is built in such way that the keys K_{NASint} and K_{NASenc} can only be renewed without change in K_{ASME} if the used algorithm is changed (which should probably be a very rare event). There are two reasons for this:

- The keys K_{NASint} and K_{NASenc} and (their mother key) K_{ASME} are all held by the same entities anyway: on the network side by the MME and on user side by the ME.
- The amount of NAS signalling is not so huge that there would be a real need for renewing keys (without running AKA) from a purely cryptographic point of view either.

7.4   Security Contexts

When two parties engage in security-related communication, for example when running an authentication protocol or exchanging encrypted data, they need an agreed set of security parameters, such as cryptographic keys and algorithm identifiers, for the communication to be successful. Such a set of security parameters is called a security context. There are different types of security context depending on the type of communication, and the state the communicating parties are in.

Note that entities may store security context data locally even when not engaged in communication. The distinction between locally stored security context data and security context shared between two communicating parties for the purpose of running a security protocol is useful in principle, but it is a bit academic and not much adhered to in practice. As the potential for confusion is low, we follow the common practice and speak only of security contexts.

Several different types of security context have been defined for EPS so as to have shorthand notations available for the various sets of security parameters used in particular situations. Their definitions are a bit tricky, and 3GPP took a while to get them right, but they are useful, and the reader will encounter them throughout the book, so we will dwell on them a little here to have a central place for security context definitions and explanations for reference. But it is true that quite a few of the parameters mentioned in this section are explained only later, notably in Chapters 8, 9 and 11.

As can be seen from the preceding section 7.3, where the EPS key hierarchy was presented, EPS security is rooted in a permanent key K. The USIM and the Authentication Centre share this key K and a set of AKA algorithms, such as MILENAGE as described in section 4.3. The key K and the set of AKA algorithms are used for UMTS AKA run over GERAN or UTRAN, and EPS AKA run over LTE. The notion of EPS security context, as defined by 3GPP, does not include the key K nor identifiers of the AKA algorithms, but only keys and related parameters particular to EPS – from K_{ASME} downwards in the EPS key hierarchy.

The following definitions draw heavily on clause 3 of [TS33.401].

EPS security context

This consists of the EPS NAS security context and, when it exists, the EPS AS security context. The EPS NAS security context is used for protecting the non-access stratum of EPS between the UE and the MME, and it may even exist when the UE is in deregistered state (see Chapter 9). The EPS AS security context is used for protecting the access stratum of EPS between the UE and the eNB, and it only exists when cryptographically protected radio bearers are established and is otherwise void. For an EPS AS security context to exist the UE needs to be in connected state.

EPS NAS security context

This context consists of K_{ASME} with the associated key set identifier eKSI (more on this later), the UE security capabilities (see below), and the NAS uplink and downlink COUNT values. These counters are relevant also for security as they are used as input parameters to key derivations in certain state and mobility transitions (see Chapters 9 and 11) and, in conjunction with integrity protection, for preventing message replay. Separate pairs of NAS COUNT values are used for each EPS NAS security context. The EPS NAS security context is called 'full' if it additionally contains the keys K_{NASint} and K_{NASenc} ('NAS keys' for short) and the identifiers of the selected NAS integrity and encryption algorithms, otherwise it is called 'partial'. An EPS security context containing a full or partial EPS NAS security context is also called full or partial, respectively. Note, however, that both K_{NASInt} and K_{NASEnc} can be derived from the K_{ASME} when the NAS integrity and encryption algorithms are known. Thus, they need not necessarily be stored in the memory.

UE security capabilities

They are the set of identifiers corresponding to the ciphering and integrity algorithms implemented in the UE. This includes capabilities for E-UTRAN, UTRAN and GERAN if these access types are supported by the UE. A network node learns the UE security capabilities from the UE, or from a neighbouring node (more on this in later chapters). The UE EPS security capabilities are a subset of the supported UE security capabilities relating to algorithms used in EPS.

EPS AS security context

This context consists of the cryptographic keys at AS level (i.e. between the UE and the eNB) with their identifiers, the Next Hop parameter NH, the Next Hop Chaining Counter parameter NCC used for next hop access key derivation (see section 9.4), the identifiers of the selected AS level cryptographic algorithms for RRC integrity protection and ciphering of RRC and user plane, and the counters used for replay protection.

Native versus mapped contexts

There are different types of EPS security context, namely 'native' and 'mapped'. These types point to the origin of the context: a native EPS security context is a context whose K_{ASME} was created during a run of EPS AKA, while a mapped EPS security context is converted from a UMTS security context when the UE moves to LTE from UTRAN or GERAN (see Chapter 11). A mapped EPS security context is always 'full', while a native EPS security context may be full or partial. A partial native EPS security context is created by an EPS AKA run, for which no corresponding successful NAS Security Mode Command (SMC) procedure has been run; in other words. a partial native EPS security context is always in state 'non-current', as explained in the next subsection. After having been put into use by running a NAS SMC procedure, a partial native EPS security context becomes full as the NAS security algorithms and the NAS keys have been agreed between the UE and the MME.

Current versus non-current contexts

There are other states in which EPS security contexts can be, namely 'current' and 'non-current'. The current security context is the one that has been activated most recently. A non-current security context is sitting on the side and waiting to replace the current one. As mentioned in the previous paragraph, a partial native EPS security context is non-current, but also a full native EPS security context can be non-current, namely when it has been pushed aside by a mapped EPS security context in a handover from UTRAN or GERAN to EPS. A mapped EPS security context never becomes non-current. The different handling of native and mapped in this respect is explained by the fact that a native context is considered of higher value as it originates from within the EPS itself. It may therefore be used later (again), while a mapped context may be discarded when no longer used as a current context. The type of a context does not change during its lifetime. The state of a context can change, but it can be only in one state at a time.

Key identification

In E-UTRAN, the NAS Key Set Identifier eKSI identifies the key K_{ASME}. It is the purpose of the eKSI to signal which K_{ASME} was used to derive the NAS keys, such as when the UE sends a NAS message in moving from idle to connected state. The use of the eKSI ensures key synchronization between the UE and the MME. The NAS Key Set Identifier information element consists of a value of three bits and a type bit. The type indicates whether an EPS security context is a native EPS security context or a mapped EPS security context ('0' denotes native, and '1' mapped). The eKSI is the EPS equivalent of the KSI in 3G. The KSI also takes three-bit values, but has no different types. The KSI points to the set of two keys, CK and IK. In mobility between E-UTRAN and UTRAN, the value of eKSI is mapped to KSI, and vice versa (see section 11.1). The eKSI is allocated by the MME, and the eKSI is therefore accompanied by the identity GUTI (see section 7.1). The GUTI tells the receiving MME at which MME the security context of the UE currently resides. The UE can also signal that no key is available by setting the eKSI value to '111'.

EPS security context storage

When the USIM is enhanced for EPS, a part of the EPS native security context is stored on the USIM under certain conditions. When the USIM is not enhanced for EPS, the non-volatile part of the ME memory takes on an equivalent role and stores that part of the EPS native security context. The idea is that, in both cases, an EPS native security context shall be kept even when the UE deregisters or is switched off. When the UE registers again and goes to connected state, the EPS native security context can be retrieved from storage and used to protect the initial NAS message. By re-using the stored context, a new run of EPS AKA can be avoided. A mapped context is never stored on the USIM. A mapped EPS security context is kept in a transition to idle state, and, if available, is used to protect the initial NAS message when the UE transitions back to connected state. A mapped EPS security context is deleted when the UE deregisters (see the definition of current and non-current security context above).

EPS security context transfer

Parts of the EPS security context may be pushed down from the MME to a base station, or may be transferred between equivalent EPS nodes (e.g. from one MME to another MME), or from one base station to another one. (Of course, base stations get to see only EPS AS security context data.) This is possible even when these nodes lie in different networks, providing the operators' security policies allow this – see section 7.2.4. EPS security context shall not, however, be transferred to an entity outside the EPS. In particular, the K_{ASME} shall never be transferred from the Evolved Packet Core (EPC) to an entity outside the EPC. In this way, K_{ASME} is not revealed to network entities handling technologies other than E-UTRAN.

8

EPS Protection for Signalling and User Data

Protecting communication over the air and inside the network is important so that confidentiality of information can be assured and attacks on the communication channels can be more easily mitigated. EPS has two layers of security for signalling: the first layer is between UE and the base stations, and the second layer is between UE and the core network (see Chapter 6). The user plane data packets are protected between UE and base stations and further in the network in hop-by-hop manner. In this chapter, we describe in detail how the communication between UE and network and inside the network is protected.

LTE has separate signalling and user planes. The signalling plane is further divided into signalling between UE and base stations (i.e. Access Stratum, AS) and between UE and core network (i.e. Non-Access Stratum, NAS). Signalling protection consists of ciphering and integrity protection with replay protection; for the user plane (data) only ciphering is provided, as explained in sections 8.1–8.3. We describe also how core network interface protection mechanisms are used within EPS (in section 8.4), how certificate enrolment to the base stations is handled (in section 8.5), and how emergency calls are handled (in section 8.6).

8.1 Security Algorithms Negotiation

Before the communication can be protected, both UE and the network need to agree on what security algorithms to use. EPS supports multiple algorithms and includes two mandatory sets of security algorithms [TS33.401], namely 128-EEA1 and 128-EIA1 based on SNOW 3G [TS35.216], and 128-EEA2 and 128-EIA2 based on AES [FIPS 197], that all implementations of UEs, eNBs and MMEs need to support. EPS can be extended to support more algorithms in the future. See Chapter 10 for more information about AES and SNOW 3G and their usage in EPS.

Algorithms are negotiated separately between UE and base stations (AS level) and between UE and the core network (i.e. MME, NAS level). The network selects the algorithms based on the UE security capabilities and the configured list of allowed security algorithms for the

---

LTE Security Dan Forsberg, Günther Horn, Wolf-Dietrich Moeller, and Valtteri Niemi
© 2010 John Wiley & Sons, Ltd

network entities (e.g. base stations and MMEs). The UE provides its security capabilities to the network during the attachment procedure and when sending TAU Request messages after intersystem handovers to E-UTRAN – see section 11.1.4. The security capabilities for AS level signalling, NAS level signalling, and user plane data are the same, except that the user plane data does not support integrity protection. However, it is possible to have different algorithms for AS and NAS active at the same time. The configured list of allowed security algorithms in the network can be used, for example, when an algorithm needs to be phased out. This list also provides the operator with a means to express a preference of certain algorithms over others.

Messages cannot be protected before algorithms have been agreed and signalling protection has been set up. The security capabilities that the UE has provided to the network are repeated in an integrity protected response message from the network in order to protect against bidding down attacks, where the attacker modifies the message carrying the UE security capabilities from the UE to the network. If UE detects a mismatch between the security capabilities it sent to the network and the ones it received from the network, the UE cancels the attach procedure. It could be argued that it would be even better if the bidding down attack protection would happen in both directions. The UE would then repeat the UE security capabilities again to the network once the integrity protection has been set up so that the network could detect bidding down attacks in case the UE failed to do so. However, EPS relies on and requires UEs to do the checking, mainly because the added security achieved by doing the check also on the network side does not justify the added complexity.

Two Security Mode Command procedures are used to indicate the selected algorithms and to start ciphering and integrity protection with replay protection. One Security Mode Command procedure exists for the AS and another one for the NAS level. The MME is responsible for selecting the NAS level algorithms, and the base station is responsible for selecting the AS level algorithms, including the user plane algorithm. It is not possible to change the AS level algorithms using the AS Security Mode Command procedure. The NAS level algorithms can be changed with the NAS Security Mode Command procedure, for example when the MME changes and the target MME supports different algorithms from the source MME.

8.1.1   Mobility Management Entities

The operator configures MMEs with a list of allowed algorithms for NAS signalling in priority order; one list for the integrity algorithms and one for the ciphering algorithms. During the security setup the MME chooses one NAS ciphering and one NAS integrity algorithm based on the configured lists and signals the decision to the UE in the NAS Security Mode Command (NAS SMC) procedure.

When MME changes (i.e. in an inter-MME mobility scenario) and the target MME wants to change the NAS algorithms, it uses the NAS Security Mode Command procedure as in initial NAS level security set-up. The target MME also includes the UE security capabilities for bidding down attack protection, similar to the initial set-up.

8.1.2   Base Stations

Similar to the MME configuration, each base station is also configured with a list of allowed algorithms in priority order, one list for integrity protection algorithms and another for ciphering algorithms. Thus, the base station decides what algorithms are used with the UE for AS signalling protection and for AS user plane data protection. The MME sends the UE security capabilities to the base station along with other UE context information like the K_{eNB} key, from which the actual protection keys are derived. The base station uses the AS Security Mode Command (AS SMC) procedure to indicate to the UE the selected algorithms and start the protection.

When the base station changes during X2 and S1 handovers (see section 9.4), the target base station can change the algorithms if the locally configured algorithm priority lists indicate algorithms different from those currently used in the source base station. Algorithms can only be changed upon handovers. In addition, during an intra-base station handover (e.g. when only the cell changes but not the base station itself) the base stations are not required to support changing of security algorithms.

In an X2 handover, the source base station provides the UE security capabilities and the currently used security algorithms in the source cell to the target base station. The target base station then checks whether the algorithms need to be changed and, if so, indicates the new algorithms by including them in the handover command message sent to the UE via the source base station [TS36.331]. In other words, the target base station creates the handover command message, sends it to the source base station, which then sends it to the UE. In this way, the UE knows the new algorithms before the actual handover happens and can configure security for the communication with the target base station. In this way, the AS level signalling and user plane data messages can be sent protected all the time, even when the algorithms are changed.

There is a security threat that a compromised source base station may lie to the target base station about the UE security capabilities. The source base station could, for example, remove some algorithms from the UE security capabilities and thus force the target base station to select a possibly weaker security algorithm. To mitigate this threat the target base station sends the UE security capabilities received from the source base station to the MME in the path switch message. The path switch message indicates to the core network that the base station has been changed for this UE. The MME can then compare the UE security capabilities it has in its memory with the UE security capabilities it received from the base station. If there is any difference, the MME needs to react to this by, for example, raising an alarm and logging the event. The standard does not require the network to cancel the handover, which would be an obvious reaction when the algorithm is changed to a possible weaker algorithm owing to the mismatch in the UE security capabilities. However, even if the UE security capabilities are different the target base station may not have to change the used algorithms when the current algorithms are listed in the UE security capabilities and also in the supported algorithms priority ordered list of the base station. Thus, it is left for operator policy to decide what happens when the source base station reports differing UE security capabilities compared to the UE security capabilities stored in the MME.

In an S1 handover, the signalling between source and target base station goes through the MME core network element. At this point also the MME may change, and in this case the source MME sends the UE security capabilities to the target MME along with other UE context information. The target MME then sends the UE security capabilities to the target base station. So, the source base station does not provide the security algorithms, but the target MME does. Thus, there is no need for the target base station to send the UE security capabilities back to the MME as it does in the X2 handover case.

8.2   NAS Signalling Protection

8.2.1   NAS Security Mode Command Procedure

In the NAS Security Mode Command procedure the MME sends the NAS Security Mode Command message to the UE, and the UE responds with a NAS Security Mode Complete (or NAS Security Mode Reject) message. The NAS Security Mode Command message contains the security capabilities of the UE (reflected back to UE) and the selected algorithms for NAS signalling protection. The message contains also a key set identifier eKSI that identifies the correct key hierarchy (i.e. the root key K_{ASME}) to be used for key derivations in the UE. Thus, it also identifies the key used to integrity-protect the message. The NAS Security Mode Command message is integrity-protected so that the UE can verify its integrity, but not ciphered, as the UE does not yet know what algorithm and key to use for deciphering. Since the network already knows which algorithms and keys have been selected, it can receive ciphered messages, and thus the UE sends the NAS Security Mode Complete message both integrity-protected and ciphered. The MME starts downlink NAS signalling ciphering after successfully verifying the Security Mode Complete message. The MME starts uplink NAS signalling deciphering after it has sent the NAS security mode command message.

For error cases the network needs to be prepared to receive unciphered messages after the NAS Security Mode Command message has been sent. If the ME is not able to verify the integrity of the NAS Security Mode Command message, it will reply with a NAS Security Mode Reject message protected with the keys used before the NAS Security Mode Command message, if any. However, during initial attachment there is no previous NAS Security Mode Command and thus the reject message cannot be protected, as there is no active security context.

There is a difference between the uplink ciphering activation stage between AS and NAS level Security Mode Command procedures (see below). On the AS level the uplink ciphering starts only after the base station has received the Security Mode Complete message and at the UE side when the UE has sent the Security Mode Complete message. But on the MME level the NAS Security Mode Complete message is ciphered. In this way the UE can send its equipment identifier, IMEISV, confidentiality-protected to the network, provided that the network asked for it in the NAS Security Mode Command message. This improves the user privacy, as the permanent user equipment identifier is not sent in plain text over the air interface and thus cannot be tracked. However, the MME then needs to differentiate between the ciphered NAS Security Mode Complete message and unciphered error message.

See Figure 8.1 for the NAS Security Mode Command procedure. This also shows nonces (NONCE$_{UE}$ and NONCE$_{MME}$) that are used in an intersystem mobility scenario. The usage of these nonces is explained in section 11.1.

8.2.2   NAS Signalling Protection

Integrity and replay protection for NAS messages is part of the NAS protocol itself. A 128-bit integrity algorithm is used with following input parameters: a 128-bit key K_{NASInt}, a 32-bit COUNT, a DIRECTION bit that indicates upstream or downstream signalling, and a constant value BEARER. The COUNT is constructed from the NAS sequence number (SQN) as follows:

$$COUNT := 0x00 \parallel NAS\ OVERFLOW \parallel NAS\ SQN$$

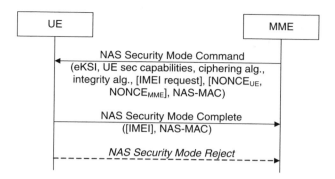

Figure 8.1 NAS Security Mode Command procedure

The leftmost 8 bits are all zero and the NAS OVERFLOW, a 16-bit value, is incremented every time the 8-bit NAS SQN overflows. Thus, the effective COUNT value has 24 bits. Note that there is no need to have a NAS level bearer identity as is the case in the AS level (see later in this chapter), because there is only one NAS level connection between UE and the MME. In other words, NAS signalling uses only one bearer with a constant bearer value. The value BEARER was only included to maximize the similarity with the algorithms on the AS level. The resulting NAS message authentication code (NAS-MAC) is 32-bits long. This full NAS-MAC is appended to all NAS messages when integrity protection applies, except for the NAS Service Request message, which uses only a 16-bit NAS-MAC owing to the space limitations on this particular message. UE sends the NAS Service Request message, for example, when it answers to paging from the MME or when uplink user data is to be sent in order to establish the radio bearers. The message has to be short so that it can be sent efficiently through the radio and to allow fast click-to-view user experience.

As a general rule, once the NAS level integrity and replay protection has been activated with the NAS level Security Mode Command procedure, messages that are not integrity-protected are discarded in the UE and the MME. Also, when the verification of integrity protection fails, the receiver will discard the message. Additionally, only certain messages can be accepted before integrity protection is activated. There are, however, some exceptions to this rule; some exceptional messages are not discarded even if they are not integrity-protected or if the integrity protection fails. All these exception cases are specified in [TS24.301]. Replay protection ensures that the receiver accepts a message with a particular incoming NAS COUNT value only once using the same NAS security context.

NAS level integrity and replay protection is active as long as the corresponding EPS security context is available in the UE and the MME. For example, the Attach Request and the Service Request messages are always integrity-protected if the EPS security context is available.

Ciphering of NAS messages is also part of the NAS protocol. The NAS ciphering algorithm uses the same input parameters as integrity protection, except for the key, which is K_{NASEnc} for ciphering, and the additional parameter LENGTH. LENGTH indicates the length of the keystream that needs to be generated. This parameter does not affect the generated keystream bitstream.

8.3   AS Signalling and User Data Protection

8.3.1   AS Security Mode Command Procedure

The base station indicates the selected algorithms and start of security in the AS Security Mode Command procedure. The base station sends the integrity-protected AS Security Command Message to the UE, which then verifies the message authentication code. Then if the code is correct the UE starts control plane signalling integrity and replay protection and prepares to receive ciphered downlink control and user plane messages. The UE does not start uplink ciphering before it has sent the AS Security Mode Complete message to the base station. This is different from the NAS Security Mode Complete, which is ciphered to allow the UE to send the device identifier, IMEISV, to the network confidentiality-protected. There is no need to provide confidential data to the network in the AS Security Mode Command message. Also, error handling is easier if the base station can activate uplink ciphering after receiving the AS Security Mode Complete message, as then the AS Security Mode Command procedure is successful. If there have been any errors in the procedure on the UE side, it sends a failure message instead (see [TS36.331] for more details). The AS security mode setup procedure is described in Figure 8.2.

8.3.2   RRC Signalling and User Plane Protection

The AS level signalling protocol is called Radio Resource Control (RRC) protocol [TS36.331]. Both the user plane data and the RRC signalling are carried over the Packet Data Convergence Protocol (PDCP) [TS36.323]. Furthermore, the security is implemented on the PDCP layer and not on the RRC layer itself nor on the user plane above PDCP. In this way both the signalling protection and user plane data protection can use the same constructs on the PDCP level. This differs from the NAS signalling protection, which is part of the NAS protocol itself. However, note here that on the NAS level no user plane data protection is needed.

For AS level integrity and replay protection a 128-bit integrity algorithm is used with the following input parameters: a 128-bit key K_{RRCint}, a 32-bit COUNT, a 5-bit BEARER identity, and a DIRECTION bit that indicates upstream or downstream. There can be multiple radio bearers on the AS level – the possible values are described in [TS36.323]. The different bearers may have different service characteristics. The 32-bit COUNT value input

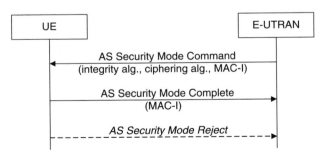

Figure 8.2   AS Security Mode Command procedure

parameter corresponds to the 32-bit PDCP protocol sequence number PDCP COUNT. The RRC integrity protection checksum (the message authentication code, MAC-I) is 32 bits long [TS36.323].

The 5-bit BEARER identity is mapped from the RRC bearer identity. RRC has three signalling radio bearers (SRBs), two for RRC control messages (SRB0 and SRB1) and one for carrying NAS messages (SRB2). Prior to the establishment of SRB2, NAS messages are sent over SRB1. SRB2 is always protected. Messages are sent over SRB1 unprotected before security activation, and protected thereafter. SRB0 is not protected. RRC can configure multiple data radio bearers (DRBs), which are all ciphered but not integrity-protected. NAS signalling is also carried over the PDCP protocol between UE and the base station, so both the AS level and NAS level protection is applied after the activation of AS and NAS level security to the NAS messages. NAS messages that do not have a valid integrity protection checksum on the AS level after the activation of security are not forwarded to the MME.

Every radio bearer has an independent COUNT variable for both uplink and downlink directions. For SRBs the same COUNT variable is used as input for ciphering, replay protection and integrity protection. The base station must take care that the same COUNT value is not used twice with a given security key and radio bearer identity to avoid keystream repetition. To avoid this reuse in large data transfer cases, for example, the base station can trigger an intra-cell handover to get fresh keys and thus also fresh keystream.

To reduce the signalling message size, the 32-bit COUNT variable is formed based on the PDCP sequence number (SQN) and an overflow counter called hyper frame number (HFN) [TS36.323]. Only the sequence number is sent in the messages and the HFN is increased every time the SQN overflows. The length of the SQN can be configured and the length of the HFN is 32 bits minus the length of the SQN (i.e. 5 bits for signalling radio bearers and 7 bits with short PDCP SQN or 12 bits with long PDCP SQN for data radio bearers).

AS level integrity and replay protection is verified both in the UE and in the base station. If the verification fails, the message is discarded. However, on the UE side a specific recovery procedure is triggered (see below; see also [TS36.331]) to allow coping with context mismatches between the UE and the base station that cause the integrity protection to fail. The procedure used for the recovery is called RRC connection re-establishment. This opens a DoS attack possibility for the attacker, as it can send messages to the UE that contain false integrity checksums in order to trigger the (in this case unnecessary) recovery procedure. However, this attack is non-persistent, and no worse than jamming (refer to the discussion in section 6.2.1 on when a potential DoS attack requires specific countermeasure). Thus 3GPP gave higher priority to the possibility to recover from this context mismatch deadlock situation.

Similar to NAS signalling, certain RRC messages have to be accepted before the AS security has been activated. But, for example, setting up bearers carrying user plane data never happens before security is activated. Moreover, the UE only accepts handover messages after security has been activated. Replay protection ensures that the receiver accepts messages with a particular incoming PDCP COUNT value only once using the same AS security context.

AS level ciphering algorithms use the same input parameters as AS level integrity algorithms, except that the ciphering key K_{RRCenc} is used instead of the integrity protection key and the keystream LENGTH input parameter is required. LENGTH indicates how many keystream blocks need to be generated.

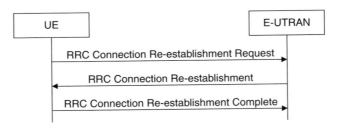

Figure 8.3   RRC Connection Re-establishment Success. Adapted with permission from © 2010, 3GPP™

8.3.3   RRC Connection Re-establishment

The RRC connection re-establishment (Figures 8.3 and 8.4) is initiated by the UE when there are multiple physical layer problems, handover failures, or possibly integrity checksum errors. The purpose of this procedure is to resume the SRB1 operation and to reactivate the security but without changing security algorithms.

The RRC Connection Re-establishment Request message from UE to the base station includes a security token parameter called shortMAC-I. This is generated by taking the 16 least significant bits of the integrity checksum (i.e. MAC-I) calculated with the RRC integrity protection key used in the source cell in case of handovers or in the cell that triggered the re-establishment procedure. The input bits of COUNT, BEARER and DIRECTION are all set to binary ones. The integrity checksum is calculated over the cell identity of the target cell, the physical cell identity of the cell the UE was connected prior to the failure, and the link layer identity of the UE called C-RNTI [TS36.331; TS33.401]. The integrity algorithm used is the same as in the source cell.

The base station sends a RRC Connection Re-establishment message including a Next hop Chaining Count (NCC) parameter to the UE. The UE uses the NCC to synchronize the current K_{eNB} key and further to derive the signalling and user plane (data) protection keys based on the previously allocated security algorithms. At this point the UE starts integrity and replay protection and ciphering for both sent and received messages.

This procedure requires that the source base station prepared the target cell in the target base station (see Figure 9.5 in Chapter 9). Both the RRC Connection Re-establishment Request and RRC Connection Re-establishment messages are sent over SRB0 (signalling radio bearer 0), but the RRC Connection Re-establishment Complete message is sent over SRB1 integrity-protected with the same algorithms as in the source cell.

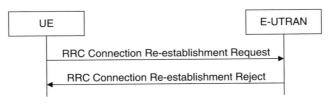

Figure 8.4   RRC Connection Re-establishment Reject. Adapted with permission from © 2010, 3GPP™

Upon failure of the RRC connection re-establishment procedure, the UE moves to idle state and may come back to connected state. This procedure includes also allocating a new C-RNTI link layer identity. Going to idle state and back to connected state involves NAS level signalling with the MME and fresh key delivery from the MME to the base station.

8.4   Security on Network Interfaces

8.4.1   Application of NDS to EPS

With the updates to the Network Domain Security (NDS) framework in Release 8, as described in Chapter 4, the framework was ready to be used within EPS. Thus clause 11 of [TS33.401] makes the application of NDS/IP [TS33.210] mandatory for all IP-based control plane signalling. This requirement is more general than the requirements for 3G (see section 4.5.3) where only the Gn, Gp and Iu reference points are mentioned explicitly.

The reference to NDS/IP implies that the provisions in [TS33.210] apply only optionally to interfaces that are inside one security domain. If the interfaces are, for example, physically protected, then no cryptographic security based on IKE and IPsec is needed.

8.4.2   Security for Network Interfaces of Base Stations

In addition to the general reference to NDS/IP for all interfaces between network elements, the specifications contain provisions for base stations owing to their potentially exposed location. A description of this special environment is given in section 6.4 on platform security for base stations. In many cases the base station location is such that the base station is not inside a (normally physically secured) security domain of the operator. On the other hand, a separate SEG for each base station is not a good solution either, as SEGs are defined for concentrated traffic between security domains, and not for the huge numbers of base stations deployed with a more mesh-like interconnection caused by the coexistence of S1 and X2 interfaces. Thus the requirements on these connections resemble the Za reference point, but without requiring a full SEG functionality at the endpoints.

The base station has connections to the EPC using the S1 reference point and to adjacent base stations via the X2 reference point. Clause 5.3 of [TS33.401] states general security requirements for these links. In particular, integrity, confidentiality and replay protection from unauthorized parties has to be provided. If these links are not considered adequately secured by other means (e.g. physical), then cryptographic means are necessary for the protection of the interface traffic. Details are given in clause 11 of [TS33.401] for control plane data and in clause 12 for user plane data. Clause 13 adds similar requirements for connections to the management system.

What is common to the security of the different planes is the reference to mandatory implementation of NDS/IP [TS33.210] with IPsec in tunnel mode and the reference to the updated IP Encapsulating Security Payload (ESP) RFC [RFC4303]. Also implementation of IKEv2-based authentication with certificates as specified and profiled in [TS33.310] is mandatory for all planes.

Some requirements on the termination points of these secured connections are discussed in section 6.4 on platform security for base stations. The discussion is mainly about the location inside the base station where the handling of integrity protection and ciphering for S1 and X2 purposes has to take place.

In contrast to the general NDS/IP requirements for the Za reference point, for all network interfaces of the base station the termination point of the secured tunnel in the core network may be in a SEG, but also other network elements are allowed for this task.

For S1 and X2 control and user plane connections transport mode IPsec is optionally allowed to be implemented and used. This is in addition to the tunnel mode which is mandatory to be implemented anyway.

8.5   Certificate Enrolment for Base Stations

The previous section explained how the backhaul link of base stations is secured by mechanisms according to Network Domain Security (NDS). Authentication for the establishment of these backhaul links is based on a Public Key Infrastructure (PKI), as specified in clauses 11, 12 and 13 of [TS33.401], which require IKEv2 certificates-based authentication according to [TS33.310].

As the LTE base stations are network elements expected to be deployed in large numbers, and as their authentication is based on operator-signed certificates, mechanisms for the automated mass enrolment of base stations to the operator PKI were specified in clause 9 of [TS33.310].

8.5.1   Enrolment Scenario

The enrolment of the base stations to the operator PKI is necessary, as NDS security mechanisms require the authentication of network elements based on an operator-based PKI (and not a vendor-based PKI). The following scenario is given as an example for the delivery of base stations.

1. The operator orders a base station from the manufacturer and receives a confirmation including the identity of the base station.
2. The manufacturer's personnel installs the base station at the intended site and connects the base station to the intended network, for example an operator virtual LAN.
3. The base station discovers its IP address via DHCP [RFC2131], and receives in the response additional information about a contact address for enrolment.
4. The base station authenticates its vendor-provided identity to the Certification Authority (CA) of the operator, and requests an operator-signed certificate. The CA generates the certificate and sends it to the base station, possibly together with an operator-defined identity.
5. The base station installs this certificate and then uses this operator-signed certificate to authenticate its identity to the Security Gateway (SEG) of the operator for operational connection to the core network.

Note that the above scenario is not usually found in the deployment of HeNBs. Thus for HeNBs enrolment to an operator PKI is not specified, but a vendor-provided device certificate is used instead for authentication. This is described in Chapter 13.

The above-mentioned CA normally consists of two logical parts: the Registration Authority (RA) and the Certification Authority (CA) proper. These are logical elements, and the exact functional split between both is not standardized, and may depend on the actual deployment scenario. This separation also allows operating only one CA in a highly secure environment, while there may be multiple RAs in front of the CA, depending on location, organizational unit, specific task, and so on. The basic functions of these elements are as follows.

- *Registration Authority*. The RA is the front end and performs all communication with the base station. It authenticates the base station, and formally checks the certificate request. In addition, it may perform the authorization check if the base station is allowed to enrol to the operator PKI, and it may assign an operator-defined identity to the base station. Also the check of the proof-of-possession of the private key may be performed in the RA, while some deployments may assign this task to be performed in the CA. After all checks are performed successfully, the RA forwards the certificate request to the CA, and in the end sends the generated certificate received from the CA to the base station.
- *Certification Authority*. The CA performs the actual generation of the certificate, based on the certificate request sent from the base station and checked and possibly augmented or modified by the RA. This is done using the private key of the CA, and thus the CA requires an environment which guarantees the long-term security of this private key. Therefore quite often a CA is operated as one central entity, serving many possible RAs for different purposes.

8.5.2 Enrolment Principles

Requirements for enrolment procedure

The main guideline for the specification of a base station enrolment procedure was to allow a plug-and-play deployment of base stations with:

- usage of an existing standardized protocol for certificate enrolment;
- minimization of manual interaction;
- no need for pre-provisioning of operator-specific data in factory;
- no need for security-relevant provisioning on installation site;
- authentication of a base station to the RA of the operator based on a vendor-signed base station certificate;
- secure provisioning of a base station certificate signed by the operator PKI;
- secure provisioning of an operator root certificate.

An extensive threat and risk analysis for the different possible solutions was carried out. As part of this analysis, the following topics were discussed and resolved. Two solution variants for the provisioning of operator root certificates (see below) were accepted in order to allow for different trade-offs between risk and complexity depending on the deployment scenarios.

Selection of enrolment protocol

Clause 7.2 of [TS33.310] requires the support of the Certificate Management Protocol CMPv2 [RFC4210] for the life-cycle management of network element certificates. Therefore the selection of CMPv2 also for base station enrolment was a natural choice.

Communication channel between base station and RA

The enrolment is done end-to-end between the base station and the RA of the operator. The CMPv2 protocol provides means for proof of origin and integrity protection of the messages, so no additional security tunnel for the communication is needed. As only public data is transferred in a CMP protocol exchange, confidentiality protection is not necessary for the communication either.

Authentication of base station by the operator network

The enrolment of the base station to the operator PKI is based on a vendor-provided base station identity. The base station authenticates itself to the operator network during the enrolment procedure using a vendor-provided public/private key pair installed in the base station before enrolment, and a certificate on the base station identity and the public key signed by a vendor CA. In the terminology of the CMPv2 specification, this enrolment is an in-band initialization using an external identity certificate according to Appendix E.7 of [RFC4210].The authenticating entity in the operator network (e.g. the RA) must be provided with the root certificate of the vendor PKI to be able to authenticate the vendor-provided base station identity. This provisioning of the vendor root certificate must occur before the enrolment procedure in a trustworthy manner.

Proof of possession of the private key

The base station must provide the RA/CA with a proof of possession for the private key which belongs to the public key to be certified. This is accomplished by usage of the Proof-of-Possession information elements within the CMPv2 messages. This private/public key pair may differ from the one used in authenticating the base station to the operator network.

Authorization of base station enrolment

Authorization of the enrolment for each base station is not in the scope of the specification [TS33.310]. Nevertheless the RA has to be informed via management means of the vendor-provided base station identities expected to be enrolled. This management data may include the identity of the base station meant to be used in the operator infrastructure (which may differ from the identity put into the base station in the factory) so that it can be provided to the base station as part of the CMP protocol run. This identity may also be provided to the base station after its installation, but before certificate enrolment – for example, in a DHCP response sent when the base station first attaches to the network.

Provisioning of the operator root certificate

The authentication of the operator network by the base station during enrolment would require a pre-provisioned operator root certificate in the base station. This, in turn, would require provisioning of the operator root certificate in the factory (contradicting the third requirement from the bulleted list above), the installation of the operator root certificate on-site at installation time (which would have some unwanted security implications), or some complex cross-signing relations between vendors and operators (where the base station would be provided with a vendor root certificate in the factory, and the operator root certificate would be cross-signed by this vendor root certificate). Solutions with cross-signing were ruled out, as they would not allow the authentication of single operators (as any cross-signed operator root certificate would be accepted by the base station), and their added benefit would not outweigh the incurred complexity of handling of the many trust relations necessary between a vendor and their many customers. Thus only provisioning of the operator root certificate before or during the CMPv2 protocol run is specified. If there is no operator root certificate provisioned at all, the base station shall assume the enrolment procedure to have failed.

Provisioning of operator root certificate during CMP run

This is the plug-and-play solution without any operator-specific pre-provisioning and security-relevant interaction on installation site. The base station extracts the operator root certificate from the CMP response message from the RA/CA. As there may be multiple certificates in the CMP response, the selection of the operator root certificate is done based on the following two criteria: (a) the certificate is a self-signed certificate (only root certificates are self-signed), and (b) the newly generated base station certificate can be validated by this root certificate, possibly via a chain of intermediate certificates also contained in the CMP response. The risk of not performing an authentication of the operator network at this time was seen as tolerable, as it is still enforced that only allowed base stations may enrol with the operator network. As a base station needs the cooperation of core network elements to be able to establish a connection with the UE, and only a base station with a certificate of the intended operator may connect to their network, a base station with a wrong certificate may never impersonate the intended operator network towards the UE. In addition, architectural provisions below the IP layer may reduce such risks, such as the deployment of virtual LANs.

Provisioning of operator root certificate before CMP run

This means that the operator root certificate is provisioned to the base station either in the factory or by service personnel on installation site. Both variants violate one of the requirements given above; that is, the operator-independent delivery of base stations from the factory, or by having the need to perform security-relevant actions at installation site. On the other hand, this allows the authentication of the operator network during enrolment if the operator is willing to accept the higher complexity of the delivery process or the additional trust into the manual installation procedure. If the base station is provisioned with the operator root certificate before the start of the enrolment procedure, it must use this root certificate for authenticating the RA/CA during enrolment.

Renewal of base station key pair or operator certificate

During the lifetime of a base station, the operator may choose to renew the private/public key pair of the base station, or he may issue certificates with a lifetime shorter than the expected device lifetime. In both cases, the key update message exchange specified in the CMPv2 protocol suite is used. The main difference to the initial enrolment of the base station is that, in this case, the base station authenticates itself to the RA/CA based on the (old) operator certificate, and not the vendor certificate. This implies that all authentications performed during the renewal process are based on the operator root certificate.

Vendor base station certificate

After successful initial enrolment of the base station, the vendor base station certificate is no longer needed according to the specification in [TS33.310]. It is left to vendor policy whether the vendor-provided certificate and the private key are deleted after initial operator enrolment, or whether they are kept to allow a return to the pristine state of the base station. Starting from this pristine state, the base station could then be enrolled to another operator's network. Also a completely new enrolment to the same operator's network could be performed, if needed.

8.5.3   Enrolment Architecture

The enrolment architecture is depicted in Figure 8.5. On the left, the communication between base station and RA/CA using the CMPv2 protocol is shown. On the right, the subsequent usage of the operator-provided base station certificate for establishing the security on the backhaul link to the operator network is given. This is not part of the enrolment proper, but shows that different paths to the operator network are used for enrolment, on the one hand, and later usage of the enrolled base station certificate, on the other hand.

It is a precondition for a successful enrolment of the base station that the RA/CA has the vendor root certificate available, as it is used during authentication of the vendor-provided base station identity. The base station authenticates to the RA using the vendor-signed base station certificate and the vendor-generated private key. After enrolment the base station only uses the operator-provided certificate for connecting to the operator network, protected by

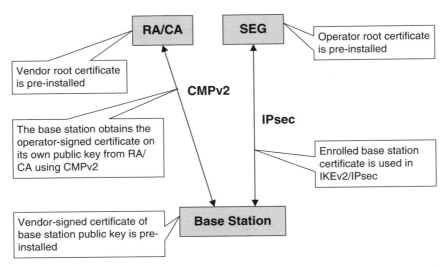

Figure 8.5   Overview of security architecture for base station enrolment. Adapted with permission from © 2010, 3GPP™

an SEG. Consequently, the SEG is not provisioned with a vendor root certificate, which, in turn, ensures that a base station with only a vendor certificate never can gain access to the core network of the operator that lies behind the SEG. The same considerations apply to X2 connections to other eNBs.

Figure 8.5 does not show the necessary security safeguards against attacks on the RA/CA from the outside. This is left to operator decision as it has no influence on the end-to-end CMPv2 interface between base station and RA/CA. Similarly, the exact separation of functionality between RA and CA is not covered in the CMPv2 specification and in [TS33.310] either. There is no need to standardize such a function split as, again, there is no impact on the CMPv2 interface. Instead the operator may choose this split according to their PKI infrastructure, and their particular security policies. An example architecture could be that the RA is located in a demilitarized zone, and then communicates with the CA located within the operator core network. A further function split of the RA is also possible where the demilitarized zone only contains an RA front end and the authorization task of the RA is performed within the operator core network.

8.5.4 CMPv2 Protocol and Certificate Profiles

The complete profile of CMPv2 [RFC4210] for usage in base station enrolment and key update is specified in clause 9.5 of [TS33.310]. It contains all requirements and preconditions, the exact definitions, which message fields are mandatory to use for each message, which entity has to sign certain messages, and how the proof-of-possession fields are to be handled. The profile also refers to the RFC on Certificate Request Message Format (CRMF) [RFC4211], which defines the content of certificate request messages used in CMP, and to the RFC on Internet usage of X.509 certificates [RFC5280].

The following gives an overview of the required message types, and some operational issues to be considered.

Supported CMPv2 messages

Based on the enrolment principles given in section 8.5.2, the CMPv2 profile only includes certificate initialization request and key update functions. Revocation processing, requests for additional certificates, PKCS#10 requests and CRL fetches are not part of the CMPv2 profile. Thus only the following CMPv2 PKI message bodies are required.

- *Initialization Request* (ir). This request allows the initialization of a base station with a certificate from the operator PKI based on a certificate from an external (i.e. vendor) PKI.
- *Initialization Response* (ip). This response to the base station contains the generated base station certificate, the operator root certificate (if provided during CMP run), the RA/CA certificate(s) and any intermediate certificates.
- *Key Update Request* (kur). This request is similar to ir, with the main difference that the request is signed with a private key whose related public key is certified by the same PKI as the new certificate will be, that is by the operator PKI.
- *Key Update Response* (kup). This response is similar to ip. It is a response to a kur message.
- *Certificate confirm* (certconf). With this message the base station signals to the RA/CA that it accepts the newly generated certificate.

- *Confirmation* (pkiconf). This response is the final (empty) response in the CMP message exchange.

Certificate and key usage in RA/CA

The RA/CA uses digital signatures for two different purposes:

- signing of base station certificates;
- signing of CMPv2 PKI messages.

The same private key could be used to sign certificates and messages, but this would require setting key usage extensions for both use cases for the same certificate. This may open up possibilities for misuse, as the part of the RA/CA responsible for signing messages may be tricked into signing a certificate instead. Such signing would not undergo the complete process of certificate generation. Thus, according to good PKI practices, it is recommended that separate private keys and certificates be used for signing certificates and CMPv2 messages.

Certificate profiles

The profiles of the different certificates used in the CMP messages and for verification of signatures are specified in clause 9.4 of [TS33.310]. They are specified based on the existing certificate profiles in clause 6 of [TS33.310].

To allow easy handling of the enrolment of eNBs from different vendors at the same RA/CA, the subject name format for the vendor-provided base station identity is clearly specified. In addition, the inclusion of a distribution point for certificate revocation information (CRL distribution point) is not mandatory in vendor-provided certificates, as the interface for distribution of such information is not in scope of the specification.

8.5.5   CMPv2 Transport

Transport of CMPv2 messages between the base station and RA/CA is done using HTTP, in accordance with [draft-ietf-pkix-cmp-transport-protocols-07]. Implementation of HTTPS is mandatory while its usage is optional.

As the CMPv2 profile given in section 8.5.4 contains only message exchanges originating from the base station, support for RA/CA-initiated HTTP requests (i.e. announcements) is not required.

8.5.6   Example Enrolment Procedure

Figure 8.6 shows the message flow for a successful initial enrolment of a base station to the operator PKI.

A short explanation of the figure is given below, but for a more extensive description see Annex G of [TS33.310].

- Step 1. The base station discovers the RA/CA address.
- Steps 2–4. The base station generates the new private/public key pair if the latter is not pre-provisioned. The Initialization Request (ir) is generated containing the new public key,

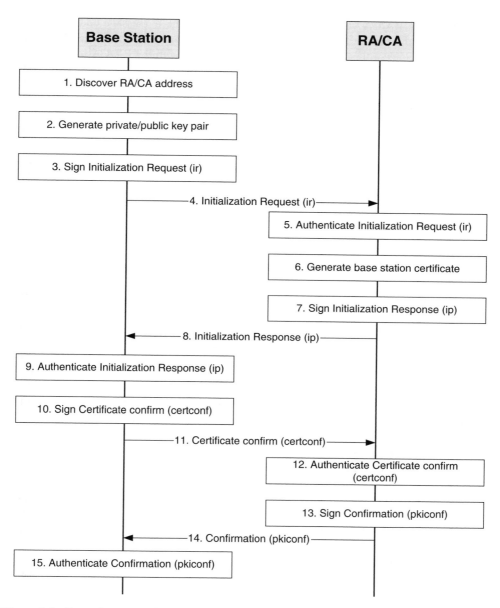

Figure 8.6 Example message flow for initial base station enrolment. Reproduced with permission from © 2010, 3GPP™

the suggested base station identity, if known, and the proof-of-possession field generated by applying a digital signature with the new private key. The ir message is signed using the vendor-provided private key. Its own vendor-signed certificate and any intermediate certificates are included in the extraCerts field of the PKIMessage carrying the ir. The signed PKIMessage is sent to the RA/CA.

- Steps 5–8. The RA/CA verifies the digital signature on the ir message and the proof of possession of the private key. The RA/CA generates the certificate for the base station with an identity according to operator policy and signs it with the RA/CA private key for certificate signing. The certificate is included into an Initialization Response (ip). The ip message is signed with the RA/CA private key for signing CMP messages. The RA/CA certificate(s) and the operator root certificate and any certificates necessary in the trust chain are included in the PKIMessage. The signed ir message is sent to the base station.
- Step 9. If the operator root certificate is not pre-provisioned to the base station, the base station extracts the operator root certificate from the PKIMessage. The base station authenticates the PKIMessage using the RA/CA certificate and installs the base station certificate on success.
- Steps 10–12. The base station creates and signs the Certificate Confirm (certconf) message and sends it to the RA/CA. The RA/CA authenticates the Certificate Confirm message.
- Steps 13–15. The RA/CA creates and signs a Confirmation message (pkiconf) and sends it to the base station. The base station authenticates the pkiconf message.

8.6   Emergency Call Handling

Although protection is typically independent of the protected content, so that all data is protected in similar manner, there is one notable exception: emergency calls and IMS emergency sessions. Regulations on emergency calls vary between different countries, such as whether unauthenticated emergency calls are permitted or not. Regulations of some countries require that it is possible to always make an emergency call with user equipment, even when there is no valid SIM or USIM inserted.

If there is no USIM in the UE then there is no way to authenticate the user in LTE. Furthermore, there is no key agreement and, consequently, neither confidentiality nor integrity protection is possible. All this implies that emergency calls become an attractive target for attackers. It is vital that the system guarantees that it is not possible to use it for anything else but emergency calls when the UE is unauthenticated.

A specific state, called limited service state, is used to describe situations in which a UE cannot obtain normal service [TS23.122]. An idle UE without a valid USIM enters limited service state. There are also situations where a UE that contains a valid USIM would enter limited service state, such as when there are no cells available from the selected PLMN. In the latter case, the UE tries automatically to re-select a PLMN but this could fail for normal service, for example in roaming situations. No other services than emergency services are provided for a UE that is in limited service state.

Specific functions for emergency call handling have been added to EPS in Release 9, whereas in Release 8 only normal procedures are available for emergency purposes. A voice solution for EPS is provided by IMS (see Chapter 12). In IMS, IMS emergency sessions are used for emergency calls [TS23.167]. On the bearer level, there are specific emergency bearers that support IMS emergency sessions. These bearers are available for normally attached UEs, and in addition to UEs in limited service state when the local regulation requires support for unauthenticated emergency calls. Altogether, there are four different ways in which a network may provide support for emergency bearers.

- Support is available only for normally attached UEs, with valid subscription and authenticated by the network. There is no support for UEs in limited service state.

- Support is provided only for authenticated UEs, with valid IMSI and subscription, but the UE may be in limited service state owing to being in a location where it is restricted from service.
- Support is provided only for UEs that have a valid IMSI but authentication may be skipped or it can fail.
- Support is provided to all UEs, even UEs without a USIM or without an IMSI. If there is no IMSI the IMEI may be used to identify the UE for emergency call purposes.

For all emergency bearer services, the MME uses a specific emergency APN to derive the correct PDN gateway for emergency purposes. Note that this gateway is always in the visited network in case of roaming UEs. The motivation for this arrangement is clear: it is also the local Public Safety Answering Point (PSAP) that is typically in a better position to handle the emergency situation than a faraway PSAP.

The PDN connection that is associated with the emergency APN is totally dedicated to IMS emergency sessions and no other services are allowed. In particular, the PDN GW blocks all traffic associated with this APN that is not to or from an IMS network entity providing emergency services.

On the IMS system, there is a specific Call Session Control Function (CSCF) devoted to emergency sessions, called Emergency CSCF (E-CSCF) [TS23.167]. One of its duties is to handle the communication towards a PSAP. The Proxy CSCSF (P-CSCF) has also a key role in handling emergency sessions. Among other duties, the P-CSCF ensures that only registrations for emergency purposes are accepted from emergency PDN connections. Moreover, the P-CSCF blocks all non-emergency session requests that are related to an emergency registration.

All of the above special arrangements guarantee, when taken together, that it is not possible to misuse emergency support to make normal but still unauthenticated calls:

- UE in limited service state can only use emergency bearers.
- Emergency bearers are limited to an emergency APN and a specific emergency-aware PDN gateway.
- This specific PDN gateway allows only traffic to/from IMS entities that handle emergency services.
- The P-CSCF on the IMS side checks that all IMS traffic to/from the specific PDN GW is indeed for emergency purposes, and selects a suitable Emergency CSCF for the further handling of the requests, including finding an appropriate PSAP for the session.

For the case of non-3GPP access to the EPC, Release 9 specifications support handover of an emergency session from E-UTRAN to HRPD (see section 11.2). The reverse direction is not supported, though. The E-UTRAN also provides an indication to the HRPD side about whether the UE has been authenticated in E-UTRAN or not [TS23.402]. The local regulations affect the handling of unauthenticated emergency sessions for the HRPD, similarly as for E-UTRAN.

8.6.1 *Emergency Calls with NAS and AS Security Contexts in Place*

Here we consider the case where the UE and the MME share a NAS security context that can be used to protect an emergency bearer.

In the most typical cases the UE making an emergency call can indeed be successfully authenticated in EPS during the establishment of the emergency call, and the NAS Security Mode Command procedure can be run, or a previously established NAS security context already exists when the emergency call is set up. In both cases, keys are available for ciphering and integrity protection purposes and can be applied normally, on both AS and NAS levels. If an integrity check fails afterwards then the handling of this situation is the same as in the case of non-emergency calls: signalling messages with a wrong message authentication code will be discarded and eventually the call can terminate.

But, even when a NAS security context already exists, the MME may decide, according to its authentication policy, to initiate an EPS AKA run at any time. As explained above, depending on its policy, the network may allow an emergency call to go forward when authentication fails. This case is handled in section 8.6.3.

8.6.2   Emergency Calls without NAS and AS Security Contexts

We next look at the case where UE cannot be authenticated in EPS during the emergency call set-up, and there is no previously established NAS security context. Then there are no established keys available, so there cannot be any ciphering or integrity protection, neither on the AS level nor on the NAS level. However, as, for normal service, integrity protection of AS and NAS signalling is mandatory; it turned out to be procedurally easier to define a 'dummy' integrity protection function for emergency services rather than define an exceptional case of not sending a NAS security mode command at all. This 'dummy' function is called NULL Integrity Algorithm and denoted by EIA0. It simply adds a constant message authentication code of 32 zeros to every message. This null integrity protection function is only allowed for UEs that are in limited service state and not successfully authenticated in EPS. Similarly, a NULL Encryption Algorithm EEA0 has been defined for emergency calls in limited service state.

8.6.3   Continuation of the Emergency Call when Authentication Fails

As already mentioned above, there is a possibility that an emergency call is allowed to proceed even when AKA was run but it ended in failure, either during the set-up of an emergency call or while the emergency call already is in progress. Now we have two different scenarios: either

• UE and MME already share a security context from a previous (successful) AKA run; or
• there is no such shared security context.

In the first case, both UE and the MME may continue using the existing EPS security context after the failed AKA. For this case, it is worth noting the contrast to a failed integrity check. Both AKA and integrity check perform authentication: the AKA for entity authentication and integrity check for message authentication (see section 2.3). However, the emergency call proceeds even when the authentication (by means of EPS AKA) fails while the call can terminate if an integrity check fails.

When there is no shared security context, the MME sends a NAS SMC message with null ciphering algorithm EEA0 and null integrity algorithm EIA0 chosen.

9

Security in Intra-LTE State Transitions and Mobility

This chapter describes security for state transitions and mobility inside LTE. These include registering to the network, moving to ECM-CONNECTED state, intra-LTE handovers, moving to idle state, idle state mobility, and de-registering from the network.

The two layers of LTE security and the key management requirements are reflected in the security of state transitions and mobility scenarios. The first layer security between the UE and the base stations, called AS security layer, is set up only when user plane data needs to be exchanged, but the second layer security between UE and the core network, called NAS security layer, is set up all the time when the UE is registered to the network. An EPS NAS security context of type native (see section 7.4) remains stored in the UE and the MME while the UE is not registered to the network and is used when the UE re-registers to the network.

The second layer (NAS) is used to bootstrap the first layer (AS) security when the UE needs to send or receive data. The first layer security is refreshed with the help of the second layer security between the UE and the core network. Running EPS AKA and a Security Mode Command procedure refreshes the second layer security itself that is, the EPS NAS security context.

Before the security has been set up between the base station and the UE there cannot be any handovers or user plane data transfers. When the UE is in ECM-IDLE state and needs to send a NAS message to the network, a Radio Resource Control (RRC) connection between the UE and the base station is established and both enter RRC_CONNECTED state. The base station needs to transfer the received NAS message to the MME and thus establishes an S1 connection with it. As a result, when the MME receives the NAS message, both the UE and the MME enter ECM-CONNECTED state. For faster transfer to the ECM-CONNECTED state, the connection initiation signalling on the RRC protocol stack will piggyback the UE-initiated NAS signalling messages (i.e. Service Request, TAU Request, Attach Request, or Detach Request).

LTE Security   Dan Forsberg, Günther Horn, Wolf-Dietrich Moeller, and Valtteri Niemi
© 2010 John Wiley & Sons, Ltd

9.1 Transitions to and from Registered State

9.1.1 Registration

When the UE initially registers to the network, the EPS Authentication and Key Agreement (AKA) protocol is run, as described in Chapter 7. As a result, both the UE and the MME share an intermediate key called K_{ASME}. This key is the root for the E-UTRAN key hierarchy (section 7.3). The MME runs the NAS level Security Mode Command procedure (section 8.2) to activate the NAS keys and security algorithms. The NAS protocol specific uplink and downlink message counters belonging to the NAS level security context, called NAS uplink COUNT and NAS downlink COUNT, are set to zero. The NAS level security context exists all the time while the UE is in registered state.

When an AS security context is established the NAS uplink COUNT value is used as a parameter to derive the K_{eNB} key delivered to the serving base station. This ensures that the K_{eNB} is always fresh as there is always NAS level signalling and thus also NAS uplink COUNT increments.[1]

When the UE registers to the network and already has a native EPS security context in the non-volatile memory of the ME or in the USIM (see below), it will use this context to integrity-protect the NAS level Attach Request message. The MME receiving the Attach Request may have this EPS security context of the UE already in the memory. If not, it needs to fetch it from the old MME or run EPS AKA. If the old and the new MME support different security algorithms, the NAS level keys need to be re-derived. Thus, the new MME sends a NAS security mode command with the new security algorithm identifiers and protects the message with the re-derived NAS keys.

If there is no NAS SMC procedure before the AS SMC procedure, the K_{eNB} is derived based on the NAS uplink COUNT in the Attach Request. Otherwise, the MME and the UE will use the start value of NAS uplink COUNT from the latest NAS Security Mode Complete message to derive the K_{eNB} as it is the latest NAS uplink message from the UE. This means that the MME cannot send the security context to the serving eNB before it knows which NAS uplink COUNT to use for deriving the K_{eNB}. See section 7.3 for more information about key hierarchies and key derivations and section 9.7 for more about concurrent security procedures.

9.1.2 Deregistration

There are different cases in which the UE enters deregistered state. The UE can itself deregister from the network, for example if it is switched off. The network can also initiate deregistration, perhaps because some procedure failed. For more information about the reasons when the network initiates deregistration, see [TS24.301].

The EPS security context handling varies depending on whether the UE goes power-off or not and whether the EPS security context is full and native or not – see section 7.4. The mapped or partial EPS security contexts are not stored in the UE or network when UE goes to deregistered state, but the full native EPS security context generally is. There are exceptions to this, such as when the network rejects the attachment request of the UE. In this case, all security context data are removed from the UE and the MME.

---

[1] However, there is an anomalous use of NAS uplink COUNT when K_{eNB} is derived in UTRAN-to-E-UTRAN handover – see section 11.1 in this book.

The UE stores the full native EPS security context to the USIM if the USIM supports storage of EPS security context; otherwise the UE stores the full native EPS security context into non-volatile memory on the ME. The latter case allows using existing USIMs used in 3G networks (i.e. Release 99 USIMs), but does not prohibit updating the USIMs to newer versions that support EPS-specific security context parameter storage.

9.2   Transitions Between Idle and Connected States

In this section we describe security context management on transitions to and from ECM-CONNECTED and RRC_CONNECTED state. UE moves to RRC_CONNECTED and ECM-CONNECTED state when it needs to send or receive data to or from the network. When there is no need to send any data, UE is moved to idle state. In idle state, the UE does not share any security context with the base station, only with the MME.

9.2.1   Connection Initiation

Service Requests and TAU Requests with the active flag set initiate the establishment of an AS security context between the UE and the base station based on the existing security context the UE shares with the MME. The NAS security is used to bootstrap the AS security. The K_{eNB} key is used to further derive RRC and UP protection keys in the base station and the UE. The K_{eNB} itself is derived based on the K_{ASME} and the NAS uplink COUNT value of the message that caused the MME to send the security context to the base station. So, the MME derives the K_{eNB} and sends it along with the UE security capabilities to the serving base station. The base station selects the AS security algorithms (see section 8.1) and sends the AS level security mode command to the UE. The UE replies with the AS security mode complete message (see section 8.3).

The MME will also derive an initial Next Hop (NH) key parameter, but does not send it to the base station in the initial security context setup because the K_{eNB} is already in the message and only one key and related NCC value is sent each time. The NH is like another K_{eNB} and has an associated NCC value. The NH key is created in an iterative fashion from K_{ASME} and the previous NH value, and the NCC indicates the number of iterations. The K_{eNB} is used as the initial NH value. An NH is never sent to the base station without NCC and thus they are denoted as an {NH, NCC} pair. These pairs are used in the handover key management for creating fresh keys in the base stations (see section 9.4).

Note that, from the security perspective, it is not useful to send also a freshly generated {NH, NCC} pair in the initial security context setup as one key only is required and there is no preceding base station that could have gained knowledge of the K_{eNB} in a horizontal X2-handover (see section 9.4).

9.2.2   Back to Idle State

The UE enters the idle state when its signalling connection to the MME has been released or broken. The release or failure is explicitly indicated by the base station to the UE or detected by the UE itself. At this point, the base station discards all AS level security context parameters related to the UE. The MME also discards the AS security context related {NH, NCC} pair when the UE enters the idle state. The UE discards the AS level security context and the {NH, NCC} pair.

If the current EPS security context is native (a result of EPS AKA), and not mapped (a result of intersystem handover from UTRAN or GERAN), the UE also updates the relevant EPS NAS security context parameters. The parameters are updated on the USIM if the USIM supports EPS security context parameter storage, otherwise on the non-volatile memory.

9.3 Idle State Mobility

In this section we describe idle state mobility inside LTE. For intersystem idle state mobility (e.g. between UMTS and LTE), refer to Chapter 11. Idle state mobility happens when the UE moves and the network broadcasted Tracking Area Identifier (TAI) changes [TS24.301]. At this point, if the TAI is not included in the list of TAIs assigned to the UE during a previous Attach procedure or Tracking Area Update (TAU) procedure, the UE needs to initiate the TAU procedure by sending a NAS level TAU Request message to the network.

In idle state the UE is not connected with any base station, but it listens periodically to the broadcasted system information messages from the network. These broadcasts also include the TAI. If the network needs to reach the UE while in idle state (e.g. due to incoming user plane data), it broadcasts a UE-specific message to the tracking area(s) in which the UE is registered (this is called paging). The UE receives and recognizes the message as it contains its identity and responds to the network by initiating a transfer from idle state to connected state with a Service Request message [TS24.301]. In this way, the network is able to reach the UE even though it does not have an actual connection with the UE through a base station in idle state. Thus, the the location accuracy of the UE in idle state is per list of tracking areas. Of course, the UE can also send the Service Request message when it needs to move to the connected state itself (e.g. when it has user plane data to send to the network).

If the UE moves and the broadcast TAI changes to a value not included in the list of TAIs assigned to the UE during a previous Attach procedure or TAU procedure, the UE needs to inform the network that it is now in a different tracking area by sending a NAS level TAU Request message. The network responds with a TAU Accept message (Figure 9.1). The TAU procedure is always UE-initiated, and there are multiple reasons why the UE sends a TAU Request message. The two main reasons are the idle state mobility and the periodic

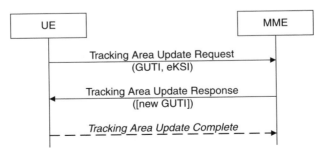

Figure 9.1 Tracking area update procedure

Figure 9.2 Distributing authentication data within one serving domain. Adapted with permission from © 2010, 3GPP™

TAU procedure. The periodic TAU procedure is needed to assure the network that the UE is still registered and has not moved out of coverage. If the UE is no longer registered, the MME can free resources and release the bearers of the UE.

As the TAU messages are protected, the TAU procedure also serves as a periodic local authentication procedure in idle state. However, the periodic TAU procedure does not replace the periodic running of EPS AKA, which is run more seldom. The EPS AKA leads to a refresh of the key hierarchy and ensures the continued presence of the USIM, which the TAU procedure does not.

Note that, in order to send the TAU Request message, the UE needs to enter connected state on the radio level – that is, RRC_CONNECTED. Also, the TAU Request itself transfers the UE and the MME into connected state from the NAS protocol point of view – that is, ECM-CONNECTED state. The UE and the base station run the connection initiation procedure to create the Signalling Radio Bearer SRB1 and enter RRC_CONNECTED state. UE uses the SRB1 to send the TAU Request message to the base station.

Multiple operators may share the same base stations. Thus, one base station can have connections to multiple MMEs owned by different operators, and for load balancing and redundancy reasons also to multiple MMEs owned by the same operator. For this reason the UE sends the network allocated Globally Unique Temporary Identity (GUTI) that contains the Public Land Mobile Network (PLMN) identity and the MME identity [TS23.122] to the base station [TS36.331]. The PLMN identity is basically the operator's network identifier. The base station then routes the NAS level message to the right MME based on the GUTI it received from the UE.

The UE integrity-protects the TAU Request message, but does not cipher it. The reason is that the MME receiving the TAU Request message needs to know the identity of the old MME from the GUTI, and it cannot decipher the message without the context of the UE that resides in the old MME.[2] Since the UE sends its temporary identity GUTI in the TAU Request message, an adversary may try to track the UE based on this identity. However, the network can allocate a fresh temporary identity and provide it to the UE in the TAU Accept message, which is always both ciphered and integrity-protected. In this case the UE responds with a TAU Complete message to acknowledge the new GUTI.

Along with the current temporary identity GUTI, the UE also inserts the current EPS security context related key set identifier eKSI into the TAU Request message. Based on these two parameters, GUTI and eKSI, the network is able to find the right old MME (MMEo, Figure 9.2) and the right security context that can be used to verify the integrity checksum of the TAU

---

[2] Partial ciphering of the message was not considered applicable.

Request message. The old MME sends the authentication data of the UE including the current EPS security context to the new MME (MMEn, Figure 9.2), which then sends a TAU Accept message to the UE. For more information on authentication data exchanged between MMEs see section 7.2.4. If the new MME receives a response from the old MME that the user could not be authenticated, it initiates an EPS AKA.

The old MME and the new MME may support different security algorithms. If the new MME wants to change the NAS level security algorithms, it needs to initiate a NAS level Security Mode Command procedure before responding with a TAU Accept message protected with the new security algorithms.

If the UE has pending data to be sent to the network and it needs to send the TAU Request at the same time (e.g. due to the tracking area change or a needed periodic TAU procedure), it can set an active flag in the TAU Request as true. In this way, the UE does not need to send a separate Service Request message in addition to the TAU Request message. As a result, the MME will send AS security context data to the base station, and the base station will establish the AS security with the UE and set up the user plane.

9.4   Handover

In this section we first take a look at the handover key management requirements and mechanisms background. Then we identify the mechanisms that are used in LTE key management. Readers who want to read only about the LTE handover key management mechanisms may skip the first two subsections and jump directly to section 9.4.3.

9.4.1   Handover Key Management Requirements Background

There are multiple definitions for key management. The Internet RFC 4949 defines it as follows: '*The process of handling keying material during its life cycle in a cryptographic system; and the supervision and control of that process*' [RFC4949]. NIST defines it as: '*The activities involving the handling of cryptographic keys and other related security parameters (e.g., IVs [i.e. Initialization Vectors], counters) during the entire life cycle of the keys, including their generation, storage, distribution, entry and use, deletion or destruction, and archiving*' [FIPS 140-2]. The Open System Interconnection Reference Model (OSI/RM) describes it in the following way: '*The generation, storage, distribution, deletion, archiving and application of keys in accordance with a security policy*' [ISO 7498-2]. Here we use the term 'key management' to mean the mechanisms and rules for creating, distributing, deriving and using cryptographical keys resulting from an authentication procedure, and we limit it to the scope of mobile networks.

There are multiple documents listing requirements for key management for mobile networks in general. For example, the Internet Engineering Task Force (IETF) has created a best current practice document [RFC4962] that describes requirements or guidance for authentication, authorization and Accounting (AAA) [Sklavos *et al.* 2007] key management [RFC2903]. IETF has also criteria for evaluating AAA protocols for network access [RFC2989]. Also, both WLAN [IEEE 802.11] and WiMAX [IEEE 802.16] and [WiMAX], follow similar guidelines in their specifications. The 3rd Generation Partnership Project (3GPP) has defined general security requirements and architectures for the cellular networks like GSM, UMTS and LTE [TS21.133; TS33.102; TR33.821 TS33.401; TS33.402]. The different standardization bodies have separate requirements documents and settings, but, at a high level, the key management security requirements are similar for mobile networks. The common requirement is to have cryptographically separate keys (see below).

The main threat for handover key management is key compromise, such as when an attacker attacks a base station to retrieve the keys. To mitigate this threat, key separation is required at many levels. Keys A and B are separate if key B cannot be derived from key A, and key A cannot be derived from key B. For the key derivation, a Key Derivation Function (KDF) is used, which must be a one-way function (e.g. a hash function like SHA256).

Partial key separation is achieved if the requirement holds only in one direction, but not in the other – when backward or forward key separation applies, but not both. For the definitions, see section 6.3.

The following security properties of the handover key management are fulfilled by the LTE key management:

- key separation between access network technologies;
- key separation between base stations;
- key separation between UEs;
- key separation between algorithms;
- key separation between control and user planes (i.e. signalling messages and user data);
- key separation between integrity protection and ciphering;
- keystream separation between bearers and stream directions;
- keystream (when using a stream cipher) always fresh, i.e. the same key stream must not be used twice to cipher the data.

In addition to these, implementation-specific security requirements are included, such as for base station security hardening purposes. LTE is the first 3GPP radio access technology that puts implementation requirements on base stations. At a high level, the requirements mean that the key derivations, integrity protection, deciphering and ciphering must happen inside a secure environment. However, the requirements do not enforce any specific mechanisms.

GSM does not follow forward or backward key separation principles as the same key is transferred between the base stations. UMTS bypasses this requirement by introducing a middle network element above the base stations called the Radio Network Controller (RNC) that terminates the signalling and data protection. The RNC is typically in a physically secure place, which makes it more resistant to physical attacks. As in GSM, the RNCs transfer the same keys to the target RNC. LTE does not have an RNC, and signalling and data protection termination happens in the base stations. However, LTE supports both backward and, with a well-defined limitation, forward key separation.

9.4.2 Handover Keying Mechanisms Background

User authentication is needed when the mobile terminal (MT) attaches to the network.[3] Then, as a result of the authentication and successful authorization for network access, the MT and

---

[3] In this section we use the terms MT and base station in a broader and network technology independent sense to mean a wireless terminal and a wireless receiver that communicates with the MT via a wireless link and with the network via wireless or wired link.

the network share a key that is used to protect the communication with the network. When the MT is in a state that corresponds to what is called 'connected state' in LTE, also the MT and the base station share a key. When the MT changes the base station in the network as a result of mobility, the new base station also needs to share a key with the MT. There are multiple ways to deliver the base station specific keys efficiently to the new base station before or during handover. Public key-based methods are traditionally not considered because asymmetric cryptographical operations (like decrypting and signing with a secret key of the public key pair) are considered computationally too heavy for mobile devices and radios, in which handovers are very time critical.

Delegated authentication

Delegated authentication means that the Authentication Server, like Home Subscriber Server (HSS) in EPS, delegates the authentication authority to a local authentication agent, which may be a local AAA server or signalling gateway close to the MT like the MME in E-UTRAN. Typically this happens deriving a key from the root key that can be distributed to the access network Key Distributor (KD). This is a common approach in many mobile network key management frameworks as the mobility happens mostly inside the access network or between the access networks, and there is no need to go back to the authentication server.

Key request

Key request is the simplest form of key delivery to the base station. The base station sends a key request to the KD when the MT hands over to it. The KD creates a fresh base station specific key and delivers it to the new base station. A modified mechanism can be used for cases where the handover signalling goes through a centralized element providing the KD functionality (e.g. a WLAN switch or the MME in EPS). In this case, the source base station sends a key request to the KD along with other mobility signalling, but the KD then sends a fresh key to the target base station, instead of to the source base station. LTE uses this modified key request scheme in S1 handovers and a normal key request mechanism in X2 handovers, except that in X2 handovers the fresh key is used in the next handover and not in the current handover (see section 9.4.3).

Pre-distribution

In a pre-distribution (or pre-emptive keying) scenario [draft-irtf-aaaarch-handoff-04; Mishra *et al.* 2003; Mishra *et al.* 2004; Kassab *et al.* 2005] the KD derives base station specific session keys and distributes them to a number of base stations when the MT has successfully attached to the access network. The specific base stations, and the number of them included in the pre-distribution scheme, can vary [Mishra *et al.* 2004]. This scenario makes the handover faster as the key is already in the target base station, provided that it was in the distribution group of the pre-distribution algorithm. The main disadvantage of a pre-distribution scheme is that it increases signalling between the KD and multiple base stations. Also, the KD needs to pre-distribute the keys to multiple neighbouring base stations, but the MT may never visit the neighbouring base stations the keys were pre-distributed to. A modified pre-distribution scheme is where the base station sends keys to the neighbouring base stations for preparing them for a possible handover. LTE does not apply pre-distribution as such, because the source base station does not prepare multiple target base stations. However, the X2 handover prepares the target base station by sending the key to it before the actual radio break. In this way the key distribution part happens before the time-critical handover break and thus makes the LTE handovers very efficient as in pre-distribution in general.

Optimistic access

Aura and Roe describe a method they call optimistic access [Aura and Roe 2005], in which the network delivers a ticket to the MT. The MT then uses the ticket as a temporary authentication key to get access before the normal authentication procedure is finished with the target base station. This resembles ticket-based methods like Kerberos [Miller *et al.* 1987; Neuman and Ts'o 1994] protocol. Ohba and Dutta describe a kerberized handover keying method, in which they also send a ticket to the target base station [Ohba *et al.* 2007]. Kamarova and Riguidel [Komarova and Riguidel 2007] continue with the same mechanism and use the tickets for fast inter-system roaming and also let the home network provide multiple tickets to multiple visited networks at the same time for the MT. This mechanism increases over-the-air signalling and is not well suited for LTE and thus not applied.

Pre-authentication

Pack and Choi introduced a pre-authentication mechanism where the MT authenticates to multiple base stations through a single base station [Pack and Choi 2002a; Pack and Choi 2002b; draft-ietf-pana-preauth-07; draft-ietf-hokey-preauth-ps-09; Dirk Balfanz Smetters *et al.* 2002]. In this way the MT can pre-establish shared keys with multiple neighbouring base stations. This makes the next handover fast as the keys are already established. However, the MT may have to run pre-authentication with multiple base stations as it is not certain to which base station the MT is handed over next. It increases signalling over-the-air (battery life) and between base stations. Pre-authentication sits well with intersystem handovers, as the source and target systems may not support the same key management or authentication mechanisms. In EPS, a form of this mechanism is applied to handovers between LTE and cdma2000 HRPD access. These two access types are sufficiently different so that the more efficient key mapping techniques applied to handovers between LTE and GSM or 3G systems cannot be used.

Session keys context

Session Keys Context (SKC) [Forsberg 2007] is a way to distribute keys to the base stations. SKC contains multiple session keys separately encrypted for each and every base station. The SKC is created in the KD, for a number of base stations and sent to the base station that the MT is currently attached to. When the MT moves, the SKC is transferred between the base stations. Transferring the SKC between base stations may use, for example, the context transfer protocol [RFC4067], or inter-access point protocol [IEEE 802.11F]. Each base station gets the session key from the SKC and creates encryption and integrity protection keys from it.

Table 9.1 Example SKC rows for three base stations

Base station identity	Encrypted key (SK)	Message authentication code (MAC)
ID_{BS1}	$E_{SA-BS1}\{SK_{MTx-BS1}\}$	$MAC_{SA-BS1}\{ID_{BS1} \parallel E_{SA-BS1}\{SK_{MTx-BS1}\}\}$
ID_{BS2}	$E_{SA-BS2}\{SK_{MTx-BS2}\}$	$MAC_{SA-BS2}\{ID_{BS2} \parallel E_{SA-BS2}\{SK_{MTx-BS2}\}\}$
ID_{BS3}	$E_{SA-BS3}\{SK_{MTx-BS3}\}$	$MAC_{SA-BS3}\{ID_{BS3} \parallel E_{SA-BS3}\{SK_{MTx-BS3}\}\}$

The session keys are encrypted and accompanied by base station identity information. The encrypted session key and base station identity information are signed using a Security Association (SA) between the KD and the base station. Each base station that receives this context finds its own encrypted SK based on its identity. Table 9.1 shows an example context row in an SKC. The row is integrity-protected with a message authentication code, such as keyed HMAC [RFC2104].

Each of the keys is derived from the root key and the target base station identifier as in formula (9.1):

$$SK_{MTx-BSi} = KDF(root\ key||ID_{BSi}||TID_{MTx}) \tag{9.1}$$

The ID_{BSi} is the access point identity. SKC assumes that the base station identity is available for the MT when attaching to it so that the MT can use formula (9.1) to derive the key from the root key that is holds. In this way the key management with SKC is simple, and the base station identity is also authenticated. The last parameter is temporary identity of the MT (TID_{MTx}) in the access network. It is assumed that the temporary identity of the ME does not change between consecutive attachments/handovers to the same base station. Also, if all the input parameters to the keystream generation are reset while attached to the base station (e.g. sequence numbers) the key must be refreshed.

Illustrative key derivation function parameters for deriving integrity protection and ciphering keys are given in formulas (9.2) and (9.3) below, where the parameters A and B are constants that make the keys different. Additionally, the selected integrity protection and ciphering algorithms should be bound to the keys as KDF input parameters (not shown in the equations).

$$K_{Int} = KDF(SK_{MTx-BSi}||TLinkID_{MTx}||A) \tag{9.2}$$

$$K_{Enc} = KDF(SK_{MTx-BSi}||TLinkID_{MTx}||B) \tag{9.3}$$

The SKC mechanism was proposed to the 3GPP security working group as a candidate mechanism at an early stage of LTE security standardization but, after intensive discussions, it was not selected. One reason was the complexity of determining the right group of base stations to be included in the SKC, and another was the lack of a base station identity on the air interface in LTE.

All these key management methods have different properties and are suitable for different mobile network architectures and deployments. Forsberg extensively compares key request, pre-distribution and pre-authentication key management methods [Forsberg 2007]. See also the LTE handover key management analysis with SKC [Forsberg 2010].

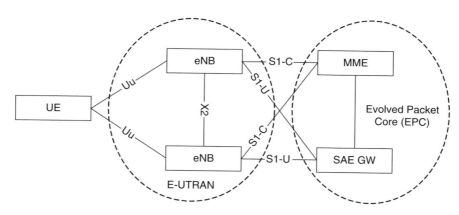

Figure 9.3   Evolved Packet System (EPS)

9.4.3   LTE Key Handling in Handover

In LTE there are two types of handover, X2 and S1, named after the interfaces over which the main handover signalling is transported (Figure 9.3). In an X2 handover, the handover preparation happens between source and target base station through a direct interface between the base stations – the X2 interface. In an S1 handover, the signalling is sent via the MME.

One of the major differences between X2 and S1 handover types is that the MME is informed of the path switching before the break in S1 handovers and after the break in X2 handovers. Path switching is a location update procedure from the target base station to the MME. So, in S1 handovers the MME can provide fresh keying material to the target base station before the UE receives the command by the source base station to hand over to the target base station. There is no need to send a path switch message in the S1 handover, as the MME already knows the target base station identity and location.

From a security perspective, these two handovers are different as the MME can provide fresh keying material for the target base station before the break (Figure 9.4), but in an X2 handover the MME can provide fresh keying material only after the handover in the path switch acknowledgement message for use in the next handover.

Fresh keying material is derived in the MME and the UE based on the Next Hop (NH) key and the local master key K_{ASME}. The key derivation steps are illustrated below showing that NH is computed in an iterative fashion. A more precise formula including length fields and constants is given in Annex A of [TS33.401].

$$NH_0 = K_{eNB-0} = KDF(K_{ASME,} \text{ NAS uplink COUNT}) \qquad (9.4)$$
$$NH_1 = KDF(K_{ASME,} K_{eNB-0}) \qquad (9.5)$$
$$\cdots$$
$$NH_{NCC+1} = KDF(K_{ASME,} NH_{NCC}) \qquad (9.6)$$

When AS security is set up, the initial K_{eNB-0} is derived from the K_{ASME} and the current NAS uplink COUNT. At the same time the initial NH_1 is also derived. For the initial NH_1 an NH

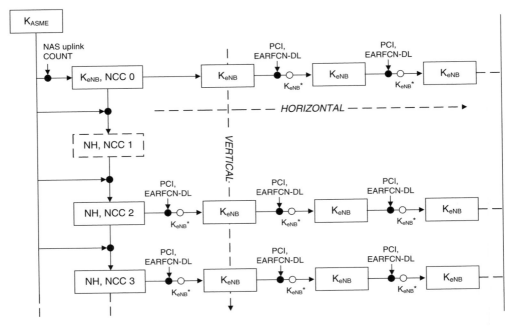

Figure 9.4 Key handling in LTE

Chaining Count value (NCC) is initialized to 1 as the initial $K_{eNB\text{-}0}$ is assumed to have NCC value equal to 0.

The NCC is a 3-bit key index (values from 0 to 7) for the NH and is sent to the UE in the handover command signalling. It never decreases from the UE perspective, as the UE cannot go backwards in the chain of deriving the NH values and does not store the old NH values. So, if the NCC that the UE receives in the handover command is greater than the NCC value for the current K_{eNB} in use, the UE will do vertical key derivation (Figure 9.4), after synchronizing the {NH, NCC} parameter corresponding to the received NCC. Otherwise, the UE will do horizontal key derivation. Thus, the vertical key derivation in Figure 9.4 happens when the NH is used, and horizontal key derivation when the current K_{eNB} is used as the basis for the $K_{eNB}*$, which is then used in the target base station to create integrity protection and ciphering keys.

Further NH values are then derived from the previous NH and K_{ASME} as in formula (9.6). Thus, for each S1 handover and path switch signalling for X2 handovers, the MME provides a fresh {NH, NCC} pair to the target base station. In X2 handovers, the fresh {NH, NCC} pair can be used only for the next X2 handover, if available in the base station. In S1 handovers, the target base station uses the fresh NH to derive a new K_{eNB}, as in formula (9.8) below.

Backward and forward key separation

LTE provides backward key separation, where the source base station uses a one-way function as a key derivation function to get the target base station specific key $K_{eNB}*$. In other words, the target base station cannot derive or deduce any keys the UE used with the source base station. Thus, the backward key separation happens after one hop. The two different possible key derivation steps are shown in formulas (9.7) and (9.8). The two parameters included in the key derivation are the Physical Cell Id (PCI) and a frequency-related parameter called EARFCN-DL.

$$\text{Horizontal} : K_{eNB*} = KDF(K_{eNB,}\, PCI, EARFCN\text{-}DL) \tag{9.7}$$

$$\text{Vertical} : K_{eNB*} = KDF(NH_{NCC}, PCI, EARFCN\text{-}DL) \tag{9.8}$$

Formula (9.7) is for cases when the source base station does not have a {NH, NNC} pair available. This happens after an S1 handover if the following handover is an X2 handover, as the source base station does not have an {NH, NCC} pair available, or for an X2 handover after an initial context set-up. It also happens after an intra-base station handover, where path switch is not required (i.e. the path to the base station does not change) and, hence, a fresh {NH, NCC} cannot be provided to the base station with the path switch signalling. But then, for an intra-base station handover, the attack scenario that motivates key separation does not apply as no key is propagated from one base station to another.

Formula (9.8) shows the case where the source base station has an {NH, NCC} pair available (X2 handover), or the target base station has an {NH, NCC} available (S1 handover) and can use it to derive a fresh $K_{eNB}*$. Note that, in the S1 handover case, the MME derives the fresh {NH, NCC} pair and provides it directly for the target base station. Thus, the source base station does not know the $K_{eNB}*$ in the S1 handover case. This is one-hop forward key separation, where the source base station does not know the target base station key after a single handover. Two-hop forward key separation happens in X2 handovers as the source base station knows the {NH, NCC} pair and thus also the $K_{eNB}*$, but after the second hop this particular source base station does not know the next $K_{eNB}*$ any more as it does not know the respective {NH, NCC} provided in the path switch signalling after the X2 handover.

9.4.4   Multiple Target Cell Preparations

In LTE the source base station may prepare multiple target cells in the target base station for a possible handover failure when the UE cannot connect to the cell that was first selected. In the handover failure case the UE either selects the original source cell again or another cell and tries to connect to it.

As can be noted from formulas (9.7) and (9.8), the $K_{eNB}*$ is different for each target cell, because the cells have different identities and frequencies. Figure 9.5 illustrates the different keys. It also shows that, even if the UE would fail handovers to multiple different target eNBs and return back to the originating cell and base station multiple times, the different target base stations have separate keys. Note that each base station may serve even 32 different cells, but for each of the cells a separate $K_{eNB}*$ is needed because the key is bound to the cell identity and frequency.

The handover failure signalling procedure is called RRC Connection Re-establishment and a token called shortMAC-I is used to authenticate the UE to the target cell. The signalling in this procedure is not protected; that is, the messages are sent in the signalling radio bearer 0 (SRB0), which is not ciphered or integrity-protected. Thus, the token is used. See section 8.3 for further details of RRC Connection Re-establishment procedure and the token generation.

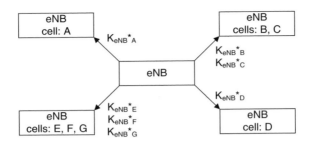

Figure 9.5   Separate keys for different target cells

9.5   Key Change on the Fly

Key change on the fly is needed to get new keys into use when AS security has already been activated. Thus, the new keys are taken into use on the fly, while sending and receiving data. There are two cases for running the key change on-the-fly procedure, namely K_{eNB} rekeying and K_{eNB} refresh. In both cases, the RRC and UP protection keys are refreshed. Also NAS level keys can be changed, but this happens with the NAS level security mode command procedure and is thus different from the AS level (i.e. RRC and UP keys) key change on-the-fly procedure.

9.5.1 K_{eNB} Rekeying

K_{eNB} rekeying happens when the whole key hierarchy has been renewed in the MME, either by activating the (partial) EPS security context generated in an EPS AKA run or by reactivating a native EPS security context after handover from UTRAN or GERAN to LTE (see Chapter 11). The MME creates a new K_{eNB} in a way similar to the procedure when AS security is set up (see section 9.2), using the key K_{ASME} and the NAS uplink COUNT from the NAS Security Mode Complete message from the UE to the MME. In other words, the MME has to run a NAS level Security Mode Command procedure to change the NAS keys before sending the new AS security context with a fresh K_{eNB} to the serving base station. In this way, the new K_{ASME} and NAS keys become the current EPS NAS security context before the key change on-the-fly procedure is triggered in the base station.

   When the base station receives the UE Context Modification request from the MME, it initiates the key change on-the-fly procedure with the UE. This procedure is based on intra-cell handover, and the same key derivation steps apply here as in the normal handover case. The intra-cell handover command includes an indication of key change on-the-fly procedure and thus the UE knows that it needs to re-derive the K_{eNB} based on the new current K_{ASME} key.

9.5.2 K_{eNB} Refresh

The K_{eNB} refresh procedure happens locally in the base station and the UE. It is based on intra-cell handover signalling. The same K_{ASME} is still in use, and no new NAS keys are derived. Only the RRC and UP keys are refreshed based on normal horizontal key derivation

(Figure 9.4), namely chaining the current K_{eNB}. The K_{eNB} refresh procedure is needed when the sequence number COUNT is about to be reused, such as due to COUNT wrap-around in a bearer that used the same bearer id when the keys derived from this K_{eNB} were taken into use. If K_{eNB} refresh did not occur the keystream used to cipher the messages would be repeated, which is a serious security flaw.

During K_{eNB} refresh, the UE gets the intra-cell handover command and notices that the current EPS NAS security context has not changed; that is, the current K_{eNB} was derived from the same K_{ASME} key as the current NAS keys. Thus, the UE uses the current K_{eNB} with the horizontal key derivation procedure.

9.5.3   NAS Key Rekeying

NAS level rekeying happens with the NAS Security Mode Command procedure as normally. The MME must initiate an EPS AKA and run the NAS Security Mode Command procedure well before the NAS uplink or downlink COUNT value overflows. This ensures that the keystream never repeats.

After EPS AKA, or when reactivating a native EPS security context after handover to LTE, the NAS level rekeying must happen before the MME sends the new K_{eNB} to the base station to initiate K_{eNB} rekeying.

9.6   Periodic Local Authentication Procedure

In some cases, the bulk data transfer – the user plane (UP) packets – may not be ciphered, perhaps because local regulations do not allow ciphering of the radio interface. However, the RRC signalling is still integrity-protected and can thus be used to locally authenticate the UE that is sending the UP packets.

The base station initiates the local authentication procedure (Figure 9.6), and sends the uplink and downlink PDCP COUNT variables of the UP bearers to the UE in the Counter Check message. UE compares the values with its own and sends the differing values to the base station in the Counter Check Response message. If the base station receives a counter check response that does not contain any COUNT values, the procedure ends and the base station knows that there has not been any packet injection or deletion attacks on the link with the link layer identity of the UE and radio configuration.

The PDCP COUNT consists of a hyper frame number (HFN) and sequence number (SQN). The SQN is visible in every data packet on the link and is increasing. The HFN is increased when the SQN overflows (i.e. the SQN value reaches its maximum and starts from zero again). So, an attacker, if able to inject packets on the unciphered link, may send UP packets so that the HFN overflows and the SQN value becomes (roughly) the same as it was before the attack. In this way the base station and UE would not notice problems in the SQN but still have differing HFN values in the memory. Since the UP is not protected, the HFN is not used as an input parameter for deciphering. So, if the UE includes PDCP COUNT values, it means that the UE has different PDCP COUNT values, or more precisely different HFN values, in memory from the base station.

The base station may release the connection with the UE if this local authentication procedure reveals differing PDCP COUNT values from the UE. The base station may also report to the MME or an Operations and Management (O&M) server for further data analysis and logging.

Normally, the user plane packets are ciphered and meaningful packet injection is much harder as the attacker does not know the ciphering keys and thus deciphering on the base station or UE results in random data and not protocol-specific headers. However, if base station and UE do not do any validity checks on the packet contents, it may cause some harm or problems on the implementations. Only integrity protection preserves the packet integrity. Even though in LTE the user plane ciphering is bound to the packet sequence numbers (i.e. the 32-bit PDCP COUNT), there is no guarantee that the receiver, when deciphering the packet, is assured that the packet contents have not been changed. However, binding the ciphering to the sequence numbers makes it much harder to replay packets, and thus packet injection with correct sequence number and some valid content after deciphering is considered hard as it would require information about the ciphering key.

9.7   Concurrent Run of Security Procedures

There are many security and mobility related procedures that both the base station and the MME can initiate, and this may happen concurrently. Whether this is a design or implementation problem does not matter if concurrency rules for the procedures help to avoid errors and reduce complexity in implementations. For example, we know that after EPS AKA there is a new key K_{ASME} in the MME, but the NAS keys are still based on the old key K_{ASME}. Thus, the MME needs to initiate a NAS Security Mode Command procedure. However, when this occurs during a service request procedure and at the same time the MME sends the K_{eNB} to the base station, there is a race condition between the AS and NAS level signalling procedures as the UE may derive the K_{eNB} based on the old or the new key K_{ASME}, dependent on whether the UE receives the AS level Security Mode Command before or after the NAS level Security Mode Command. Next, we describe the ten concurrency rules listed in [TS33.401].

1. The first rule states that the MME must not initiate any procedure that includes sending a new K_{eNB} to the base station during an ongoing NAS level security mode command procedure. The reason is that, otherwise, the UE and MME may derive the new K_{eNB} from the different root keys. Note that the new K_{eNB} will be taken into use with either the AS

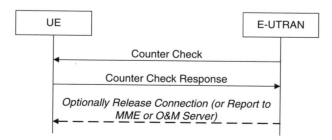

Figure 9.6   eNB Periodic Local Authentication Procedure. Reproduced with permission from © 2010, 3GPP™

level security mode command procedure or with the key change on the fly procedure, but not with inter-cell handovers.

2. Similarly, the second rule states that, if there is a procedure ongoing that includes sending a new K_{eNB} to the base station, the MME must not initiate a NAS security mode command procedure. This rule is similar to the first one, and is required to make sure that the UE and MME derive the K_{eNB} from the same root key.

3. The third rule is also about deriving the K_{eNB} from the right {NH, NCC} pair or K_{eNB} key during handovers. If the MME has an ongoing NAS security mode command procedure, the {NH, NCC} pair sent to the target base station during handovers must still be based on the old K_{ASME} root key. The reason is that, at the AS level, the K_{eNB} based on the old root key K_{ASME} is still used. Only after the MME has sent the new K_{eNB} to the base station, and the base station has successfully taken it into use, the MME can send fresh {NH, NCC} pairs based on the new K_{ASME} root key.

4. This is the counterpart rule to the third rule and is for the UE. The UE must continue to use the AS level parameters based on the old root key K_{ASME}, even if the UE has received the NAS security mode command message. Only after the base station has successfully run the RRC Connection Re-configuration procedure for taking the new K_{eNB} from the new K_{ASME} root key into use, the UE must use the fresh AS level parameters for handovers and discard the old AS parameters.

5. The fifth rule says that while there is an ongoing inter-base station handover procedure, the base station shall reject any S1 UE Context Modification procedures from the MME (i.e. the procedure that delivers a new K_{eNB} based on a new K_{ASME}). This is simply to avoid key synchronization problems. Also, the base station must not initiate any new handover procedures before the current RRC Connection Re-configuration procedure is finished.

6. The sixth rule is similar to rule five, but is for the MME. It says that the MME must not proceed with an inter-MME handover or inter-RAT handover signalling while there is an ongoing NAS security mode command procedure. The reason is that the NAS level security context changes as a result of the NAS security mode command procedure. In the inter-MME handover procedure, the source MME sends the NAS level security context to the target MME. Thus, during the NAS security mode command procedure the NAS level security context is not well defined. Similar problems would occur in an inter-RAT handover.

7. Similarly to rule six, the MME must not continue with inter-MME handover signalling before it has finished an ongoing S1 UE Context Modification procedure. The reason is the AS level parameter synchronization among the base station, the UE and the MME. If during this synchronization the source MME sent new or old NAS level security context, including the {NH, NCC} pair, to the target MME, the target MME might send different K_{ASME} root key based parameters to the target base station from what UE and the base station are currently using.

8. Rule eight describes the case when the MME has taken new NAS keys into use based on the new K_{ASME} root key but has not yet initiated the S1 UE Context Modification procedure to synchronize the new K_{eNB} with the base station and the UE. If at this point there is an inter-MME handover, the source MME must send both the old and the new K_{ASME} root key security contexts with corresponding key set identifier (eKSI) to the target MME.

9. As a response to rule eight, the target MME needs to know what to do with the two K_{ASME} during inter-MME handover. Rule nine says that the target MME must use the new K_{ASME} root key based NAS level security context for NAS protection, but the old K_{ASME} root key based parameters for the AS level operations. Then the target MME must at some point initiate the S1 UE Context Modification procedure to synchronize the new K_{eNB} based on the new K_{ASME} root key with the base station and the UE.

10. During inter-MME mobility, the source MME must not send any NAS messages to the UE after it has sent the UE context to the target MME. Only if the handover is cancelled or fails, the source MME can start sending NAS messages to the UE.

These rules are self-evident after the right situations or concurrency cases have been identified. However, it is not easy to come up with a complete list of possible error cases and concurrent run of procedures. These ten rules help to understand the possible race conditions and error cases better, but also provide guidance on how to implement the system to better avoid them.

10

EPS Cryptographic Algorithms

In this chapter we discuss in detail the cryptographic algorithms that are used in EPS. One principle that has been used in the design of EPS security is that of algorithm agility: the system should be flexible in the sense that new algorithms can be introduced and outdated ones can be removed, both without major hassle. Therefore, it is expected that in the future new algorithms would appear in EPS but they are potentially not even invented at the time of writing and hence naturally not yet discussed in this chapter. The need for better algorithm agility has stemmed from experiences with 2G and 3G systems where new algorithms have been introduced and one algorithm (A5/2) has also been removed from the 3GPP system.

On the other hand, we are here discussing standardized algorithms. A general principle for any standardized mechanisms (including non-security related ones) is that options should only be introduced if they serve a clear benefit for the system as a whole. If the difference between one option and another is more like a matter of taste, or if the benefit of each option over the others materializes only in a small minority of all circumstances, options should not be introduced because they complicate the system, add development cost, and put the interoperability at risk. Hence, the number of different algorithms should be kept small and introduction or removal of algorithms should be done only after it is clear that such action adds value to the system as a whole.

As explained in Chapter 2, cryptographic algorithms – at least the ones that are usable for mass-market products – share the same (from a deployment point of view, nasty) characteristic that they can be broken. Sometimes it can even be the case that theoretical breaks are followed fairly rapidly by practical exploitations. This is the reason why we need to have options in the choice of algorithms: in case one algorithm breaks, we still have others standing. Another consequence of this kind of reasoning is that the design of the algorithms in use should differ from each other as much as possible. Then it is less likely that even a major breakthrough in cryptanalysis would affect many of them.

Cryptographic algorithms are needed for both AS and NAS level protection, and the same algorithms may be used for both purposes. In principle, completely different algorithm sets could have been specified for these two purposes. One advantage of such an approach would have been, once again, greater diversity of the algorithms and, consequently, a smaller effect of a single broken algorithm. However, the big disadvantage would have been that, especially on the UE side, independent implementations would have been needed, one set for AS protection and another set for NAS protection, consuming double the implementation effort.

Note that the message authentication code is denoted by MAC-I for AS level integrity protection in both [TS33.401] and [TS24.301], while message authentication code for NAS level protection is denoted by NAS-MAC in [TS33.401] and simply by MAC in [TS24.301]. Throughout this chapter we use the abbreviation MAC for both AS level and NAS level integrity protection.

10.1   Null Algorithms

Although protection of communications is needed, in some circumstances it is not possible to provide cryptographic protection. One such situation is an unauthenticated emergency call, as discussed in section 8.6. It is always tricky to take care of exceptional situations where protection is lifted: there have to be guarantees that the exception of no protection does not, by accident or intentionally, spill over to cases where protection could have been provided. For this reason, it is typically better to design systems so that the case of no protection is explicitly triggered by actions of some of the communicating parties. In other words, the protection needs to be explicitly turned 'off' instead of just *not* turning it 'on'. Of course, this kind of explicit triggering of no protection does not alone guarantee that the triggering is done only in appropriate situations, other measures are needed also.

It is now easier to understand why a concept of 'Null algorithm' makes sense. Because the start of no-protection has to be done explicitly, it is simplest from the system point of view to use procedures for starting no-protection similar to those that are used for starting protection. Bear in mind here that the start of protection needs to be done explicitly as well, mainly for synchronization reasons. Thus, instead of choosing a proper algorithm to be put in place in order to start protection, we choose a Null algorithm to be put in place to start no-protection.

The flip side of the coin is that the concept of 'Null algorithm' may be confusing to some people who are not familiar with security issues. Indeed, a Null algorithm is not a cryptographic algorithm; in fact it is not really an algorithm at all. Whereas the question of whether a Null algorithm should be called an algorithm or something else may be interesting from a semantic point of view, the important fact is that the use of the Null algorithm provides no protection.

There are a couple of different ways in which a Null algorithm may be realized. One obvious choice is that the Null algorithm does not do anything. This is the option that has been chosen for the Null ciphering algorithm in EPS; mathematically speaking it is an identity function: ciphertext is identical to plaintext. It is also called EPS encryption algorithm number 0 (EEA0). In Annex B of [TS33.401] the Null algorithm is described slightly differently: instead of doing nothing, it uses a keystream of all zeros. Because the ciphertext is obtained by a bitwise xor operation from the plaintext and the keystream (see section 2.3.3), this also results in the ciphertext being equal to the plaintext.

Another way of realizing a Null algorithm is to do some very simple operation, just in order to make it explicit that a Null algorithm has indeed been in use. This is the option that has been chosen for the Null integrity algorithm in EPS: regardless of the message content or key or any other parameter, a 32-bit string of all zeros is appended to the message as the result of applying the Null integrity algorithm. The reasoning behind choosing this option for integrity rather than the other option of doing nothing is similar to the reasoning described earlier in for

Null algorithms in general: (a) it becomes explicit that no integrity protection is intended to be provided; and (b) from a procedural point of view, the protected case and the non-protected case become as similar as possible.

A couple of remarks are needed here. Regarding (a), note that a MAC of all zeros may occur also in the case of a proper integrity algorithm but only for one message out of 2^{32} (on average). Regarding (b), all processing of integrity check values (MAC or expected MAC) is quite similar when using the Null algorithm or proper algorithms. On the sending side, a MAC of all zeros is appended to a message in the same way in which a (proper) MAC is appended to a message when using a proper integrity algorithm. On the receiving side, as an example, if a message that requires a MAC to be included is received without any MAC, then it is simply discarded even when a Null integrity algorithm is in use.

10.2 Ciphering Algorithms

The encryption mechanisms used in EPS are very similar to those used in 3G. There are many differences between EPS and 3G in how keys are generated and managed but, once the correct key is in place, the usage of the key is very similar in these systems. This is fortunate in the sense that it allows terminals to use some internal components for both LTE and 3G. Bear in mind that it would be quite natural to support both 3G and LTE in the same terminal, similar to the way in which most 3G terminals support also GSM.

In EPS there are independent instances of confidentiality protection mechanisms, one on the AS level and another on the NAS level. However, both mechanisms are very similar to each other; and, with regard to the encryption algorithm itself, there is no difference: an algorithm suitable for the AS level is also suitable for the NAS level, and vice versa.

It was clear from the start of EPS security design that a sufficient amount of cryptographic diversity would be useful. Therefore, it was decided that two ciphering algorithms would be supported from the start of EPS. In addition, these two algorithms should be as different from each other as possible, in order to minimize the chance that both would be broken by the same breakthrough in cryptanalysis.

From what has been written so far in this section it would be easy to draw the conclusion that the same set of algorithms that is in use for 3G would also be a good choice for EPS. However, starting a completely new system always presents the possibility of seeking new approaches and even better solutions than those that would be most natural from a legacy system point of view.

The history of the selection process for the 3G ciphering algorithms is explained in Chapter 4. The two 3G algorithms are, at the time of writing, UEA1 based on KASUMI and UEA2 based on SNOW 3G. It is notable that the leading general-purpose algorithm AES [FIPS 197] is not among the two. The reasons for this were explained in section 4.3. In short, AES was not ready yet when KASUMI-based UEA1 was chosen, while SNOW 3G-based UEA2 was the preferred choice as the base algorithm, over AES, because its design was more different from that of KASUMI.

It was expected that LTE terminals would probably need to support AES anyway, for various application layer protection purposes. Therefore, choosing AES as a ciphering algorithm for EPS would probably also offer some re-usability benefits, although of a type different from those offered by choosing 3G algorithms also for LTE.

Altogether, designers of the EPS security architecture faced a kind of positive problem: there were three good choices available for the algorithms while only two were needed in the beginning. Referring back to the different design strategies that were discussed in section 4.3, it was seen that the strategy of choosing an off-the-shelf algorithm was a better choice for EPS than the other two more complicated strategies: inviting submissions and commissioning a design task force.

Although there were also many other potential off-the-shelf algorithms than the three mentioned above, the selection was restricted soon to these three, mainly because of the re-usability aspect. Taking into account the requirement of cryptographic diversity, it seemed natural to include SNOW 3G into the final two. Between the other two, a decision was finally made to choose AES over KASUMI as the base for the other algorithm that would be supported from the beginning. Because of the general principle of avoiding unnecessary options in the standardized system, the number of mandatory algorithms was limited to two.

For the case of SNOW 3G, the specification work needed to adapt it to EPS security architecture was minimal, and was carried out by references to the existing 3G specifications. The algorithm is called 128-EEA1, to explicitly denote that the algorithm uses a 128-bit key and to distinguish it from a possible future 256-bit version of this algorithm. Note that 3GPP has decided that, if at some point in the future 128-bit keys are no longer seen as long enough, it is better to double the key size instead of introducing, for example, 192-bit keys.

For AES, the situation was slightly more complicated. Several modes of operations for AES had already been defined in NIST specifications but, for obvious reasons, none of them had been produced for this particular purpose. However, it was soon found that there were already existing modes of operation that could be adapted to the EPS environment. The tasks of choosing the most appropriate existing mode of operation and creating the necessary specifications was, once again, delegated to the ETSI SAGE group. Since the needed effort was much smaller than in the earlier design projects, no special task force was established. Also, instead of creating stand-alone specifications, the needed definitions were appended to [TS33.401]: Annex B contains extensions needed to existing NIST standards and Annex C contains the necessary new test data.

The Counter mode [NIST800-38A 2001] was chosen for the EPS purposes. The needed adaptation was simple. The initial 128-bit counter value was defined to contain the ciphering algorithm input parameters COUNT, BEARER and DIRECTION in the most significant part while the least significant part was defined to be all zeros. Then the counter would be incremented by the normal integer addition as long as new keystream blocks of 128 bits would be needed.

The EPS algorithm based on AES in Counter mode is called 128-EEA2.

10.3 Integrity Algorithms

Many of the facts explained for the background of EPS ciphering algorithms also apply to integrity algorithms. The integrity protection mechanisms are similar in both 3G and LTE, although there are big differences in key management. Each integrity algorithm applies as such to both AS level and NAS level protection. In order to have a good security margin against progress in cryptanalysis, two different algorithms are in place from the beginning of EPS.

From an implementation point of view, especially for terminals, it would be good to have algorithms that are usable also for some other purposes.

There is a typical practice of using the same core cryptographic functions for both ciphering and integrity purposes. This practice is also mainly due to re-usability benefits, and there are no cryptographic reasons behind it. However, no heavy arguments were found that would have spoken against such a practice, so it was decided that the two integrity algorithms that are supported from the start are based on AES and SNOW 3G.

As was the case for ciphering, UIA2 could be adapted in a straightforward manner from 3G specifications. The algorithm is called 128-EIA1. Again, the numerical value of 128 refers to the possibility that a 256-bit version is needed in later releases.

For AES, similar to the case of ciphering, some more adaptation work was needed, and it was carried out by ETSI SAGE. The Cipher-based MAC (CMAC) mode was chosen [NIST800-38B 2005]. The additional definitions can be found in Annex B of [TS33.401] and the necessary new test data is in Annex C of the same specification. Similar to ciphering, the input parameters needed for EPS integrity protection are mapped to the CMAC initialization parameters, using all zeros for filling in the rest of the parameters.

The new EPS algorithm based on AES in CMAC mode is called 128-EIA2.

10.4   Key Derivation Algorithms

As explained in earlier chapters, the EPS key hierarchy is significantly more complex than that of 3G or GSM. One consequence is that there has to be a standardized way to derive keys from each other. From the security point of view, it is crucial that the derivation is one-way: it should not be possible to use physically less protected keys on the lower layers of the hierarchy to get information about the physically more protected keys that are higher up in the hierarchy. In addition, two keys derived from the same key should be independent. Note here that the difference in the physical protection refers rather to the network side; on the UE side there are fewer differences.

Although 3G access security did not require defining a standardized Key Derivation Function (KDF), it has been needed for other 3GPP features. Most notably, the Generic Bootstrapping Architecture (GBA) ([TS33.220; Holtmanns *et al.* 2008] includes the derivation of new keys as one of its core features. EPS key derivation re-uses the standard key derivation function of GBA. The core of the KDF is the cryptographic hash function SHA-256 [FIPS 180-2]. It is used in the keyed HMAC mode [RFC2104], where the key for HMAC is the 'mother' key from which the lower layer key is derived. The other input parameter for HMAC is called the 'message', a name motivated by the primary use of HMAC for message integrity purposes. In the case of 3GPP key derivation, the message is a bit string S with a clearly defined structure:

$$S = FC||P0||L0||P1||L1||P2||L2||\ldots||Pn||Ln$$

Here $||$ denotes the concatenation operation. The parameter FC is a single octet that is used to differentiate between various purposes that the KDF is used for in the 3GPP system. The parameters P0, P1, P2, $\ldots$, Pn are the additional input parameters that are needed in the key derivation. The parameter Li is a two-octet encoding of the length of the parameter Pi (counted in octets). Using length values explicitly as part of the input guarantees that the string S can be unambiguosly parsed.

Annex A of [TS33.401] contains descriptions of all the instances of the KDF that are needed for EPS purposes. For instance, it is explained in section 7.3 that, for the case of deriving K_{eNB} from K_{ASME}, the only additional input parameter is the NAS uplink COUNT value. For this key derivation purpose FC $= 0 \times 11$ and P0 $=$ NAS uplink COUNT value. The length of NAS uplink COUNT is 4 octets, so L0 $= 0 \times 00 \ 0 \times 04$.

As a slightly more complex example, let us take a look at the derivation of the RRC encryption algorithm key K_{RRCenc} from K_{eNB}. It would appear that the only additional input parameter required was the algorithm identifier, as explained in section 7.3. It was decided, however, to use the same FC for all key derivations leading to a leaf key in the key hierarchy. Therefore, a further additional input parameter is needed to separate these leaf keys from each other. Now FC $= 0 \times 05$ for all these key derivations while, for example, P0 $= 0 \times 03$ as coding for 'algorithm type distinguisher' in the case of RRC encryption key. P1 is the parameter for algorithm identity. Both the algorithm type distinguisher and the algorithm identity are single octets long, so L0 $=$ L1$= 0 \times 01$. The algorithm identifiers have been defined in clause 5 of [TS33.401]. For example, using AES as the encryption algorithm corresponds to value P1 $= 0 \times 02$ (in clause 5 of [TS33.401], the binary representation 0010 is given).

11

Interworking Security Between EPS and Other Systems

In this chapter, we describe how the EPC can interwork with other 3GPP access technologies like GERAN and UTRAN, but also how EPC supports interworking with non-3GPP access technologies like cdma2000®HRPD, WiMAX and WLAN. Interworking is important as LTE allows different deployment options for the operators. Clients supporting multiple-access technologies, including LTE, need to have continuous access to the network for good user experience. For example, a client using 3G data may enter an LTE hotspot and thus be handed over to it for a higher data throughput.

We start by describing the interworking with GSM and 3G in section 11.1 and move then to the non-3GPP interworking part in section 11.2.

11.1 Interworking with GSM and 3G Networks

Here we describe intersystem idle state mobility and handovers between 3G or GSM and EPS. This is particularly interesting as there are multiple cases for handling the security context depending on the originating system and mobility mode (handover or idle state mobility). Refer to section 7.4 for detailed definitions relating to security contexts. A process called mapping of security contexts from 3G or GSM to EPS, and vice versa, is applied in intersystem mobility. This mapping means that a 3G or GSM originated security context is used to derive an EPS security context, and vice versa. In handovers, the mapping of security contexts is always applied for efficiency reasons even if a native security context (one created by an authentication in the respective system) is available in the target system. In idle state procedures, existing security context in the target system is used if available.

Security context mapping satisfies the security requirement of backward key separation. Backward key separation means that the target system cannot derive the keys used in the source system. This requirement is realized by deriving new ciphering and integrity protection keys for 3G from the existing K_{ASME} during EPS to 3G or GSM mobility. In the other direction, from 3G or GSM to EPS, it is the MME, which does the mapping and thus the backward key

LTE Security   Dan Forsberg, Günther Horn, Wolf-Dietrich Moeller, and Valtteri Niemi
© 2010 John Wiley & Sons, Ltd

separation does not hold. Also, the security context mapping does not provide forward key separation; that is, the source system knows the mapped keys used in the target system.

When an intersystem mobility event results in the use of mapped keys it is therefore advisable to establish native security context soon after the event in order to minimize the trust required of the target system in the source system. But even if there is unconditional trust between the systems, as when they are part of the same trust domain of an operator, when moving from 3G or GSM to EPS it is recommended in the specifications to establish a native EPS security context as soon as possible because native EPS security is considered stronger for the reasons explained in Chapter 7. In practice this is, of course, up to the operator's security policy resulting from his risk analysis. After a handover from 3G or GSM to EPS, an EPS AKA authentication can be run, and all keys in the key hierarchy can be renewed, even while the UE is in connected state, by using the key change on-the-fly procedure as described in Chapter 9.

Idle state signalling reduction

EPS specifies Idle state Signalling Reduction (ISR) [TS23.401], which implies that the UE can be registered simultaneously with multiple systems using different Radio Access Technologies (RATs). With ISR the UE can be registered in E-UTRAN and UTRAN or GERAN at the same time while in idle state and listen to paging messages from the RAT where it is currently camping. While re-selecting cells between these RATs the UE does not need to send any location update signalling messages provided that the respective routing or tracking areas do not change. As a result, the idle state signalling is reduced if the UE switches back and forth between E-UTRAN and UTRAN or GERAN. On the other hand, the network pages the UE on both technologies when the UE receives incoming data, as the network does not know which RAT the UE is actually listening to (see Figure J.4-1 Downlink data transfer with ISR active in [TS23.401]). When ISR is switched on the UE can also go to connected state in one of the systems while remaining registered in the other system.

The ISR is activated when the network responds to a Tracking Area Update (TAU) Request or Routing Area Update (RAU) Request with a TAU Accept or RAU Accept message that indicates ISR activation. When the ISR is activated and the UE has multiple temporary identities available, it sets a 'Temporary Identity used in Next update' (TIN) parameter value to 'RAT-related TMSI'. This means that the UE will use the RAT-specific temporary identity when it sends a RAU Request or TAU Request message to the network. For example, with a TAU Request the UE will use a Globally Unique Temporary Identity (GUTI) and with the RAU Request a P-TMSI. Table 11.1 (see also [TS23.401], Table 4.3.5.6-1) shows how UE sets the TIN value when receiving Attach Accept, TAU Accept, or RAU Accept.

ISR can be deactivated upon a number of conditions, for example through timer expiry. Whenever the ISR is not active the TIN is set to be the currently used temporary identity allocated in the currently used RAT (i.e. GUTI in E-UTRAN and P-TMSI in UTRAN). When the TIN has values of 'GUTI' or 'P-TMSI', it means that the corresponding temporary identity is used in the next TAU or RAU Request message. If the TIN value is 'GUTI', the GUTI is mapped to a P-TMSI and P-TMSI signature in the RAU Request message. GUTI is used as is in the TAU Request. If the TIN value is 'P-TMSI', the P-TMSI is used without any modification in a RAU Request while a GUTI mapped from a P-TMSI is used in TAU Requests. Table 11.2 (see also [TS23.401], Table 4.3.5.6-2) shows how UE sets the temporary identity in the Attach Request, TAU Request, or RAU Request messages depending on the currently set TIN value.

Table 11.1 Setting of the TIN in the UE

Message received by UE	Current UE TIN value	New UE TIN value
E-UTRAN Attach Accept (never indicates 'ISR Activated')	Any value	GUTI
GERAN/UTRAN Attach Accept (never indicates 'ISR Activated')	Any value	P-TMSI
TAU Accept	Any value	GUTI
TAU Accept (indicates 'ISR Activated')	GUTI	GUTI
	P-TMSI or RAT-related TMSI	RAT-related TMSI
RAU Accept	Any value	P-TMSI
RAU Accept (indicates 'ISR Activated')	P-TMSI	P-TMSI
	GUTI or RAT-related TMSI	RAT-related TMSI

For more details, see the informative Annex J in [TS23.401] and the next two subsections on cases when the UE may include two GUTI or P-TMSI values into a message.

11.1.1   Routing Area Update Procedure in UTRAN

When the UE is moving in idle state into a UMTS routing area from an EPS tracking area,[1] and is not registered on the UMTS side, it needs to send a RAU Request message over UTRAN. When a UE is already registered in UMTS it needs to send a RAU Request when the routing area changes, or periodic RAU Requests when remaining in the same routing area. There are two cases for the UE to select the UMTS security context for protecting the RAU procedure: using an existing UMTS security context, or obtaining the UMTS context through a mapping from the EPS security context in the MME (see section 7.4 for more information about security contexts).

Table 11.2 Temporary identity of the UE in Attach/TAU/RAU Request messages

Message of the UE	TIN: P-TMSI	TIN: GUTI	TIN: RAT-related TMSI
TAU Request	GUTI mapped from P-TMSI/RAI	GUTI	GUTI
RAU Request	P-TMSI/RAI	P-TMSI/RAI mapped from GUTI	P-TMSI/RAI
E-UTRAN Attach Request	GUTI mapped from P-TMSI/RAI	GUTI	GUTI
GERAN/UTRAN Attach Request	P-TMSI/RAI	P-TMSI/RAI mapped from GUTI	P-TMSI/RAI

---

[1] UMTS Routing Area and EPS Tracking Area are similar concepts. They both allow the network to find the UE when an incoming call needs to be delivered by paging it in a defined area.

UMTS Routing Area Update with existing UMTS security context

When a UE needs to send a Routing Area Update Request in one of the above-mentioned cases, the UE uses an existing UMTS security context to protect the RAU procedure if the temporary identity used in the RAU Request – according to Table 11.2 – is a P-TMSI that is not mapped. This is the case when ISR is activated and the TIN indicates 'RAT related TMSI', or when ISR is deactivated and the TIN indicates 'P-TMSI'.

Along with the temporary identity P-TMSI, the UE includes the Key Set Identifier (KSI) into the RAU Request to allow the SGSN to identify the keys. The previous SGSN may have assigned a P-TMSI signature to the UE earlier. If it was allocated, the UE will include it into the RAU Request so that it can be used to authenticate the RAU Request. If the SGSN does not have the corresponding security context indicated with the KSI and P-TMSI, it fetches it from the old SGSN indicated in the P-TMSI/RAI. If this is unsuccessful the SGSN runs a UMTS AKA authentication.

UMTS Routing Area Update with mapped UMTS security context

This case occurs if the temporary identity used in the RAU Request – according to Table 11.2 – is a P-TMSI that is mapped from a GUTI. When a UE in idle state moves from an E-UTRAN tracking area to a UTRAN routing area with ISR deactivated, it always uses a UMTS security context mapped from the EPS security context to protect the RAU procedure. The value of the EPS Key Set Identifier (eKSI) associated with the current EPS security context is mapped to the UTRAN KSI information field of the RAU Request.

The UE will create a so-called NAS-token (see more below), based on the NAS integrity protection key in the EPS security context, and include it into the P-TMSI signature field of the RAU Request. In this way, when the SGSN requests the UE security context from the MME, the MME can authenticate the request based on the EPS security context of the UE. If the MME can verify the NAS-token it creates a mapped UMTS security context by mapping it from the EPS security context and transfers it to the new SGSN. This mapped UMTS security context includes the UE security capabilities and ciphering and integrity keys CK' and IK' derived from K_{ASME}, and the KSI mapped from the eKSI corresponding to K_{ASME}. The MME has the UTRAN and GERAN security capabilities as the UE provided them along with the EPS security capabilities for the MME while registering in EPS (see section 11.1.2 in EPS below).

The NAS-token is created based on the NAS integrity protection key and the NAS uplink COUNT. CK' and IK' are derived from the K_{ASME} using the same NAS uplink COUNT value. For the exact formulas see Annex A of [TS33.401]. The current NAS uplink COUNT value in the UE and in the MME may be different owing to lost or pending uplink NAS messages. For this reason, the MME will calculate the NAS-token with a range of NAS uplink COUNT values and compare the bits with the received P-TMSI signature.[2] If match is found the NAS-token is verified and the MME identifies the NAS uplink COUNT value that was used to calculate the NAS-token and marks it as used. Based on this NAS uplink COUNT value, the MME will also derive the CK' and IK'. Thus, the same NAS-token cannot be used twice in the MME, unless the NAS-token is retransmitted (perhaps because of a lost message during the same mobility event).

---

[2] The NAS-token is actually truncated to at least 16 bits and included into the P-TMSI signature field. The P-TMSI signature field is longer than 16 bits, but a part of the remaining bits is used for other purposes [TS24.301].

The UE stores the CK', IK' and KSI on the USIM as the new UMTS security context. This is a mapped UMTS security context as it was derived from the EPS security context in the MME. UE also computes a Kc from CK' and IK' using the conversion function c3 specified in [TS33.102] and updates the Kc value in the ME and the USIM. This is necessary as any previously stored Kc was calculated from a previous UMTS security context CK and IK and thus is not synchronized with CK' and IK'. The GPRS Ciphering Key Sequence Number (CKSN) is set to the KSI. Operators wanting to create fresh keys and a native UMTS security context can always run UMTS AKA on the UTRAN side, perhaps next time when the UE changes from idle to connected mode.

The RAU procedure in GERAN using a GSM security context mapped from the EPS security context in the UE and the MME is similar to the RAU procedure in UTRAN using a UMTS security context mapped from the EPS security context in the UE and the MME, except that the UE and the SGSN will derive the cipher key Kc or K_{c128} [TS33.102] from CK' and IK' transferred from the MME to the SGSN, and use the GERAN specific security algorithms. An SGSN that supports interworking between E-UTRAN and GERAN must be able to handle UMTS security contexts. Thus, the MME provides the same security context to the new SGSN as it provides for the SGSN supporting interworking between E-UTRAN and UTRAN described above.

11.1.2 Tracking Area Update Procedure in EPS

When the UE is moving in idle state into a new EPS tracking area from a UTRAN routing area, and is not registered on the EPS side to that tracking area, it needs to send a Tracking Area Update (TAU) Request message to the MME. (To be precise: for sending the TAU Request the UE has to move to connected state. The UE may fall back to idle state after the completion of the TAU procedure.) There are two cases for the UE to choose the security context for protecting the TAU Request message: the UE can use a current EPS security context (either native or previously mapped EPS security context) if it is available in the UE, or map the current UMTS security context to an EPS security context.

Moving in idle state into an EPS tracking area from an UTRAN routing area is not the only case when a UE sends a TAU Request. When a UE is already registered in EPS it needs to send a TAU Request when the tracking area changes, or periodic TAU Requests when remaining in a tracking area.

EPS specifies a flag in the TAU Request called 'active flag'. When this flag is set, the MME will create an AS security context for the UE, including K_{eNB}, and send it to the serving base station. As a result, the UE and the base station establish AS level security with the AS level security mode command (see section 8.3), and can start sending and receiving user plane data.

EPS Tracking Area Update with current EPS security context

The UE uses a current EPS security context to protect the TAU procedure if the temporary identity used in the TAU Request – according to Table 11.2 – is a GUTI that is not mapped. This is the normal case for TAU procedures inside E-UTRAN. It is also used when ISR is activated, and the tracking area changes compared to the currently registered tracking area with the UE (meaning that the UE needs to send a TAU Request, even if the ISR is activated). This case has already been described in section 9.3 on idle state mobility.

UE includes its GUTI and eKSI into the TAU Request message. The UE will only integrity-protect the TAU Request message, so that if the MME changes, the new MME is able to find out what the old MME identity is based on the GUTI as it cannot decipher the message without the security context. If the old MME does not have the security context of the UE and keys indexed with the eKSI, or the integrity protection validation fails, the new MME will run EPS AKA.

Since the EPS supports multiple algorithms, it may be that the new MME supports different algorithms from the old MME, and thus the NAS level security algorithms need to be changed. To do this, the MME will run the NAS Security Mode Command procedure as described in section 8.2 before sending the TAU Accept message to the UE protected with the new NAS keys and NAS algorithms.

EPS Tracking Area Update with mapped EPS security context

If the TIN value in the UE is set to 'P-TMSI' – according to Table 11.2 – the UE includes a GUTI mapped from a P-TMSI and the old Routing Area Identifier (RAI) in the TAU Request along with the KSI that identifies the keys in the UMTS security context. If the SGSN allocated a P-TMSI signature it is also included into the message.

If the UE has a current EPS security context, it will additionally include the GUTI (and eKSI) that is associated with this context as an additional GUTI information element. This GUTI is then different from the GUTI mapped from the P-TMSI. The UE then also integrity-protects the TAU Request with the keys and algorithms in the current EPS security context, but does not cipher it. The new MME can then fetch the current EPS security context from its own memory if the EPS security context indicated by the additional GUTI and the eKSI is still available.

If the UE does not have a current EPS security context, it will not integrity-protect nor cipher the TAU Request message.

If the new MME does not have the current EPS security context indicated by the received eKSI, or the TAU Request was received without integrity protection, the new MME will request the UMTS security context from the old SGSN and convert the security context to a mapped EPS security context. The MME finds the old SGSN on the basis of the GUTI mapped from a P-TMSI, and the old SGSN identifies the UMTS security context using the GPRS ciphering key sequence number that was sent along by the UE together with the mapped GUTI. This mapped EPS security context then becomes the current EPS security context. The next paragraph explains how the mapped EPS security context is created.

The UE always includes a 32-bit $NONCE_{UE}$ into the TAU Request message. The nonce is used only when a mapped EPS security context needs to be created. But as the UE cannot know when it sends the TAU Request whether this will be the case, it always includes the $NONCE_{UE}$. A nonce is, by definition, a number used only once. In this case, $NONCE_{UE}$ even has to be a random number. (For more information about the randomness requirements on $NONCE_{UE}$, see Annex A in [TS33.401]).

When creating the mapped EPS security context from the CK $\|$ IK received from the SGSN, the MME will also create a nonce, called $NONCE_{MME}$ (which has to satisfy similar randomness requirements as for $NONCE_{UE}$), and use both nonces to derive a fresh mapped K'_{ASME}. The reason for using the nonces for creating the K'_{ASME} is that the same CK $\|$ IK may be delivered to the new MME in idle state mobility procedures repeatedly, for example when the UE moves back and forth between UTRAN and E-UTRAN several times, and no new keys CK and IK are created on the UMTS side during this period. If now the mapped K'_{ASME} was created only from CK and IK, without further fresh input the same mapped K'_{ASME} would be created each time, and consequently also the same NAS encryption and integrity keys (see Chapter 8). This would violate the security requirement that the same keys with the same sequence numbers (COUNT values) must not be used twice as they are used as input values for integrity protection and ciphering. But when the mapped EPS Security context is created, the respective NAS uplink and downlink COUNT values are set to start values. For this reason, the K'_{ASME} must be always fresh and thus the use of the nonces is required. The new MME will take the new current EPS security context into use with the NAS security mode command procedure (see section 8.2), and include both of the nonces into the NAS Security Mode Command message. The MME includes the $NONCE_{UE}$ so that the UE can verify that its own $NONCE_{UE}$ was not modified while sent to the MME. Then the UE is also able to derive the fresh K'_{ASME} by using the nonces as input values to the key derivation function. Annex A in [TS33.401] shows the exact key derivation parameters and formulas. As normally, the MME may also change the algorithms at the same time (e.g. if the new MME supports different algorithms from the old MME).

This description applies also to idle state mobility from GERAN to E-UTRAN as the requirement is that the source SGSN shares a UMTS security context with the UE.

11.1.3 Handover from EPS to 3G or GSM

Here we describe the key management during handovers from EPS to 3G or GSM. Note that, before the handover can happen, both the NAS and AS security must be set up on the EPS side. In other words, the base station cannot send handover commands without AS security being active. This protects against attacks where the attacker sends unprotected commands to hand over to other less secure RATs before the UE and base station have set up AS level security.

When the EPS network decides to do a handover to UTRAN/GERAN, it will create UMTS keys and send them to the target SGSN along with UE security capabilities and the KSI. The target system will then create handover command parameters, including security algorithms for use in the target system, that are delivered to the MME and finally to the serving base station in E-UTRAN that commands the UE to do handover to UTRAN.

We now first explain the handover to UTRAN. We then explain the (small) differences from the case of handover to GERAN.

As already mentioned above, in handovers always mapped keys are used in the target system for efficiency reasons, irrespective of the availability of security contexts in the target system. Both the UE and the MME need to create fresh UMTS keys CK′ and IK′ from the current K_{ASME}. The KSI identifying CK′ and IK′ equals the value of eKSI identifying the current K_{ASME}. To ensure key freshness, the UE and the MME use the current NAS downlink COUNT value as input parameter to the CK′ and IK′ derivation and then increase its value. In this way, the same NAS downlink COUNT value is never used twice to derive CK′ and IK′ with the same K_{ASME}. To ensure that both the MME and the UE use the same NAS downlink COUNT value, the MME includes four least significant bits of the current 32-bit NAS downlink COUNT value into the message delivered to the eNB that commands the UE to do handover to UTRAN. The eNB forwards these bits to the UE. The UE will then synchronize its NAS downlink COUNT value with the one used by the MME, perhaps by increasing the NAS downlink COUNT value until the four least significant bits match. The UE also checks that the same NAS downlink COUNT value is not used twice to derive the CK' and IK' with the same K_{ASME}. Note that the NAS downlink and uplink COUNT values must never decrease in the UE or MME.

Along with the CK' and IK', the MME provides also UE security capabilities to the target SGSN. The target system then decides what algorithms to use.

The new mapped UMTS security context replaces all the stored values in the USIM, in the ME, and in the target SGSN. In this way, the mapped context remains available to both UE and SGSN after the UE has gone to idle state. Note further that the KSI mapped from eKSI may be identical to a previously established KSI. It is therefore important that previously stored key and KSI values be overwritten in all places where they may have been stored so as to avoid future key synchronization problems between UE and SGSN. As in the idle state mobility case described in section 11.1.1, the UE also derives Kc from the CK' and IK' and stores it on the USIM, together with the CKSN set to KSI.

If the handover fails, the new mapped UMTS security context is deleted. If the target SGSN had a security context with the same KSI as the new mapped security context, the SGSN will delete it. This is needed to avoid possible security context synchronization problems.

This description applies also for handovers from E-UTRAN to GERAN as the requirement is that the target SGSN shares a UMTS security context with the UE. However, the UE and SGSN will derive Kc from the CK' and IK', and K_{c128} when the new encryption algorithm requires a longer key. Also, the target SGSN and UE assign the eKSI value associated with the CK' and IK' to the GPRS CKSN associated with the GPRS Kc or K_{c128}. The UE updates also Kc on the USIM.

11.1.4   Handover from 3G or GSM to EPS

In handover from UTRAN or GERAN to EPS, the mapped EPS security context is always used in the target system after handover. Only after the handover can the EPS take an available native EPS security context into use if it decides to do so by running a Security Mode Command procedure or run EPS AKA to create a fresh native EPS security context.

The source system SGSN delivers the CK, IK and KSI to the target MME. If the user has only a GSM subscription, and hence the security context in the SGSN was derived from a SIM instead of a USIM, the SGSN delivers a Kc to the target MME, which then aborts the procedure. Remember that the user is allowed to access UMTS, but not EPS, when using a GSM subscription. To make the system more efficient, the source RNC may check whether the UE was authenticated with UMTS AKA (see more details in [TS33.401]). If the UE was not authenticated with UMTS AKA, the source RNC may decide not to perform a handover to E-UTRAN. (Actually, it does not make much sense for the RNC to go ahead with the handover to E-UTRAN at this point as the MME will block it anyway.) Additionally, the source RNC may choose another target system for the UE for the handover.

The target MME uses the CK and IK along with a $NONCE_{MME}$ to create a fresh mapped K'_{ASME}. The MME creates the $NONCE_{MME}$ to make sure that the K'_{ASME} is fresh as the same CK and IK may be delivered to the MME repeatedly, such as during handover procedures when the UE is moving back and forth between UTRAN and E-UTRAN. $NONCE_{MME}$ will satisfy the same randomness requirements as the nonces used in idle state mobility procedures.

The K_{eNB} is derived from the K'_{ASME} and sent to the target base station (i.e. eNB). As usual, the K_{eNB} derivation parameters include the NAS uplink COUNT value. However, in the case of handover from UTRAN to E-UTRAN the specifications state that the NAS uplink COUNT value used in the K_{eNB} derivation has to be $2^{32} - 1$, while the NAS uplink COUNT value used in the NAS protocol as a message counter is set to zero after handover as it should be with a fresh key K'_{ASME} and, hence, fresh NAS encryption and integrity keys. The rationale is that 3GPP discovered a particular scenario where the same K_{eNB} would be derived twice, leading to a keystream repetition and potential security vulnerability. 3GPP decided to address this vulnerability even if it seemed quite difficult to exploit it.

The scenario is as follows. Assume that a UE is handed over to E-UTRAN, and a K_{eNB} is derived from the K'_{ASME} using a NAS uplink COUNT value of zero. Assume further that no TAU Request is sent after the handover, due to ISR being on, and no other NAS message is sent on the uplink, so the NAS uplink COUNT remains at zero. Then the UE goes to idle state. When the UE comes back to connected state and sends a Service Request, then this request uses the current NAS uplink COUNT value, which is zero. But according to the general rule for deriving K_{eNB}, described in section 7.3, the current NAS uplink COUNT value has to be used in the K_{eNB} derivation. As the K_{ASME} has not changed, this results in the derivation of the same K_{eNB} as the one created after the handover. The next Service Request will increase the NAS uplink COUNT, so the problem cannot occur again, and the use of the value $2^{32} - 1$ in the K_{eNB} derivation right after handover solves the problem as in the NAS signalling the NAS uplink and downlink COUNT values are used as 24-bit values and the most significant eight bits are always set to zero [TS24.301]. Thus, the NAS uplink or downlink COUNT value never reaches 32-bit maximum value ($2^{32} - 1$).

Note that the KSI is marked as KSI_{SGSN} when stored in the eKSI information element in EPS. In this way, the EPS can distinguish between KSIs allocated by an MME and KSIs coming from the SGSN as both network entities may have assigned the same value (KSI is only three bits, but eKSI uses a fourth bit to distinguish between KSI_{SGSN} and KSI_{ASME} – see section 7.4). This works as the newly mapped EPS security context overwrites any existing current mapped EPS security context. Note that, in the other handover direction, the mapped UMTS security context overwrites the current UMTS security context to, for example, avoid the KSI overlap.

The target MME selects the NAS security algorithms and indicates them to the target base station (eNB), together with KSI_{SGSN}, UE security capabilities and the $NONCE_{MME}$. The target base station then selects the AS level security algorithms and includes all these parameters into the handover command message. The handover command message is then delivered to the source system, which sends it to the UE. When the UE receives the handover command it will activate the AS and NAS level security for the EPS side. Similarly, when the target eNB receives the handover complete message from the UE, it activates the AS level security. The target MME activates the NAS level security when it receives the Handover Notify message from the target eNB.

The source system SGSN sends the UE security capabilities to the target MME. The UE security capabilities, including the UE EPS security capabilities, were sent by the UE to the SGSN via the UE Network Capability information element, which includes also UE EPS security capabilities, in Attach Request and RAU Request. It is possible that an SGSN of a previous release does not forward the UE EPS security capabilities to the MME. When the MME does not receive UE EPS security capabilities from the SGSN, the MME will assume that the default set of EPS security algorithms, which is the set of algorithms defined for 3GPP Release 8, is supported by the UE (and will set the UE EPS security capabilities in the mapped EPS NAS security context according to this default set).

To protect against bidding down attacks from the source system, the UE includes its security capabilities in the following TAU Request message after the handover so that the MME can check them and change the NAS and AS security algorithms if needed. It is possible that UE does not send the TAU Request within a certain period, perhaps because the tracking area does not change and ISR is on, but this can only happen when the UE previously registered with the MME, and the MME should then still have the UE EPS security capabilities from that previous registration. If the MME has already deleted that context, then at most bidding down to the default set of capabilities is possible.

If the handover fails, the target MME deletes the new mapped EPS security context to avoid possible security context synchronization problems.

If the tracking area changes, the UE sends a TAU Request (also in the case the UE was not at all registered in the EPS before) protected with the mapped EPS security context. If the UE has a native EPS security context, it will include a GUTI into the message, either in the Old GUTI information element or in the Additional GUTI information element. The UE will include also an eKSI that is a KSI_{ASME} (i.e. key set identifier for a native EPS security context). In this way the MME is able to search the native EPS security context from its memory and activate it, if available, on the NAS level after completion of the handover procedure. The MME sends the new K_{eNB} derived from the native EPS security context to the target eNB. The target eNB uses the key change on-the-fly procedure to activate the new keys on AS level, as described in section 9.5. In this way, the EPS does not have to run EPS AKA for getting forward key separation from the source system keys. However, if the UE does not have a native EPS security context, it is strongly recommended to run an EPS AKA and perform a key change on-the-fly of the entire key hierarchy as soon as possible after the handover. The reason is that the security target is to always have separate keys in the EPS.

This description applies also for handovers from GERAN to E-UTRAN as the requirement is that the source SGSN shares a UMTS security context with the UE.

11.2   Interworking with Non-3GPP Networks

This section has a close relationship with Chapter 5 on 3G–WLAN interworking. The reader interested in this section is advised to first have a look at the principles of 3G–WLAN interworking laid out in section 5.1 as they will be re-used here. The procedures here are also quite similar to the security mechanisms described in section 5.2, but they will be described in detail as there are some significant differences that would be difficult to explain with only a wholesale reference to section 5.2.

11.2.1   Principles of Interworking with Non-3GPP Networks

Scope

While Chapter 5 dealt with the security for accessing a third-generation core network via a non-3GPP access network, using the example of WLAN, this chapter deals with the security for accessing the Evolved Packet Core (EPC) via a non-3GPP access network. A non-3GPP access technology is a technology not defined by 3GPP; that is, any access technology other than E-UTRAN (LTE), UTRAN (3G) or GERAN (2G). The security procedures for non-3GPP access to the EPC specified in [TS33.402] are not specific for any particular non-3GPP access technology, but, for one particular access technology, namely cdma2000® HRPD [TS33.402] provides a mapping of the procedures to the network entities. This reflects the fact that access to the EPC via a cdma2000® HRPD access network represents probably the most important use case at the time of writing this book. It permits operators using the cdma2000® technology in their second- or third-generation networks to smoothly migrate to LTE. The cdma2000® access technology as such is specified in [C.S0024-A v2.0] and is widely used in the Americas and parts of Asia. Other access technologies explicitly mentioned in the 3GPP specifications [TS23.402] and [TS24.302] are WiMAX (Worldwide Interoperability for Microwave Access) [WiMAX], WLAN and Ethernet.

The present treatment differs somewhat in scope from the preceding section 11.1 on interworking with GSM and 3G networks, because the latter dealt only with the security aspects of movements between different types of network while this section deals also with situations where the user remains stationary. Furthermore, user movements in non-3GPP access to the EPC, considered in the context of the present section, are supported in such a way that the EPC network remains the same; in contrast, in the context of section 11.1, the type of core network would, in general, also change for a moving user.

Trusted versus untrusted access networks

A crucial concept in the context of non-3GPP access to the EPC is the distinction between trusted and untrusted access networks. The intuitive meaning of 'trusted access network' is that both the security measures anyway present in the access network and the security of the links between the access network and the EPC are good enough from the EPC operator's point of view. Consequently, for trusted access networks, no additional security measures need to be defined to protect the communication between the terminal and the EPC, while for untrusted access networks such additional measures are needed. Therefore, the procedures vary substantially depending on the trust status of the access network.

The 3GPP specifications do not give precise criteria when an access network should be considered trusted or not. The reason for this is that specifications are meant to capture the technical behaviour of a system, while the question whether somebody trusts in something goes beyond technology, and has to do also with organizational, commercial and legal considerations. Consequently [TS23.402] states:

> Whether a Non-3GPP IP access network is Trusted or Untrusted is not a characteristic of the access network.[3]

It is therefore up to operators and users whether they consider an access network as trusted or not. In most practical cases, a user's home operator will take the decision for the user, and the user's terminal will learn of the trust status of the access network by configuration or by an explicit indication in a protected signalling message sent during 3GPP-based access authentication – see clause 6.2.3 of [TS24.302]. The explicit indication takes precedence over the data configured in the terminal in case of conflict as it is likely to be more up to date. The home operator may be assisted in this decision by information received from a visited operator in roaming situations.

---

[3] Extract reproduced with permission from © 2010, 3GPP™.

When an access network is untrusted, an IPsec tunnel must be established between the User Equipment and a node in the EPC, called the evolved Packet Data Gateway (ePDG). All traffic, apart from the initial signalling for setting up the tunnel, must then travel across the access network inside this secure tunnel. In this way, the security level of the communication between the user equipment and the EPC becomes independent of the security properties of the access network, and the overall security level is high even if the access network provides no security at all. When the access network is trusted there is no need to use an ePDG and a secured tunnel between the UE and the ePDG. As we will see later, however, even for trusted access networks, IPsec protection is required between the UE and the Home Agent in certain cases for protecting Mobile IPv6 signalling.

In the view of the authors, in practical deployments cdma2000® HRPD access networks would often be considered as trusted, while public WLAN hotspots with no air interface protection would be candidates for the status 'untrusted'.

Mobility concepts for non-3GPP access to the EPC

When users are attached via a 3GPP-defined access network – E-UTRAN, UTRAN or GERAN – their mobility is supported using mobility mechanisms specific to these access networks. Examples are the mechanisms for handover in E-UTRAN described in Chapter 9. For non-3GPP access to the EPC, these mechanisms are not available, and mobility mechanisms specific to these non-3GPP access networks (not specified as part of the Evolved Packet System) and IP mobility mechanisms are used instead. There are three such IP mobility mechanisms relevant here:

- Proxy Mobile IP (PMIP) [RFC5213];
- Mobile IPv4 (MIPv4) in Foreign Agent mode [RFC3344];
- Dual Stack Mobile IPv6 (DSMIPv6) [RFC5555].

It is specified in [TS23.402] how these mechanisms are used in the context of non-3GPP access to the EPC. As this book focuses on security, and many other sources are available to find more information on these IP mobility mechanisms, we do not attempt to describe here how they work in any detail.

The use of these IP mobility mechanisms depends on the trust status of the non-3GPP access network as follows (for details see [TS23.402]).

- *PMIP*. The salient property of PMIP is that the UE is unaware of any IP mobility handling and therefore assumes itself to be connected to its home network all the time. The handling of mobility is performed on behalf of the UE by a so-called Mobile Access Gateway (MAG). Its counterpart on the core network side is the Local Mobility Anchor (LMA). When PMIP is used with a trusted access network, the MAG resides in the non-3GPP access network; when PMIP is used with an untrusted access network, the MAG resides on the ePDG. In both cases, the LMA resides on a gateway in the EPC. ([TS23.402] also treats a case where a MAG resides in a Serving GW, but this case is not further described in this chapter as it is not particular to non-3GPP access to the EPC.)

- *MIPv4*. This is used only with trusted access networks. The Mobile Node resides on the UE, the Foreign Agent resides in the non-3GPP access network, and the Home Agent resides on a gateway in the EPC.
- *DSMIPv6*. This can be used with both trusted and untrusted access networks. The Mobile Node resides on the UE, and the Home Agent resides on a gateway in the EPC. There is no Foreign Agent.

The security procedures relating to these mobility mechanisms are described in section 11.2.4.

The EAP framework

The EAP framework was described in Chapter 5. 3GPP decided to apply the EAP framework for authentication and key agreement for non-3GPP access to the EPC for the same reasons it had already decided to use the EAP framework for 3G–WLAN interworking. These reasons are that the EAP framework provides a means for using the same authentication and key agreement method across different types of access networks and generating and distributing keys in a uniform manner. An important reason is also that possibly existing credential infrastructures can be re-used. The only aspect that is specific to the type of access network is the way in which the transport of EAP messages is supported.

The peer resides on the UE, and the EAP server resides on the 3GPP AAA server. The allocation of the authenticator varies with the scenarios as we will see below.

EAP methods used with non-3GPP access to the EPC

The following two EAP methods are allowed to be used for non-3GPP access to the EPC:

- EAP-AKA as specified in [RFC4187];
- EAP-AKA′ as specified in [RFC5448].

Both the UE and the 3GPP AAA server must implement both EAP-AKA and EAP-AKA′. The same USIM may be used for both EAP-AKA and EAP-AKA′. The use of a USIM is required only if the terminal supports also 3GPP access capabilities. Note that the USIM always resides, by definition, on a smart card, the UICC. If the terminal does not support 3GPP access capabilities, 3GPP does not require that a UICC be present, and 3GPP does not specify where the credentials used with EAP-AKA and EAP-AKA′ reside. But the mobile terminal must support equivalent functionality as provided by a USIM also in the latter case for the two aforementioned EAP methods to work. This rule was introduced so as to ease the re-use of terminal types in legacy environments where UICCs have not been used.

In contrast to 3G-WLAN interworking, EAP-SIM is no longer allowed. This is in line with the general decision to deny access to the Evolved Packet System based on a SIM (see also Chapter 6).

Overview of EAP-AKA′

An overview of EAP-AKA has already been given in Chapter 5. Here we explain the additional features of EAP-AKA′ not present in EAP-AKA.

Both EAP-AKA' and EAP-AKA allow using USIMs (or equivalent functionality) and UMTS authentication vectors and UMTS cryptographic functions, as described in sections 4.2 and 7.2, within the framework of EAP. In addition, EAP-AKA' provides a binding of derived keys to the access network identity. The relationship between EAP-AKA and EAP-AKA' is therefore quite similar to that between UMTS AKA (Chapter 4) and EPS AKA (Chapter 7).

The differences between EAP-AKA and EAP-AKA' are as follows.

- For each EAP-AKA' full authentication, UMTS authentication vectors, including the keys CK and IK, are generated. However, the keys CK and IK are not directly used in EAP-AKA'. Rather, a further key pair (CK', IK') is derived from (CK, IK) by including the access network identity in the derivation. This provides the desired binding of the keys to the access network identity.
- Any further EAP keys are then derived from (CK', IK') rather than from (CK, IK).
- The way in which the keys of type MSK, EMSK, and the TEK keys K_aut and K_encr are computed is also substantially different compared to EAP-AKA.
- The key derivation function is based on the hash function SHA-256 rather than the weaker SHA-1 (see section 2.3).
- In order to allow both sides to unequivocally derive the same keys, the access network identity (called 'access network name' in [RFC5448]) is sent from the EAP server to the EAP peer in an appropriate attribute carried in the EAP-Request/AKA'-Challenge message. The access network identity must be constructed as defined, for each access network type separately, in [TS24.302].
- The EAP peer and the EAP server can find out about their mutual support for EAP-AKA' by using the normal EAP method negotiation procedures, based on the different EAP type codes associated with each EAP method. As we will see in this chapter further below, 3GPP specifies under which conditions EAP-AKA and EAP-AKA' respectively are to be applied. The 3GPP AAA server enforces the use of the correct EAP method on the core network side.
- EAP-AKA' provides a means for preventing bidding down to EAP-AKA. This is necessary as the security properties of EAP-AKA' are stronger, and hence a bidding down to EAP-AKA would unnecessarily weaken security in situations where both sides, peer and server, would be able to support the stronger method, but would allow falling back to the weaker method if otherwise communication was not possible.

EAP-AKA' may be used for full authentications or fast re-authentications, just like EAP-AKA. EAP-AKA' also uses pseudonyms and re-authentication identities in the same way as EAP-AKA.

We now discuss the advantages of being able to bind the key MSK to the access network identity. Assuming the link between the peer and the authenticator to be protected by keys derived from MSK, the key binding can eliminate some forms of the 'lying authenticator' problem, which has already been mentioned in section 5.1. The AAA procedures between the authenticator and the EAP server may be assumed to provide strong authentication so that the authenticator cannot lie about its identity to the EAP server. This enables the EAP server to ensure that a key MSK bound to a particular access network identity is delivered only to an authenticator associated with this access network identity. The EAP server then informs the EAP peer about this access network identity in the EAP message EAP-Request/AKA'-Challenge, which is integrity-protected with the key K_aut. Therefore, the authenticator can no longer lie to the peer about the access network identity, with which it is associated. This prevents in particular an attack mentioned in [RFC5448]:

> A roaming partner, R, might claim that it is the home network H in an effort to lure peers to connect to itself. Such an attack would be beneficial for the roaming partner if it can attract more users, and damaging for the users if their access costs in R are higher than those in other alternative networks, such as H.

Note that for EAP-AKA such an attack would be, in principle, possible. Whether this attack constitutes a risk depends on the circumstances.

The benefits obtained from the key binding to the access network identity depend on the granularity, with which 'access network identity' is defined. The access network identity could identify an individual authenticator in an access network, or all authenticators in an access network (as the name suggests), or even refer only to the access technology. The latter approach was chosen by 3GPP2 and WiMAX for the cdma2000® HRPD and WiMAX access technologies. 3GPP also specified general access network identities for Ethernet and WLAN in [TS24.302] for potential future use. This approach prevents breaches in one access technology to spill over to another access technology, while breaches inside each technology may still occur. More fine-grained definitions would be possible in the future if desired.

The precautions to be taken with the authentication vectors are the same as for the case of EPS AKA. Anybody getting hold of the keys CK and IK can also compute all the keys derived from them and, consequently, perform the key binding to the access network identity and impersonate any authenticator. It is therefore crucial for the security of EAP-AKA' that the keys CK and IK never leave the 3GPP HSS. Authentication vectors for use with EAP-AKA' must therefore have the AMF separation bit set to '1', and the terminal side must check this, just as for authentication vectors used with EPS AKA (see Chapter 7).

The development of EAP-AKA' is a nice example of a smooth cooperation between two standardization organizations, 3GPP and the IETF. While 3GPP first discovered the need for an extension of EAP-AKA providing a binding of keys to the access network identity, people at the IETF then took up this requirement and produced [RFC5448] in time for 3GPP Release 8, in constant discussion with the relevant Working Groups in 3GPP, WG SA3 and WG CT1.

Conditions for applying EAP-AKA and EAP-AKA', respectively

The conditions are slightly complicated and relate to the trust status of the access network and the IP mobility scheme used.

- *Trusted access networks.* As a general rule, EAP-AKA' must be used with trusted access networks. The procedure described in section 11.2.2 always applies when PMIP (with the MAG in the non-3GPP access network) or MIPv4 are used as mobility schemes. The authenticator then resides in the non-3GPP access network, and there is no protected tunnel to the EPC; it therefore is a good idea to benefit from the EAP-AKA' property of binding keys to the access network identity. There is, however, an exception from the general rule when DSMIPv6 is used as described below in this chapter.

- *Untrusted access networks.* For untrusted access networks, the establishment of an IPsec tunnel between the UE and the ePDG is required, as described in section 11.2.3. This IPsec tunnel is established via IKEv2, in a manner quite similar to 3GPP IP access in 3G–WLAN interworking (see section 5.2). The UE, which is the IKEv2 initiator, is authenticated using EAP-AKA inside IKEv2. One may wonder why EAP-AKA′ is not used here. Indeed, the use of EAP-AKA′ would be perfectly possible for this purpose. But EAP-AKA′ would provide no security advantage over EAP-AKA in this case because IKEv2 mandates the use of certificates for responder authentication (i.e. authentication of the ePDG to the UE); and this certificate-based authentication also guarantees a binding of the IPsec security association used to protect the message in the tunnel to the identity in the certificate. 3GPP opted for maximum commonality with the case of 3GPP IP access and therefore decided in favour of using EAP-AKA here. But, as reported in Chapter 5, efforts are under way in the IETF to remove the requirement of certificate-based authentication from IKEv2. Removing this requirement without adding further protocol elements would, however, be only possible with EAP-AKA′.

  Prior to tunnel establishment, access authentication may be required to gain IP connectivity over the untrusted access network. This access authentication may, or may not, involve the 3GPP AAA server and is independent of the EAP-AKA run inside IKEv2. We quote from [TS33.402]:

> This additional access authentication and key agreement is not required for the security of the Evolved Packet Core. However, it may be required for the security of the untrusted non-3GPP access network. Any authentication and key agreement procedure deemed appropriate by the access network provider, including EAP-AKA′, may be used.[4]

In particular, the access authentication need not even be an EAP method.

As mentioned earlier, DSMIPv6 can be used with both trusted and untrusted access networks. The use of DSMIPv6 requires the establishment of an IPsec tunnel between the UE and the PDN GW, which acts as the Home Agent, to protect the Mobile IP signalling. Again, IKEv2, with EAP-AKA for authenticating the UE to the network and certificate-based network authentication, is used for this purpose. Note that the PDN GW and the ePDG are two different functional entities, and, when using DSMIPv6 over untrusted access, there are two IPsec tunnels whereby the tunnel between the UE and the PDN GW runs inside the tunnel between the UE and the ePDG. When DSMIPv6 is used over trusted access, there is only one tunnel, the one between the UE and the PDN GW.

Similar to the case of untrusted access, access authentication may be additionally required to gain IP connectivity over the trusted access network. As only the Mobile IP signalling, and not all traffic, is protected by the IPsec tunnel in this case, 3GPP recommends using EAP-AKA′ for access authentication, as described in section 11.2.2. But it is also possible to use some other strong authentication method, documented in a standard covering the non-3GPP access network, if the following conditions are fulfilled [TS33.402].

1. The trusted access network authenticates the UE and provides a secure link for the data to be transferred from the UE to the trusted access network.

---

[4] Extract reproduced with permission from © 2010, 3GPP™.

2. The trusted access network protects against source IP address spoofing.
3. The trusted access network and the PDN GW have a secure link between them to transfer the user data.
4. The trusted access network and the EPC need to co-ordinate when the UE detaches from the trusted access network in order to ensure that the IP address that was assigned to the UE is not used by another UE without the EPC being aware of the change. If such an IP address change happened the PDN GW would have to remove the CoA address binding for the old UE.

This sounds a bit complicated, and needs more explanation. The idea underlying these four conditions is that origin authentication of IP packets from the UE can be achieved using a form of IP address binding to the IMSI as follows. Condition 2 ensures that a user cannot illegally use somebody else's IP address, and conditions 1 and 3 ensure that no attacker can alter the IP address while the IP packet is in transit from the UE to the PDN GW. So, these three conditions together ensure that packets with the same source IP address always originate from the same user, at least as long as this user is assigned that IP address.

But how can the operator of the EPC know which user is behind that IP address? The required binding of the source IP address to an IMSI is provided by IKEv2 with EAP-AKA authentication between the UE and the PDN GW because the authenticated user identity is a NAI based on the IMSI. Condition 4 takes care of the additional problem that an IP address may be reassigned by the access network without the EPC noticing. If this happened the EPC would wrongly belief that the new user, to which the IP address was reassigned, had the IMSI that in fact belongs to the old user previously authenticated by means of EAP-AKA.

Condition 4 may not be easy to fulfil in practice as it asks for a coordination of IP address assignment in the access network with functions in the core network. Examples of how condition 4 may be achieved are mentioned in [TS33.402]. One example is a coordination of timers for IP address reallocation in the access network with timers for Mobile IP binding expiry or IKE Dead Peer Detection in the PDN GW. Another example is the use of a gateway control session defined in the context of the policy control (PCC) mechanism; for details see [TS33.402].

All in all, the conditions listed above ensure that the EPC can verify that a certain IP packet originates from an authenticated user with a particular IMSI. But as it is easy to get things wrong with this kind of coordination, the specification [TS33.402] cautions that EAP-AKA' should be used if there is any doubt about the four conditions above being fulfilled in practice.

11.2.2   Authentication and Key Agreement for Trusted Access

This chapter presents the procedure using EAP-AKA' for trusted access networks. The precise conditions, under which this procedure applies, were explained in the preceding Chapter 11.2.1. The procedure is depicted in Figure 11.1.

The numbering of the steps in Figure 11.1 is the same as that in Figure 6.2-1 of [TS33.402], to make it easier for the reader to compare the text explaining the figure here with the text in the 3GPP specification. Figure 6.2-1 of [TS33.402] shows an additional network element, a AAA proxy, but as such a proxy takes no active role in the procedure and simply passes all messages through we have omitted the proxy in our figure. The textual description in this book is shortened in some places compared to [TS33.402] as not all details presented there are

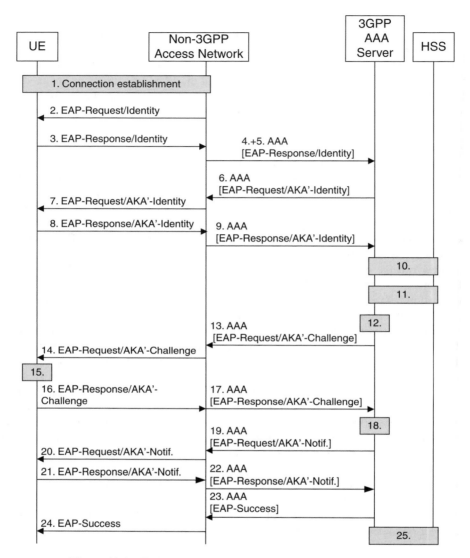

Figure 11.1   EAP AKA authentication for trusted non-3GPP access

essential for the understanding of authentication for trusted access networks. It is expanded in other places so as to explain the rationale for certain steps.

We limit ourselves to presenting the full authentication procedure as the fast re-authentication procedure is very similar. (See section 5.2 for a high-level explanation of how the fast re-authentication differs from the full authentication.)

1. A link layer connection is established between the UE and the authenticator in the non-3GPP access network. This establishment procedure is specific to the access technology, such as cdma2000® HRPD.

2. The authenticator in the non-3GPP access network starts the EAP procedure by sending the EAP-Request/Identity message to the UE. According to the EAP framework, this message is not specific to EAP-AKA′.
3. The UE sends the EAP-Response/Identity message, which is not specific to EAP-AKA′ either. The identity included by the UE is either a pseudonym received in a previous protocol run, or it is derived from the IMSI. In either case, the identity includes a leading digit hinting that the UE supports EAP-AKA′ [TS23.003].
4. The authenticator encapsulates the EAP Response/Identity message in a suitable AAA message, using the DIAMETER protocol, and forwards it towards the 3GPP AAA server. The AAA message also includes the access network identity. For a discussion of the access network identity, see section 11.2.1.
5. The 3GPP AAA server receives the AAA message. When it includes a pseudonym the server derives the IMSI from it. If this fails, the server goes to step 6. For the security of EAP-AKA′ it is important that the 3GPP AAA server can authenticate the origin of this message at least to the extent that it can verify whether the sender of the message – the authenticator – is indeed authorized to use the included access network identity. If the authenticator could lie about the access network identity, the binding of keys to this identity would be no longer meaningful. As mentioned in section 11.2.1, for the most important access technologies the access network identity just refers to the access network type; for example it equals the string 'HRPD' or 'WIMAX'. This means that the 3GPP AAA server must at least be able to verify (for the purposes of EAP-AKA′) whether the authenticator resides in an access network of type 'HRPD' or 'WIMAX'.

Steps 6 to 9, are optional. They contain a message exchange for the 3GPP AAA server obtaining the user identity in messages specific to EAP-AKA′. [RFC5448] on EAP-AKA′ strongly recommends using these steps in a general setting. They serve two purposes. First, intermediate nodes between the authenticator and the 3GPP AAA server may, in general, modify the identity sent as part of the EAP-Response/Identity message in step 4, while they never do that for EAP method-specific messages. However, when the intermediate nodes are all under operator control it can be ensured by configuration that this does not happen. Second, the 3GPP AAA server may have been unable to recognize a pseudonym sent by the UE in steps 3 and 4, in which case the server would request the UE to send the permanent identity. But when a pseudonym is constructed by encrypting the IMSI with a long-term key, as described in section 5.1, it is unlikely to happen that the server cannot recognize it. So, if it can be ruled out with sufficient probability that the 3GPP AAA server can correctly process the identity received in step 5, then the server can be configured to skip steps 6 to 9.

6. This consists in the 3GPP AAA server sending the EAP-Request/AKA′-Identity message encapsulated in a AAA message.
7. The authenticator retrieves the EAP-Request/AKA′-Identity message from the AAA message and forwards it to the UE.
8. The UE responds with the type of identity requested in the EAP-Request/AKA′-Identity message.
9. The authenticator encapsulates the EAP-Response/AKA′-Identity message in an AAA message and forwards it to the 3GPP AAA server. The AAA message again includes the access network identity in the same way as in step 4. The 3GPP AAA server then performs

the same checks on the access network identity as in step 5. The server uses the received user identity in the remaining protocol steps.

10. The 3GPP AAA server knows from the origin of the AAA message, and the information elements contained in the AAA message ([TS29.273]), that the current protocol run is for trusted access, and hence EAP-AKA′ must be used. As stated earlier, all UEs accessing the EPC must support EAP-AKA′. So, if the user identity received in steps 5 or 9 does not contain the hint that the UE supports EAP-AKA′ there must be an error case, and the server abandons the procedure. Otherwise, the server sends a request for an authentication vector to the HSS, together with the user's IMSI and the access network identity. The HSS sees from the inclusion of the latter that this is a request for EAP-AKA′ and performs the transformation of the keys CK and IK to the keys CK′ and IK′, as described in section 11.2.1. Note that the specification allows the 3GPP AAA server to fetch an entire batch of authentication vectors in one go, and then the server could have a suitable authentication vector already locally available at the beginning of step 10, and would not need to contact the HSS. But the gain in performance and reliability of doing so is limited as the 3GPP AAA server and the HSS both reside in the home network and can be assumed to have a fast and reliable link between them. Furthermore, when the 3GPP AAA server always requests only one new authentication vector and then consumes it immediately, the likelihood for synchronization errors in the USIM (see Chapter 4) is minimized irrespective of the sequence number management scheme.

11. The 3GPP AAA server fetches the user profile from the HSS if not yet available. The user profile tells the server that the user is authorized to access the EPC.

12. The 3GPP AAA server derives the keys MSK and EMSK and the TEK keys – see [RFC5448] and sections 5.1 and 11.2.1. In the context of the procedures described in this book, EMSK is used only to derive a root key for the purposes of Mobile IPv4 (see section 11.2.4).

13. The 3GPP AAA server encapsulates the EAP-Request/AKA′-Challenge message, which includes the access network identity and, optionally, an attribute indicating the trust status of the access network, in a AAA message and sends it to the authenticator.

14. The authenticator retrieves the EAP-Request/AKA′-Challenge message from the AAA message and forwards it to the UE.

15. The UE processes the received message according to the rules for EAP-AKA′. In particular, the UE must check whether the authentication vector is allowed to be used with EAP-AKA′ – whether the AMF separation bit (Chapter 7 and section 11.2.1) is set to '1'. Furthermore, the UE compares the access network identity in the received message with the locally observed one in all cases where [TS24.302] specifies how to construct an access network identity from local observations, such as on a link layer broadcast channel. [RFC5448] contains detailed rules how to perform the comparison. These rules are made such that defining more fine-grained access network identities in the future would be backwards compatible, so legacy UEs not understanding the more fine-grained access network identities would still be able to perform the comparison successfully. The UE – or the human user – may use the network name as a basis for an authorization decision. For example, the UE may compare the network name against a list of preferred or barred network names. If any of these checks does not succeed the UE abandons the procedure.

    The UE also derives the keys MSK and EMSK and the TEK keys at this point.

16. The UE sends the EAP-Response/AKA′-Challenge message to the authenticator.

17. The authenticator encapsulates the EAP-Response/AKA'-Challenge message in a AAA message and forwards it to the 3GPP AAA server.
18. The 3GPP AAA server performs the checks on the response required by EAP-AKA'; that is, it uses the key K_aut to check the message integrity, and compares the RES received from the UE with the XRES received from the HSS.

Steps 19 to 22 are conditional. They are only peformed if the 3GPP AAA Server and the UE have indicated in steps 13 and 16, respectively, that they want to use protected result indications. Otherwise the procedure continues from step 23 onwards.

19. This step consists in the 3GPP AAA server sending the EAP-Request/AKA'-otification message encapsulated in a AAA message.
20. The authenticator retrieves the EAP-Request/AKA'-Notification message from the AAA message and forwards it to the UE.
21. The UE sends the EAP Response/AKA'-Notification message.
22. The authenticator encapsulates the EAP-Response/AKA'-Notification message in a AAA message and forwards it to the 3GPP AAA server.
23. The 3GPP AAA server sends the EAP-Success message encapsulated in a AAA message. The latter also contains the key MSK.
24. The authenticator retrieves the EAP-Success message from the AAA message and forwards it to the UE. The authenticator stores the MSK and does not forward it to the UE; but, as the UE has already derived the MSK in step 15, UE and authenticator now share the MSK. The authenticator and the UE use the MSK according to the security procedures specific to the non-3GPP access technology. For example, they use MSK to derive further keys, which are then used to protect the radio access link.
25. The 3GPP AAA server registers the user with the HSS and maintains session state.

11.2.3 Authentication and Key Agreement for Untrusted Access

This section presents the procedure using IKEv2 with EAP-AKA for untrusted non-3GPP access networks. The precise conditions under which this procedure applies were explained in section 11.2.1. The procedure is depicted in Figure 11.2.

The numbering of the steps in Figure 11.2 is the same as that in Figure 8.2.2-1 of [TS33.402], to make it easier for the reader to compare the text explaining the figure here with the text in the 3GPP specification. The numbering seems a little odd in places as, apparently, some steps were added to the figure later. The textual description in this book is shortened in some places compared to [TS33.402] as not all details presented there are essential for the understanding of authentication for trusted access networks. It is expanded in other places so as to explain the rationale for certain steps. We limit ourselves to presenting the full authentication procedure as the fast re-authentication procedure is very similar.

The procedural steps are almost identical to those for 3GPP IP access in 3G–WLAN interworking, as described in section 5.2.

1. The UE and the ePDG exchange the first pair of messages, known as IKE_SA_INIT, in which the ePDG and the UE negotiate cryptographic algorithms, exchange nonces and perform a Diffie–Hellman exchange.

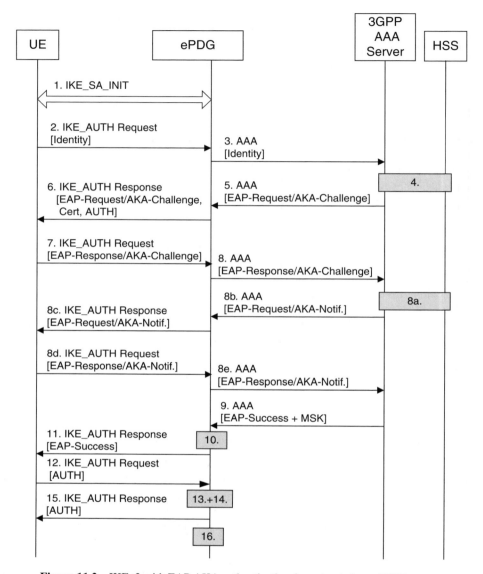

Figure 11.2   IKEv2 with EAP AKA authentication for untrusted non-3GPP access

2. The UE sends the user identity in the form required for EAP-AKA in this first message of an IKE_AUTH exchange. In accordance with [RFC4306], the UE omits the AUTH parameter in order to indicate to the ePDG that it wants to use EAP over IKEv2.
3. The ePDG sends an appropriate AAA message to the 3GPP AAA Server, containing the user identity.
4. The 3GPP AAA Server sees from the information elements contained in the AAA message ([TS29.273]) that this is a request for authentication and authorization in the context

of untrusted access to the EPC, and not trusted access to the EPC, or 3G–WLAN inter-working. As the 3GPP AAA Server trusts the sender (the ePDG) to include the correct information elements, the server knows that EAP-AKA must be applied. The 3GPP AAA Server then deduces the IMSI from the received user identity and fetches a fresh authentication vector and the user profile from the HSS (unless already available). The authentication vector has the AMF separation bit set to '0' as it must be for EAP-AKA. The user profile tells the server that the user is authorized to access the EPC.

5. The 3GPP AAA server encapsulates the EAP-Request/AKA-Challenge message in a AAA message and sends it to the ePDG. The user identity is not requested again by using the EAP-AKA-specific identity request/response messages as the user identity received in step 3 could not have been modified or replaced by any intermediate node.

6. The ePDG sends its identity, a certificate, and an AUTH parameter to the UE. The ePDG generates this AUTH parameter by computing a digital signature over parameters in the first message it sent to the UE (in step 1). The ePDG also includes the EAP-Request/AKA-Challenge message received in step 5.

7. The UE verifies AUTH using the public key in the certificate received in step 6 and sends the EAP-Response/AKA-Challenge message towards the ePDG.

8. The ePDG forwards the EAP-Response/AKA-Challenge message to the 3GPP AAA Server, encapsulated in a AAA message.

8a. The 3GPP AAA server performs the checks on the response required by EAP-AKA (i.e. it uses the key K_aut to check the message integrity), and compares the RES received from the UE with the XRES received from the HSS. At this point the UE is authenticated from an EAP point of view.

Steps 8b to 8e are conditional. They are used only if dynamic IP mobility selection (IPMS) is applied embedded in the EAP-AKA run. IPMS consists essentially in selecting one of the IP mobility schemes, PMIP, MIPv4, or DSMIPv6, that may be used with non-3GPP access to the EPC as described in section 11.2.1. For details on the IPMS, see [TS24.302]. Step 8b consists in the 3GPP AAA server sending the EAP-Request/AKA-Notification message including the selected mobility mode, encapsulated in a AAA message. Steps 8c, 8d and 8e, consist in the forwarding of this message by the ePDG to the UE, and the corresponding response from the UE, forwarded by the ePDG to the 3GPP AAA server.

9. The 3GPP AAA server sends the EAP-Success message to the ePDG, encapsulated in a AAA message. The latter also contains the key MSK.

10. The ePDG generates two additional AUTH parameters by computing message authentication codes over parameters in the two messages exchanged in step 1 using the shared key MSK. Note that the ePDG could defer the generation of these two AUTH parameters until receiving the message in step 12.

11. The ePDG forwards the EAP-Success message to the UE over IKEv2.

12. The UE generates two AUTH parameters in the same way as the ePDG in step 10 and then sends the AUTH parameter protecting the first message from the UE to the ePDG (sent in step 1).

13. This includes also step 14. The ePDG verifies the AUTH parameter received in step 12 by comparing it with the corresponding value computed in step 10. At this point the UE is authenticated also from an IKEv2 point of view.

14. The ePDG then sends the other AUTH parameter it computed in step 10 to the UE. The UE verifies the received AUTH parameter by comparing it with the corresponding value computed in step 12.

15. The ePDG deletes any old IKE SAs relating to the same Access Point Name ([TS23.402]) and informs the UE of this deletion in an informational exchange (not shown in the figure).

Handling of IPsec tunnels in case of UE mobility

IPsec was originally designed without mobility in mind. In order to allow for terminal mobility while keeping the IPsec tunnel alive, the IETF developed MOBIKE [RFC4555]. With MOBIKE, the terminal (the initiator in IKE terms) may change its IP address while maintaining the IPsec tunnel and inform the responder of the new IP address. But MOBIKE still has to assume that the same, stationary, responder is used.

The procedures for untrusted non-3GPP access to the Evolved Packet Core take advantage of MOBIKE using the following rules.

- When the UE moves from a source access where the UE is connected to an ePDG to a target access that involves the same ePDG, the UE uses MOBIKE.
- When the UE moves from a source access where the UE is connected to an ePDG to a target access that involves a different ePDG, the UE establishes a new IPsec tunnel with the new ePDG using the procedures described in this subsection.
- When the UE is connected to the EPS without being connected to an ePDG and then moves to a target access which involves the UE and an ePDG, the UE establishes a new IPsec tunnel with the new ePDG again using the procedures described in this subsection.

11.2.4   Security for Mobile IP Signalling

This subsection has three parts, corresponding to the three variants of Mobile IP used with non-3GPP access to the EPC: Proxy Mobile IP, Mobile IPv4, and Dual Stack Mobile IPv6.

Each of these Mobile IP variants has its own way of securing the mobility signalling. In each case, the main threat is that Binding Updates may be tampered with by an attacker. A Binding Update is sent by the Mobile IP client (Mobile Node or MAG) to the Home Agent (HA) to inform the latter about the client's new IP address (the so-called care-of-address), under which the client can be reached. The HA then knows to where it must forward incoming IP packets, destined to the client's home address. If such tampering with Binding Updates was possible an attacker could register a wrong care-of-address with the HA, and the client would be unreachable until the next Binding Update. Binding Updates therefore need to be at least integrity-protected. Confidentiality protection is not possible with all Mobile IP variants. It may be desirable, however, for protecting the client's privacy. As always, cryptographic integrity protection requires two elements:

- the availability of cryptographic keys;
- an integrity protection mechanism using the keys.

It is advantageous to derive the cryptographic root keys required for the purposes of Mobile IP from other keys present in the system anyway, such as an authentication key available in the

Mobile Node. The process of deriving root keys for one use case of cryptography from other security parameters already present for other purposes is often referred to as 'bootstrapping' the security of that use case. Here we describe how this bootstrapping is performed in our setting.

The integrity protection mechanisms used by the three Mobile IP variants in the context of non-3GPP access to the EPC are quite different: PMIP and DSMIPv6 rely on IPsec while MIPv4 uses a mechanism specific to MIPv4. IPsec used with PMIP and DSMIPv6 can provide confidentiality, if desired, while the MIPv4-specific mechanism cannot provide confidentiality.

We now look at the three Mobile IP variants one by one.

Proxy Mobile IP

The MAG is a network node, and not a user terminal. This makes life easy as far as key distribution is concerned because the number of network nodes is quite small compared to the number of terminals and users. While network operators have so far shied away from distributing public key certificates to potentially hundreds of millions of their customers' terminals, they do not see a major problem with supplying certificates to network nodes.

The PMIP signalling messages are exchanged between the MAG, either the authenticator in a trusted access network or an ePDG in the case of an untrusted access network, and the PMIP Home Agent (LMA) – either the Serving GW or the PDN GW (see section 11.2.1). So, the task to be solved is the distribution of keys to these nodes, and to implement an integrity protection mechanism in these nodes. Fortunately, when it comes to protecting IP-based signalling traffic between network nodes, 3GPP has a panacea called Network Domain Security (NDS/IP) – see section 4.5. Consequently, [TS33.402] requires the use of NDS/IP for protecting PMIP signalling. NDS/IP implies that cryptographic protection need not be used if the traffic between the two nodes in question travels entirely inside one security domain. However, when the traffic crosses security domain boundaries, the use of IPsec with integrity protection (message authentication) becomes mandatory. The use of confidentiality protection is optional. Protection may be provided either by a chain of security associations in a hop-by-hop fashion, or directly end-to-end.

There is another threat to consider in the context of PMIP: not only may an attacker modify signalling messages between the MAG and the LMA, but the MAG may be compromised itself. If this happens then, of course, the security of PMIP is entirely broken for all users that may be potentially served by this MAG. (They need not even be actually served by it at the time of the attack as the compromised MAG can anyway send false Binding Updates on their behalf.) 3GPP discussed whether more elaborate protection schemes, involving the UE, would be required to contain the damage potentially caused by a compromised MAG, but finally concluded that the use of NDS/IP was sufficient as the MAGs reside on nodes in a trusted access network. UE involvement would anyhow have defeated the main purpose of PMIP, namely to leave the UE unaffected by the mobility scheme.

PMIP is further based on the assumption that a MAG can securely identify which user is attached to the access network served by the MAG. If access authentication was weak then an attacker could impersonate a user in the access network. If this happened a MAG would report in good faith to the LMA that a certain user was present in the access network, while in fact the attacker was present. But, fortunately, when the MAG resides in a trusted non-3GPP access, the use of EAP-AKA' is required (see section 11.2.1), thus providing strong authentication.

Similarly, when the MAG resides on the ePDG strong authentication is provided by EAP-AKA with IKEv2.

Mobile IPv4

In the context of the EPC, MIPv4 is used only with trusted non-3GPP access networks and always comes with EAP-AKA′ access authentication according to the procedure described in section 11.2.2. The integrity protection of MIPv4 signalling messages uses a mechanism specific to MIPv4, the so-called authentication extensions as defined in [RFC3344]. In our context, two such extensions are used:

- the mandatory MN-HA authentication extension, applied between the Mobile Node (MN) and the Home Agent;
- the optional MN-FA authentication extension, applied between the Mobile Node and the Foreign Agent (FA).

In our setting, the MN resides on the UE, the HA resides on the PDN GW, and the FA resides in the trusted access network. The FA need not coincide with the authenticator in that trusted access network. Both authentication extensions contain message authentication codes computed over suitable parts of the protected messages using the MN-HA key and MN-FA key, respectively. We now describe how these two keys are generated and distributed.

MIPv4 key generation
We only explain the principles and refer to [TS33.402] for the key generation formulas and the handling of special cases, such as dynamic HA assignment, and EAP-AKA′ re-authentication. The key generation proceeds in the following steps.

- As a result of EAP-AKA′ access authentication, the UE and the 3GPP AAA server share the key EMSK.
- The UE and the 3GPP AAA server derive a Mobile IP Root Key (MIP-RK) from the EMSK. The MIP-RK never leaves the 3GPP AAA server.
- The UE and the 3GPP AAA server derive a Foreign Agent Root Key (FA-RK) from the MIP-RK. The 3GPP AAA server sends the FA-RK to the authenticator.
- The UE and the 3GPP AAA server derive the MN-HA key from the MIP-RK. The 3GPP AAA server sends the MN-HA key to the PDN GW.
- The UE and the authenticator derive the MN-FA key from the FA-RK. The authenticator sends the MN-FA key to the FA.
- No keys are sent to the UE as they are derived in the UE locally. No keys derived in the UE leave the UE.

MIPv4 message protection
The MIPv4 message protection is described with the help of Figure 11.3. It proceeds in the following steps.

1. During EAP-AKA′ access authentication (see section 11.2.2) the key EMSK is generated in the UE and the 3GPP AAA server. The UE and the 3GPP AAA server then derive the

keys MIP-RK and FA-RK as described above. The 3GPP AAA server sends FA-RK to the authenticator.

2. The UE sends a Registration Request (RRQ) message to the FA [TS23.402]. The UE includes the MN-HA authentication extension and optionally the MN-FA authentication extension.

3. The FA processes the RRQ message according to [RFC3344] and, in particular, validates the MN-FA authentication extension if present, using the MN-FA key it obtained from the authenticator. The FA then forwards the RRQ message to the PDN GW. The RRQ message is protected between the FA and the PDN GW using NDS/IP; that is, 3GPP does not make use of the Foreign–Home Authentication Extension defined in [RFC3344].

4. The PDN GW contacts the 3GPP AAA server to learn whether the UE has been authenticated and authorized and obtain the MN-HA key.

5. The PDN GW validates the MN-HA authentication extension. If the check is successful the PDN GW sends a Registration Reply (RRP) to the UE through the FA. As in step 3, the RRP message is protected between the PDN GW and the FA using NDS/IP.

6. The FA processes the RRP message according to [RFC3344] and then forwards it to the UE. The FA includes an MN-FA authentication extension if the FA received an MN-FA authentication extension in the RRQ message.

7. The UE validates the MN-FA authentication extension, if present, and the MN-HA authentication extension.

Dual Stack Mobile IPv6

In our setting, the Mobile Node resides on the UE and the Home Agent resides on the PDN GW. There is no Foreign Agent in DSMIPv6. An IPsec tunnel is set up between the UE and the PDN GW (acting as a Home Agent) using IKEv2 with EAP-AKA for the purpose of

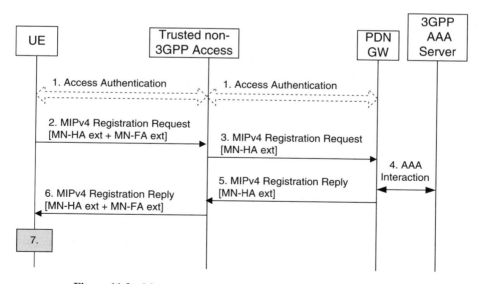

Figure 11.3 Message protection for mobile IPv4 with foreign agent

protecting the Mobile IP signalling between these entities. As required by IKEv2, the PDN GW is authenticated using a public key certificate. While the purpose of this tunnel set-up is different from tunnel set-up for untrusted access shown in section 11.2.3, the information flow is almost identical. As in the case of tunnel set-up for untrusted access, an EAP-AKA full authentication procedure and an EAP-AKA fast re-authentication procedure may be used. Corresponding information flows and their textual descriptions can be found in [TS33.402]. Once the IPsec tunnel is established, the UE and the PDN GW can securely exchange DSMIPv6 signalling messages sent through this tunnel.

There is one additional security consideration to be taken into account. It needs to be ensured that the UE can only send Binding Updates for its own Home Address and not for other Home Addresses of other Mobile Nodes through this tunnel. This is achieved by binding the Home Address to the IPsec security association as follows. The PDN GW allocates a Home Network Prefix during the IKEv2 run and sends it to the UE. The UE then auto-configures a Home Address from the IPv6 prefix received from the HA. This Home Address is then bound to the IPsec security association.

11.2.5   Mobility between 3GPP and non-3GPP Access Networks

The preceding sections 11.2.1 to 11.2.4 dealt with the security procedures applied when accessing the Evolved Packet Core via a non-3GPP access network. These procedures would also apply while the User Equipment was stationary. Here we deal with the additional procedures that apply when the UE moves between E-UTRAN and a non-3GPP access network in idle state or connected state.

For a UE moving between E-UTRAN and a non-3GPP access network while in connected state (i.e. a UE performing handover), 3GPP defines two types of procedure:

- handover without optimizations between E-UTRAN and a general non-3GPP access;
- handover with optimizations between E-UTRAN and a cdma2000® HRPD access.

The description of the information flows for these two types of procedure takes more than forty pages in clauses 8 and 9 of [TS23.402] and considers the many different combinations of interfaces that may occur. To provide a level of detail similar to this description would be far beyond the reach of this book, and would give little insight on the security aspects. We therefore limit ourselves to describing the new security concept used in these procedures.

For the sake of completeness, we mention that [TS23.402] also contains a brief clause on general principles for optimized network-controlled dual radio handover between E-UTRAN and Mobile WiMAX. That clause does not, however, contain any description of detailed procedures.

The security procedures are embedded in the descriptions of the overall handover procedures in [TS23.402]. For the case of handover without optimizations, there is nothing new in terms of security. When the UE moves to the target access network the UE first attaches to that access network and then performs the security procedures defined for that access network. So, when the UE moves, for example, from a non-3GPP access network to E-UTRAN the UE attaches to E-UTRAN, performs EPS AKA as described in Chapter 7, and establishes confidentiality and integrity protection as described in Chapter 8.

In Chapter 9 on mobility between two E-UTRAN access networks the central concept is security context transfer between the source and the target network. In section 11.1 on mobility between an E-UTRAN access network and a GSM or 3G access network the central concept is security context mapping from the source to the target network. The use of either concept obviates the need for another round of authentication and key agreement in the target network and therefore enhances performance. Neither security context transfer nor security context mapping are, however, applicable to mobility between a 3GPP access and a non-3GPP access network because the security architectures of the involved network are too different.

In the general case, nothing much can be done to improve performance from a security point of view. For the case of handover with optimizations between E-UTRAN and a cdma2000® HRPD access, however, the concept of pre-registration can be used to improve handover performance. Pre-registration includes pre-authentication. The concept is particularly useful for single-radio terminals that can attach only to one radio access technology at a time. We explain the idea of pre-registration in the following.

Pre-registration

The basic idea of pre-registration is that a UE can register in the target network, using procedures specific to the target network, while still being attached to the source network. The UE communicates with the target network through a series of tunnels spanning across the source network to a defined exit point in the source network, and further on to the target network. For E-UTRAN access, this exit point towards a cdma2000® HRPD access network is the MME. The MME communicates with the HRPD Serving Gateway (HS-GW) in the HRPD access network.

Once pre-registration has been completed the actual handover phase can start. The handover messages are tunnelled across the source access network using the same series of tunnels that was established in the pre-registration phase, thereby speeding up the handover phase significantly. Only later in the handover procedure, namely when the UE receives the Handover Command message, does the UE have to attach to the target network; the UE can remain attached to the source network, and send and receive data there, while the registration and part of the handover procedure are already ongoing.

For illustration, we show a simplified Figure 11.4 of a pre-registration procedure in a trusted cdma2000® HRPD access network being performed while the UE is still attached to E-UTRAN, and explain the steps in the procedure.

1. The UE is attached to E-UTRAN and registered with the MME. It may be in idle state or connected state.
2. Based on a radio layer trigger, the UE decides to initiate a pre-registration procedure with the target HRPD access. The pre-registration procedure allows the UE to establish and maintain a dormant session in the target HRPD access while attached to the E-UTRAN.
3. Registration to the HRPD is achieved by exchanging a series of HRPD messages between the UE and the HRPD Access Network. The HRPD signalling that is tunnelled transparently over the E-UTRAN and EPC creates an HRPD session context between the UE and the HRPD Access Network.

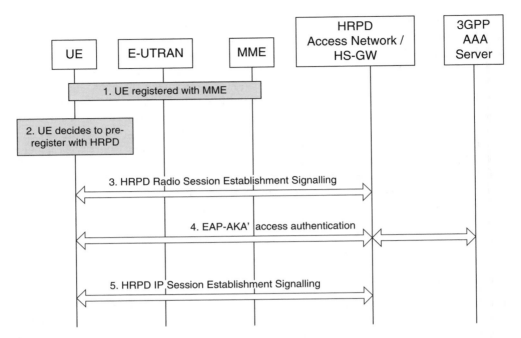

Figure 11.4   Pre-registration procedure

4. The UE, HS-GW and 3GPP AAA server exchange EAP-AKA′ signalling to authenticate the UE on the HRPD system, in accordance with the procedure described in section 11.2.2 on trusted access.
5. The UE and HS-GW exchange signalling to establish context to support the bearer traffic environment in use over the E-UTRAN.

12

Security for Voice over LTE

Voice has historically been the first application of mobile communication networks, and the success of GSM has been primarily based on voice. While it is true that data applications have considerably gained in importance over the years, voice is still a major source of revenue for mobile operators. It is expected that voice will remain an important application even in the era of LTE, so there has been a lot of discussion about the best way to provide voice in an LTE environment. Owing to this importance, we include this chapter on security of voice over LTE in this book although, as we will show in the following, the corresponding security mechanisms are largely orthogonal to the LTE security mechanisms discussed in the rest of the book.

The nature of this chapter is therefore somewhat different from the rest of the book in that it describes all the relevant mechanisms, but does not go to a similar level of detail. It includes the necessary references for readers who want to delve into this subject more deeply.

In section 12.1 we briefly introduce the methods standardized by 3GPP for providing voice over LTE. Then in section 12.2 we discuss the security mechanisms used with these methods.

12.1 Methods for Providing Voice over LTE

There are two standardized methods for providing voice over LTE:

- *IMS over LTE*. IMS is a largely access-independent service control architecture that enables various types of multimedia services using IP connectivity.
- *Circuit Switched Fallback* (CSFB). This provides voice service by fallback from LTE to the circuit-switched infrastructure offered in UTRAN, GERAN or a 3GPP2-defined network.

These two methods are complemented by the following feature:

- *Single Radio Voice Call Continuity* (SRVCC). This provides a means to hand over a call between IMS over LTE and the circuit-switched domain of UTRAN, GERAN or a 3GPP2-defined network.

LTE Security   Dan Forsberg, Günther Horn, Wolf-Dietrich Moeller, and Valtteri Niemi
© 2010 John Wiley & Sons, Ltd

12.1.1   IMS over LTE

What is IMS?

IMS stands for Internet-protocol Multimedia Subsystem. It is a subsystem or a domain within a mobile communications system. IMS may be used for providing voice services over different network technologies offering IP connectivity. In particular, it can be used for providing voice services over LTE.

IMS is a huge knowledge area in itself and it is far beyond the scope of this book to attempt to give an overview. Therefore, this section introduces the key IMS concepts (in case the reader does not know them already). A detailed insight into IMS is given in other books [Poikselkä and Mayer 2009; Gonzalo Camarillo and García-Martín 2008]. The former provides the following definition of IMS:

> IMS is a global, access-independent and standard-based IP connectivity and service control architecture that enables various types of multimedia services to end-users using common Internet-based protocols.

We discuss below some of the key words in this definition.

- The access-independence of IMS means that, in principle, IMS services and procedures may be provided and executed over different access technologies in the same way. So, in principle, there should be nothing particular to IMS over LTE compared to IMS over other access network types, such as over DSL-type. And, to a large extent, this is indeed true. There are, however, some dependencies of IMS on the access technology. These dependencies are partly due to the inevitable influence of the nature of the bearers available from the access technology on the service that can be provided over those bearers. But they are also partly due to the fact that IMS has grown out of the originally disjoint efforts of several standardization organizations, each with their own legacy environments. These efforts were then unified in the so-called 'Common IMS' defined from 3GPP Release 7 onwards. The second reason applies, in particular, to security as we will see in section 12.2.1.
- IMS is standards-based in contrast to proprietary Voice-over-IP solutions present in the market today. IMS can be deployed globally based on the Common IMS specifications, which provide a global standard.
- The IMS multimedia services suite enables voice, together with supplementary services known from traditional telephony services such as communication barring or call forwarding, as well as video, presence, group management, conferencing, messaging, and other services.
- IMS enables these services through its service control architecture, which provides users with a means to set up sessions between them and exchange media using the Internet Protocol (IP).
- The signalling protocol used for setting up sessions in IMS is the Session Initiation Protocol (SIP) in conjunction with the Session Description Protocol (SDP), as defined by the Internet Engineering Task Force [IETF]. The SIP core specification can be found in [RFC3261], and SDP is specified in [RFC4566]. Most of the effort spent by 3GPP on IMS security was related to securing SIP signalling in IMS.

IMS functional entities

For a full description of the IMS architecture and its functional entities, we again refer the reader to one of the books cited above. The relevant 3GPP specification describing the IMS

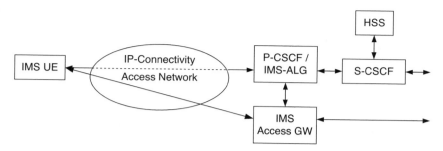

Figure 12.1   Partial view of the IMS architecture

architecture is [TS23.228]. We shall describe a few key functional entities that are essential for the understanding of IMS security, the UE, the P-CSCF/IMS-ALG, the IMS Access GW, the S-CSCF, and the HSS. The relationships among these functional entities are depicted in Figure 12.1. The dotted lines in the figure show signalling paths while the continuous lines show media paths.

- *IMS UE* (IMS User Equipment). This contains an IMS client and, in general, the user's security credentials. It communicates with the P-CSCF and the IMS Access GW over the IP-Connectivity Access Network, such as 3GPP, xDSL, cdma2000® or packet cable access networks.
- *P-CSCF / IMS-ALG* (Proxy Call Session Control Function / IMS Application Level Gateway). A P-CSCF is always present in the IMS architecture, but a P-CSCF does not always include IMS-ALG functionality. The P-CSCF is the first contact point for IMS UEs in the IMS. This means that all SIP signalling traffic from and to an IMS UE will be sent through the P-CSCF. Depending on the signalling security mechanism applied, the P-CSCF is the termination point for confidentiality and integrity protection of signalling traffic towards the user. The P-CSCF may act as an IMS-ALG, for example in support of end-to-access edge IMS media plane security (see section 12.2.1). The P-CSCF / IMS-ALG acts as a controller of the IMS Access GW in the media path. The general functions of an IMS-ALG and its interaction with the IMS Access GW are described in [TS23.228] and [TS23.334], while the particular functions related to IMS media plane security are described in [TS33.328] and [TS24.229].
- *IMS Access GW* (IMS Access Gateway). This need not be present in the media path. When it is present it may support end-to-access edge IMS media plane security (see section 12.2.1). The same references as for the IMS-ALG apply.
- *S-CSCF* (Serving Call Session Control Function). This handles the registrations of subscribers to the IMS, makes routing decisions, maintains session states, and stores user profiles. A user is able to initiate and receive services only after a successful registration with the S-CSCF. The S-CSCF forwards and receives SIP signalling messages to and from entities in other networks (rightmost arrow in Figure 12.1). It is responsible for handling subscriber authentication during registrations and, for certain security mechanisms, the distribution of session keys for signalling security. For this purpose, it fetches authentication information and service profiles from the HSS.

- *HSS* (Home Subscriber Server). The HSS is a database storing all data relevant to subscription and service use. In particular, the HSS stores the security credentials tied to the private user identities and computes authentication information from the credentials upon request from the S-CSCF.

12.1.2   Circuit Switched Fallback (CSFB)

According to the 3GPP specification [TS23.272], CS fallback in EPS enables the provisioning of voice and other CS-domain services by re-using the circuit-switched infrastructure when the UE is served by E-UTRAN. A CS fallback-enabled terminal connected to E-UTRAN may use GERAN, or UTRAN, or a 3GPP2-defined 1xRTT network, to connect to the CS domain for originating and terminating voice services. This function is only available when E-UTRAN coverage is overlapped by either GERAN, or UTRAN, or 1xRTT coverage. In other words, voice service is not provided via LTE, only via 2G or 3G networks.

For CSFB to work, it needs to be supported by signalling in EPS. In particular, the UE needs to be registered in the CS domain once it attaches to LTE. This is achieved through an interaction between the MME and the MSC/VLR in the CS domain. When the UE originates a call it first switches over to the CS domain; when there is an incoming call for the UE the CS domain tells the MME to initiate paging for the UE over LTE. Upon receiving the paging message, the UE then switches over to the CS domain and attaches to it to receive the call.

For GERAN and UTRAN, the reader is referred to Chapters 3 and 4 of this book. References for 3GPP2-defined networks can be found in [TS23.272], and, more generally, under [3GPP2].

12.1.3   Single Radio Voice Call Continuity (SRVCC)

Single Radio Voice Call Continuity is designed to ensure that a call started using IMS over LTE or IMS over the 3G High Speed Packet Access (HSPA) can continue even when the radio conditions become inadequate for the call to proceed with IMS over LTE or HSPA. This may be the case, for example, when the user moves out of LTE or HSPA coverage, or the quality of service has become inadequate. When this happens, and the user is within coverage of another radio network offering circuit-switched services, the SRVCC makes it possible for the call to continue in the circuit-switched domain of that other radio network. The call remains anchored in an IMS application server, the Service Centralization and Continuity Application Server (SCC AS) while the user is being served by the circuit-switched domain.

SRVCC currently supports the following types of handover for voice calls from IMS over a packet domain to a circuit-switched domain:

- from LTE to UTRAN;
- from LTE to GERAN;
- from LTE to 3GPP2 1xCS;
- from HSPA to UTRAN;
- from HSPA to GERAN.

SRVCC in the converse directions to the ones in the above list has not been defined in the first 3GPP release specifying SRVCC because it is assumed that the typical situation will be to have wide coverage for UTRAN, GERAN or 3GPP2 1xCS, while, at least initially, LTE and HSPA may be more likely to be provided in limited geographical areas. However, efforts are under way to enhance SRVCC in later releases (see Chapter 14). We will only treat the SRVCC handovers from LTE to UTRAN or GERAN in this book.

For SRVCC handovers from LTE to UTRAN or GERAN, a Mobile Switching Centre (MSC) server enhanced for SRVCC is required in the target circuit-switched domain. The enhanced MSC server communicates with the MME and the SCC AS.

The term SRVCC refers to 'single radio' because typical terminals cannot connect to more than one of the radio networks in the above list at a time. This makes the task of ensuring the continuity of a voice call more difficult as the handover has to be performed in a very short time so that the user experience is not negatively impacted. In order to improve the efficiency of such handovers, procedures for mapping security contexts from the MME to the enhanced MSC server have been defined for SRVCC. These procedures are presented in section 12.2.3.

SRVCC may also involve the handover of packet-switched non-voice services to the packet-switched domain of the target network. For this type of handover from LTE to UTRAN or GERAN, the security procedures described in section 11.1 apply.

SRVCC is defined in [TS23.216]. The IMS service continuity aspects and the SCC AS are defined in [TS23.237] and [TS23.292]. The 3GPP2-specific aspects are defined in [X.S0042-0 v1.0].

For the sake of completeness we mention that there is also a dual radio VCC, which applies when a terminal can connect to source and target radio network simultaneously. This is often the case when one of the radio technologies is UTRAN or GERAN (over which circuit-switched services would be provided), or LTE (over which IMS-based services would be provided), and the other is WLAN (over which IMS-based services would be provided).

12.2   Security Mechanisms for Voice over LTE

In this section we address the security aspects of the three methods for providing voice over LTE briefly described in section 12.1.

12.2.1   Security for IMS over LTE

We first give a brief overview of IMS security and then explain which of the mechanisms defined by 3GPP for IMS security apply to IMS over LTE.

One book [Poikselkä and Mayer 2009] describes the IMS signalling security procedures in some detail, but there presently is no book describing IMS media plane security mechanisms, so the reader is referred to [TS33.328].

IMS signalling security

For many years, IMS security, as defined by 3GPP, was solely concerned with securing SIP signalling in IMS. IMS signalling security provides subscriber authentication as well as integrity and confidentiality of signalling messages in registration and session set-up procedures. In particular, IMS signalling security ensures that only authorised subscribers have access to

IMS resources and can set up multimedia sessions, and that charges are attributed to the right subscribers.

IMS signalling security is provided in a hop-by-hop fashion. The difficult part is securing the first hop from the IMS UE to the P-CSCF because of the key management involving a large number of subscribers. The 3GPP specification that defines the IMS access signalling security mechanisms is [TS33.203] whose first version was approved in 2002. IMS signalling sent between IMS core network nodes is secured using Network Domain Security as described in section 4.5 of this book.

Subscriber authentication in IMS registrations

In order to cater for the different needs of the various IMS deployment scenarios, and the legacy of the terminals and networks that use IMS, the IMS signalling security specification [TS33.203] offers a variety of subscriber authentication mechanisms. We distinguish three types of IMS subscriber authentication mechanisms in [TS33.203]:

- SIP-layer authentication;
- access-network bundled authentication;
- trusted-node authentication.

SIP layer authentication
[TS33.203] specifies two SIP-layer authentication mechanisms:

- IMS AKA;
- SIP Digest.

SIP Digest is based on HTTP Digest, while IMS AKA is based on an extension of HTTP Digest called HTTP Digest AKA. We therefore first briefly describe HTTP Digest and its extension.

HTTP Digest. The 'Digest Access Authentication Scheme' for HTTP defined in [RFC2617] is often simply referred to as 'HTTP Digest'. HTTP Digest uses the username and a password shared between the user and an HTTP server as authentication credentials. The password has to be distributed by administrative means before authentication can start. HTTP Digest is based on a simple challenge–response paradigm. The server sends a challenge in the form of a nonce value. (A nonce is a number used only once.) A valid response by the user contains a checksum of the username, the password, the given nonce value, the HTTP method, and the requested URI. In this way, the password is never sent in the clear.

HTTP Digest AKA. For 3G networks, subscriber credentials are contained in the USIM, which, by definition, resides on a smart card, the UICC. The USIM is used in the UMTS AKA protocol to authenticate the subscriber (see Chapter 4). This form of credential is stronger than a mere username/password combination. This observation motivated work extending HTTP Digest by combining it in a particular way with UMTS AKA. This work resulted in HTTP Digest AKA [RFC3310]. The main advantage of HTTP Digest AKA over plain HTTP Digest is that the former provides a one-time password for HTTP Digest. This is achieved as follows. As we know from Chapter 4, in UMTS AKA the VLR or SGSN retrieves an authentication vector from the Authentication Centre in the HLR and then sends a challenge RAND, AUTN to the UE. The USIM generates a response RES and session keys CK and IK, and sends them to the ME. The ME stores the session keys and sends the response RES back to the VLR or SGSN. In HTTP Digest AKA, it is the HTTP server that fetches authentication vectors and sends the challenges. The appropriately encoded parameters RAND, AUTN are used in HTTP Digest AKA as the nonce required by the HTTP Digest scheme. HTTP Digest AKA uses the parameter RES as the password required by the HTTP Digest scheme. Because every authentication run generates a different RAND and, hence, a different parameter RES, HTTP Digest AKA indeed produces a one-time password for HTTP Digest. Note also that [RFC3310] consistently refers to the ISIM, and does not mention the USIM; the relationship between the two terms is discussed further below in this section.

HTTP Digest AKA was later enhanced to HTTP Digest AKAv2 (see [RFC4169]) in order to counter certain man-in-the-middle attacks in tunnelled authentication scenarios. HTTP Digest AKA and HTTP Digest AKAv2 differ in the way the HTTP Digest response is created: HTTP Digest AKAv2 computes the password from RES, CK and IK using a pseudorandom function.

SIP Digest. [RFC3261] describes the modifications and clarifications required to apply the HTTP Digest authentication scheme to SIP. The SIP scheme usage is almost completely identical to that for HTTP described in [RFC2617]. We refer the reader interested in the differences to [RFC3261]. Starting from [RFC3261], 3GPP specified in [TS33.203] how to apply the HTTP Digest authentication scheme to the usage of SIP in IMS. 3GPP called the resulting scheme 'SIP Digest'. Like HTTP Digest, SIP Digest uses username and password as authentication credentials. In SIP Digest, the S-CSCF takes the role of the server challenging the user. The S-CSCF retrieves the password from the HSS when a subscriber registers to the S-CSCF. The IMPI (IP Multimedia Private Identity), which can be seen as the equivalent in IMS of the IMSI in GSM, 3G or EPS, is used as the username. An IMSI can be converted into an IMPI in a canonical way [TS23.003]. The challenge and response are carried in specific headers of the messages in the IMS registration procedure.

IMS AKA. The IMS Authentication and Key Agreement scheme IMS AKA is specified in [TS33.203]. The subscriber authentication part of IMS AKA is an application of HTTP Digest AKA to the usage of SIP in IMS. HTTP Digest AKAv2 is not required for the purpose of IMS as the attack scenarios motivating the creation of HTTP Digest AKAv2 do not apply. The IMS AKA authentication credentials are the functional equivalent of a USIM. They may be derived from a USIM, or may be a separate replica of USIM functions and/or data. When IMS is accessed over a 3GPP-defined network, the IMS AKA authentication credentials must reside on a UICC. When they reside on a UICC they are, according to [TS33.203], called ISIM (IP Multimedia Services Identity Module). For a more precise definition of ISIM and a warning about a slightly inconsistent use of the term 'ISIM' across 3GPP specifications, we refer the reader to clause 8 of [TS33.203], as well as to the definition of an ISIM application on a UICC in [TS31.103]. When IMS is accessed over a non-3GPP-defined network the IMS AKA authentication credentials need not reside on a smart card. In IMS AKA the S-CSCF takes the role of the server challenging the user. The S-CSCF retrieves authentication vectors from the HSS when a subscriber registers to the S-CSCF. The IMPI (possibly converted from an IMSI), which is included in the ISIM (possibly derived from a USIM), is used as the username. The challenge and response are carried in specific headers of the messages in the IMS registration procedure. IMS AKA also has a key agreement part which is used for creating IPsec security associations (see further below in this chapter). An information flow for a successful registration of an unregistered subscriber using IMS AKA is shown later in this chapter.

Applicability of IMS AKA and SIP Digest. 3GPP allows the use of SIP Digest only when IMS is accessed over access networks that are not defined in 3GPP specifications [TS33.203]. Correspondingly, 3GPP decided that UICC-based credentials are required for subscriber authentication when accessing IMS over 3GPP-defined access networks. The reason for this decision was that 3GPP wanted to ensure that the credentials for access-level authentication and IMS-level authentication had the same strength. Only two of the IMS subscriber authentication mechanisms presented in this chapter offer UICC-based credentials: GIBA (see below) and IMS AKA. GIBA allows the use of a SIM or USIM, but is defined only for IMS access over GERAN or UTRAN. Therefore, the only subscriber authentication mechanism defined by 3GPP that is applicable to IMS access over LTE is IMS AKA. We can note, though, that the specification of IMS AKA in [TS33.203] does not mention LTE or EPS explicitly; but then it does not really have to do this as IMS is access-independent. We would further point to the fact that the S-CSCF needs to retrieve UMTS authentication vectors, not EPS authentication vectors, from the HSS even when IMS is accessed over LTE. Therefore, when receiving a request from an S-CSCF, an HSS used also for EPS needs to instruct the Authentication Centre to generate authentication vectors with the AMF separation bit set to '0' (see section 7.2). An Authentication Centre in the HSS used for EPS is always capable of generating UMTS authentication vectors.

Access-network bundled authentication

In access-network bundled authentication, IMS subscriber authentication is coupled to the authentication in the access network over which IMS is carried. 3GPP has defined two such bundled authentication mechanisms: GPRS-IMS-Bundled Authentication (GIBA) and NASS-IMS-Bundled Authentication (NBA). Both schemes are specific to the access network technologies that gave them their names: GIBA applies only when IMS is accessed over GPRS (Chapter 3) or the 3G packet domain (Chapter 4). NBA applies only when IMS is accessed over a Network Access Subsystem (NASS) defined by ETSI TISPAN [ETSI ES 282 004], which is an xDSL-based access network.

For both GIBA and NBA, the idea is to bind the IP address to the private IMS user identity, the IMPI. The idea exploits the fact that, in access authentication, the dynamically allocated IP address is bound to the identifier used at the access level – the IMSI in the case of GPRS and the Line Identifier in the case of NBA. Furthermore, the access level identifier is assumed to have a long-term binding to the IMPI in the HSS. No access-network bundled authentication mechanism has been standardized for LTE. Therefore these mechanisms are not considered any further in this book.

Trusted-node authentication

Trusted-node authentication allows a subscriber to gain access to IMS based on successful access-level authentication being provided by a trusted node in the network which provides an interworking function towards the IMS. In practice this is achieved by having this trusted node take on the role of both the UE and the P-CSCF from an IMS perspective. One example of such a scenario is the MSC Server enhanced for IMS Centralized Services (ICS) as described in [TS 23.292]. Trusted-node authentication is not relevant for IMS over LTE; we therefore do not consider it any further in this book.

Confidentiality and integrity protection for SIP signalling in IMS

3GPP defines two mechanisms in [TS33.203] for providing confidentiality and integrity protection for SIP signalling between the UE and the P-CSCF, namely IPsec ESP and Transport Layer Security (TLS). We briefly describe their use in IMS in the following.

For the sake of completeness, we mention that [TS33.203] defines two further mechanisms providing a limited form of SIP message origin authentication, namely an IP address check mechanism performed in the P-CSCF, and SIP Digest proxy-authentication performed in the S-CSCF. While these methods have their merits in particular environments, neither of them provides confidentiality nor full integrity protection. As 3GPP ruled out the use of these methods with 3GPP-defined access networks, in particular with LTE, we do not dwell on them any further in this book. The reader interested in the strengths and boundary conditions for the use of these two mechanisms compared to IPsec and TLS is referred to the discussion in Annex Q of [TS33.203].

IPsec. IPsec is a very well-known mechanism, and many security textbooks are available describing it. We therefore do not explain IPsec any further here. The usual means for setting up an IPsec security association is the Internet Key Exchange protocol IKE [RFC2409], or its successor, IKEv2 [RFC4306]. However, the use of IKE or IKEv2 is not mandated; it is allowed to use other means for setting up IPsec security associations. This is what 3GPP does: it uses the keys (CK, IK) agreed by means of IMS AKA as the ciphering and integrity keys required for IPsec Encapsulating Security Payload (ESP), possibly after a suitable key expansion (depending on the cryptographic algorithm). Note that [TS33.203] refers to the version of IPsec ESP defined in [RFC2406], not the updated version of IPsec ESP in [RFC4303]. The other parameters required for setting up IPsec security associations, including Security Parameter Index (SPI), cryptographic algorithms, IP addresses and ports, are either established by means of the SIP Security Mechanism Agreement protocol, also known as Sip-Sec-Agree protocol (see below), or set to pre-determined values. For the details of the establishment of IPsec security associations by means of IMS AKA the reader is referred to the specification [TS33.203] or the book [Poikselkä and Mayer 2009].

3GPP took the decision to use IMS AKA in conjunction with IPsec ESP in 2002. TLS was not considered a viable alternative at the time as TLS requires TCP as the transport protocol; so SIP over UDP, whose support was considered essential by 3GPP, could not be protected by TLS. Note that the work on Datagram TLS (DTLS) [RFC4347], which provides Transport Layer Security over UDP, was completed by the IETF only in 2006. An extension to SIP Digest providing better integrity protection than SIP Digest was also considered at the time, but discarded largely because it was unable to provide confidentiality, which was already known at the time to become a requirement in 3GPP Release 6. In the context of the work on Common IMS in Release 7, 3GPP discussed (starting in the year 2005) an extension to the confidentiality and integrity protection mechanism for SIP signalling to accommodate scenarios with Network Address Translation (NAT), which do not usually occur with cellular access networks, but are common with fixed access networks. During these discussions, the continued use of IMS AKA as a subscriber authentication mechanism was not contentious, but (D)TLS was proposed by some as an alternative confidentiality and integrity protection mechanism. 3GPP finally decided to stick with IPsec, and enhance it with UDP encapsulation to enable NAT traversal, mainly for the reason to have the same type of solution for the cases with and without NAT, not because there would have been any security concerns with the alternative.

TLS. TLS also is a very well-known mechanism, and many security textbooks are available describing it. We therefore do not explain TLS any further here. The introduction of TLS as an additional mechanism for confidentiality and integrity protection of SIP signalling over non-3GPP access networks was motivated by the following observation. Terminals in a non-3GPP environment often do not have the functional equivalent of a USIM, whether residing on a UICC or not. They therefore have to rely on other types of authentication credentials. For this reason, 3GPP introduced SIP Digest as a subscriber authentication mechanism, as explained above. 3GPP defined the use of TLS for confidentiality and integrity protection for SIP signalling in conjunction with SIP Digest. The set-up of IPsec security associations in IMS is not possible with SIP Digest because this set-up, described in the previous paragraph, is tightly coupled with IMS AKA, which requires the use of the functional equivalent of a USIM. Therefore, IPsec for IMS access signalling protection is not a feasible alternative in many environments. TLS is used with server authentication by means of server certificates where the TLS server is the P-CSCF; client authentication is provided by SIP Digest, not TLS.

Sip-Sec-Agree. The SIP Security Mechanism Agreement protocol is defined in [RFC3329]. It allows negotiating the security mechanisms used between a Session Initiation Protocol (SIP) user agent and its next-hop SIP entity. The mechanisms that can be negotiated according to Sip-Sec-Agree are: Digest, TLS, IPsec with IKE, IPsec with manual keying, and IPsec-3GPP (i.e. IPsec with IMS AKA as described above). In the context of the 3GPP IMS authentication the Sip-Sec-Agree mechanism is used to negotiate the security mechanisms applied between the UE and the P-CSCF. Only TLS and IPsec-3GPP are supported in 3GPP. Note that the Digest mechanism that can be negotiated by means of Sip-Sec-Agree has to be run between the UE and the next-hop SIP entity, which in IMS would be the P-CSCF, while SIP Digest, as described for IMS above, is run between the UE and the S-CSCF. The Sip-Sec-Agree protocol is integrated into the initial registration procedure as shown in the information flow for IMS AKA below.

Applicability of IPsec and TLS in IMS. 3GPP specifications strictly tie the choice between IPsec and TLS to the choice of the subscriber authentication mechanism: IPsec is always used in conjunction with IMS AKA, and TLS is always used in conjunction with SIP Digest. For access to IMS over a 3GPP-defined network, an ISIM – in the sense of [TS33.203], i.e. possibly derived from a USIM – is required for the reasons explained above. This implies that, according to the 3GPP IMS specifications, IMS AKA with IPsec shall be used for SIP signalling security when accessing IMS over LTE. The same remark as in the discussion of IMS AKA above applies, namely that LTE is not explicitly mentioned in the 3GPP IMS security specifications.

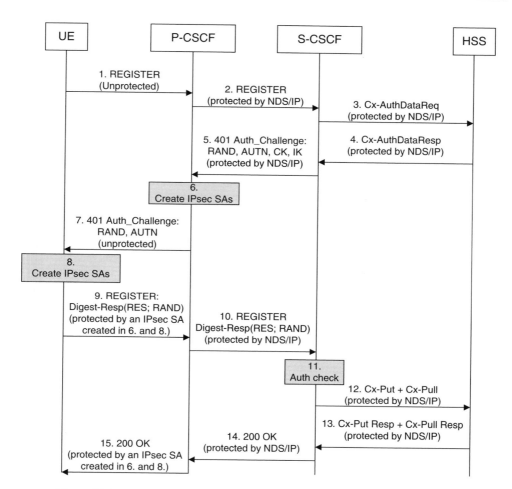

Figure 12.2   Successful registration of an unregistered subscriber using IMS AKA

Information flow for a successful registration with IMS AKA

Figure 12.2 shows the information flow for a successful registration of an unregistered subscriber using IMS AKA. This information flow is explained in brief to demonstrate the similarities and differences between UMTS AKA, EPS AKA and IMS AKA. The stage 2 specification can be found in [TS33.203]; the stage 3 specifications can be found in [TS24.229] for the messages between the UE and the S-CSCF, and in [TS29.228] and [TS29.229] for the messages between the S-CSCF and the HSS.

1. The UE sends a REGISTER request including the IMPI and the appropriate Sip-Sec-Agree header.
2. The P-CSCF processes the Sip-Sec-Agree header according to [RFC3329], strips it off, and forwards the message to the S-CSCF. (To be more precise: the message is sent via an intermediate node called I-CSCF, which first contacts the HSS to find a suitable S-CSCF.

A description of the I-CSCF is omitted in this chapter as it does not play an important role in the security procedures. The interested reader is referred to [TS23.228].)

3. The S-CSCF requests authentication vectors from the HSS.
4. The HSS returns authentication vectors of the form (RAND, XRES, CK, IK, AUTN) known from UMTS AKA – see section 4.2 and Figure 7.2.
5. The S-CSCF sends a so-called 401 Unauthorized message [RFC3261] to the P-CSCF containing the Authentication Challenge (RAND, AUTN) and also the keys (CK, IK).
6. The P-CSCF creates IPsec security associations (SAs) from CK, IK, the Security Parameter Index, and parameters received in message 1 and to be sent in message 7 (IP addresses and ports, cryptographic algorithms).
7. The P-CSCF forwards the 401 Unauthorized message with (RAND, AUTN), but it does not forward the keys (CK, IK). The P-CSCF also includes the appropriate Sip-Sec-Agree header.
8. The UE sends RAND and AUTN to the USIM or ISIM and gets RES, CK, and IK back. The UE creates IPsec SAs in the same way as the P-CSCF did in step 6. The UE computes the Digest-Response over RAND and further parameters using RES as the password as described for HTTP Digest AKA earlier in this chapter.
9. The UE sends another REGISTER request to the P-CSCF. This request includes the Digest-Response and the appropriate Sip-Sec-Agree headers. The request is protected by an IPsec SA created in step 8.
10. The P-CSCF strips off the Sip-Sec-Agree headers and forwards the message to the S-CSCF. Note that the message is discarded if it cannot be successfully processed by IPsec at the P-CSCF using the appropriate IPsec SA created in step 6.
11. The S-CSCF computes the Digest-Response in the same way as the UE did in step 8 using XRES as the password and checks whether it matches the Digest-Response received in message 10. If it does the UE is successfully authenticated.
12. The S-CSCF registers the subscriber with the HSS.
13. The HSS returns the subscriber profile to the S-CSCF.
14. The S-CSCF checks the subscriber's authorization using the received profile. If this check is successful the S-CSCF sends a so-called 200 OK message [RFC3261] to the P-CSCF indicating the success of the registration.
15. The P-CSCF forwards the 200 OK message to the UE.

IMS media plane security

The primary motivation for IMS media plane security was protecting the confidentiality of the IMS media in transit, for example in order to prevent eavesdropping on voice calls. In addition, IMS media integrity protection is supported. At the time of writing this book, 3GPP has specified IMS media plane security only for real-time services in IMS that use the Real-Time Transport protocol RTP [RFC3550]. These real-time services include voice. The specification defining the IMS media plane security mechanisms is [TS33.328], which was approved by 3GPP in late 2009. The detailed protocol work on IMS media plane security was still not fully completed at the time of writing.

For confidentiality protection, 3GPP originally relied on the security of the underlying bearer networks, provided either by cryptographic means, such as link layer protection in cellular access networks, or by assumed inherent physical properties of, for example, xDSL access links. But with the more widespread adoption of the Common IMS applicable to all sorts of access network types (e.g. unencrypted public WLAN hotspots), a uniform protection method for media realized above the transport layer seemed desirable. This led to the definition of the end-to-access edge security mechanism where IMS media plane traffic is secured between the IMS UE and the IMS Access GW at the edge of the access network. Furthermore, uninterrupted end-to-end security between terminals gained in importance. This led to the definition of IMS media plane end-to-end security mechanisms.

End-to-end media plane security comes in two variants, which differ in the key establishment protocols and cater to different use cases. All the mechanisms defined by 3GPP for IMS media plane security so far can be implemented when accessing IMS over LTE. As opposed to the case of IMS signalling security, there are no restrictions on the use of any of these IMS media plane security mechanisms in the context of IMS access over LTE in 3GPP specifications. But the need for end-to-access edge security when accessing IMS over LTE may be considered not all too pressing, given that LTE provides strong access security at the link layer, both between the UE and the base station (see section 8.3) and between the base station and the edge of the core network (see section 8.4). Still, if a uniform handling of all traffic, irrespective of the access network type, is desired then end-to-access edge media plane security may be applied also to IMS access over LTE.

12.2.2   Security for Circuit Switched Fallback

When Circuit Switched Fallback is used, voice services are not provided over LTE but over the circuit-switched domains of GERAN, UTRAN or 3GPP2 1xRTT. Therefore, voice services using CSFB are of no concern to LTE security; and the security mechanisms applied to voice services are those generally applied to GERAN (Chapter 3), UTRAN (Chapter 4) or 3GPP2 1xRTT. The signalling in the EPS required to support CSFB is protected by the LTE security mechanisms that are the main subject of this book.

12.2.3   Security for Single Radio Voice Call Continuity

Here we describe the security mechanisms for SRVCC handover from LTE to UTRAN or GERAN. The corresponding security mechanisms for SRVCC handover from HSPA to UTRAN or GERAN are not handled as they are not in the scope of this book. The interested reader can find them in [TS33.102].

For SRVCC handover from LTE to UTRAN or GERAN, a security context mapping from the MME to an MSC server enhanced for SRVCC is provided in order to improve the efficiency of the handover. The main task is the mapping of the keys in use before and after the handover. Before the handover, the UE and the MME share the current EPS security context containing the key K_{ASME} (see Chapter 7). The idea is therefore to use the key K_{ASME} and further parameters, to derive the keys required in the target network. The derived keys are then transferred from the MME to the MSC server enhanced for SRVCC as part of the SRVCC handover procedure.

As the keys used in UTRAN and GERAN are different, we treat the two cases separately.

SRVCC handover from LTE to UTRAN

The UTRAN target network requires a cipher key CK and an integrity key IK, as described in Chapter 4. In order to distinguish the keys derived in the SRVCC procedure from other keys

(CK, IK) possibly already present in the UE and the VLR from a previous visit to UTRAN, the keys derived for SRVCC carry the subscript 'SRVCC' for the purposes of the description. The keys CK_{SRVCC} and IK_{SRVCC} are obtained by applying a specific key derivation function KDF to the key K_{ASME} in the current EPS security context and a freshness parameter. The freshness parameter was chosen to be the current value of the NAS downlink COUNT (see Chapter 8). It is to ensure that two different SRVCC handovers do not result in the same keys CK_{SRVCC} and IK_{SRVCC} in the target network. The key derivation for SRVCC is specified in Annex A of [TS33.401]. It uses a framework for key derivation common to various 3GPP features (see section 10.4). This framework also ensures that the derived keys are only usable for SRVCC purposes.

After key derivation, the MME increases the value of the NAS downlink COUNT by 1 in order to ensure continued freshness of this parameter.

The MME also sends the four least significant bits of the NAS downlink COUNT to the eNB, which forwards them to the UE in the Handover Command. This is done so as to allow for synchronization of the NAS downlink COUNT values used by the MME and the UE. The NAS downlink COUNT values could be out of synch, perhaps due to a NAS message sent by the MME down to the UE, which caused the MME to increase the NAS downlink COUNT value, but got lost and was never received by the UE, so that the UE did not increase the NAS downlink COUNT value correspondingly. The algorithm for synchronizing the NAS downlink COUNT values in the UE is implementation-specific. Once this task has been performed successfully in the UE, the UE updates the NAS downlink COUNT value accordingly.

SRVCC handover from LTE to GERAN

The GERAN target network requires a cipher key of type Kc (64 bits) or Kc_{128} (128 bits) used with algorithms A5/1, A5/3 or A5/4, respectively (see section 3.4). It depends on the ciphering algorithm selected by the Base Station Subsystem (BSS) in the target network which of the two types of keys is required. The keys Kc and Kc_{128} are derived in a two-step procedure. The first step consists in deriving CK_{SRVCC} and IK_{SRVCC} from K_{ASME} in the UE and the MME in exactly the same way as described above for LTE to UTRAN SRVCC handover, and transferring them from the MME to the enhanced MSC server. In the second step, the key conversion function c3, known from UTRAN to GERAN interworking (see section 4.4), is applied to the keys CK_{SRVCC} and IK_{SRVCC} in the UE and the enhanced MSC server to obtain Kc; and the key derivation function KDF defined in Annex B.5 of [TS33.102] is applied to the keys CK_{SRVCC} and IK_{SRVCC} in the UE and the enhanced MSC server to obtain Kc_{128}.

13

Security for Home Base Station Deployment

To allow for a more efficient usage of the available spectrum, and to allow customer specific deployments (e.g. closed subscriber groups managed by the hosting party of the base station), an extension to UTRAN was specified for base stations serving very small cells. Their coverage is comparable to a WLAN access point and they are deployed similarly within customer premises. These 'femto' base stations serving 'femto cells' are called Home NodeB, as they are a home version of macro NodeBs. Similarly, an extension to E-UTRAN was specified with femto base stations called Home eNodeB. The service requirements for both types of home base station are specified in [TS22.220]. The technical report [TR23.830] handles architectural aspects of home base stations. The normative text derived from this report is not contained in a separate document, but distributed over the applicable EPS-related specifications.

Support of the standardization work in 3GPP comes from the Femto Forum [FF], a non-profit organization to promote femto cell deployment worldwide within the residential and small and medium enterprise markets. It is comprised of mobile operators, telecoms hardware and software vendors, content providers and others. The Femto Forum is not a standards-defining organization but works in the forefront of standardization, gathering and harmonizing requirements from the stakeholders.

As the definition of the security features for Home NodeBs happened in parallel with the definition of EPS in 3GPP, the security for both types of home base stations was specified in a common specification in 3GPP Release 9 [TS33.320]. As the security measures for home base station deployment are more governed by the deployment scenario in customer premises, and less so by the actual radio and core network technology, there are only a few differences between the security for 3G Home NodeBs and for EPS Home eNodeBs.

This chapter always refers to the deployment of home base stations in EPS – that is, the Home eNodeB (HeNB) – but the differences to Home NodeBs (HNB) are mentioned where applicable.

LTE Security   Dan Forsberg, Günther Horn, Wolf-Dietrich Moeller, and Valtteri Niemi
© 2010 John Wiley & Sons, Ltd

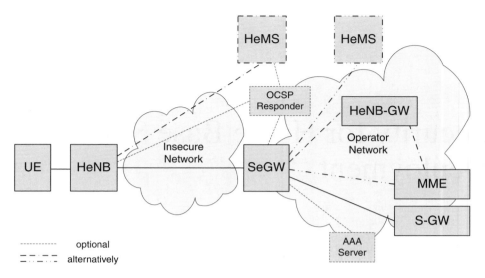

Figure 13.1   Deployment scenario and architecture for HeNBs

13.1   Security Architecture, Threats and Requirements

13.1.1   Scenario

The concept of HeNBs was introduced to provide small-area or indoor coverage for mobile communications based on the radio technology that is also used in the macro range. This allows using the same user equipment for global and local access. The HeNB is located within the customer premises and connected to the operator core network via the existing broadband access line, such as DSL or broadband cable. Figure 13.1 gives an overview of the deployment architecture.

The following paragraphs give short descriptions of the elements shown in Figure 13.1, and their roles in HeNB deployment.

Home eNodeB. The HeNB is a base station located on the customer premises and radiating in licensed spectrum. As the licensed spectrum is owned by the operator, and the regulator holds the operator responsible for the usage of this spectrum, the HeNB is subject to the same regulatory requirements as any other base station. As a consequence of this responsibility, the deployment of a HeNB by a customer is based on a contract with the mobile operator. In addition, specific security requirements exist that do not allow the customer to have full control over the HeNB. Consequently, certain configuration settings may be managed by the operator only. In the context of HeNBs the customer is called Hosting Party (HP) to differentiate this customer from an ordinary subscriber of mobile networks.

User Equipment (UE). The user equipment is an ordinary UE as used for macro cells in EPS. All EPS-capable UEs are also aware of the special HeNB functionality of Closed Subscriber Groups (CSGs), which is described further below and in section 13.6. This is different from

3G networks, where CSG-unaware UEs have to be served as well, so the HNB architecture is required to have a separate treatment for such legacy UEs.

Backhaul Link. The backhaul link is the link between the HeNB and the security gateway (see below). It carries the S1 traffic, and also the management traffic when routed via the security gateway. The backhaul link, and possibly also the link to the management system, extends across the public Internet in the general case. It is assumed that the hosting party has an existing broadband connection to the Internet, such as via DSL or broadband cable. As this connection is routed through the public domain, it is seen as insecure, and many of the HeNB security features address the threats related to a connection over an insecure network.

Security Gateway (SeGW). The security gateway is the door to the operator core network for all traffic originating and terminating in the HeNB. It is the only additional mandatory network element introduced owing to the security requirements. The acronym SeGW was intentionally chosen to be different from the SEG as used in NDS/IP [TS33.210]. While the SEG connects two different security domains, the SeGW connects an 'outlying' element belonging logically to the same security domain as the core network of the operator it is connected to. The specific functionality of the security gateway used for macro base stations (eNBs) is outlined in section 8.4, while the SeGW used for HeNBs is described in section 13.4.

HeNB Management System (HeMS). The HeNB management system is responsible for the management of the HeNB. As a HeMS must be able to manage HeNBs of different manufacturers, the so-called Type 1 interface between management system and HeNB has been specified in [TS32.591] and [TS32.593] to allow for vendor interoperability. This specification builds heavily on the management protocol for Customer Premises Equipment (CPE) specified by the Broadband Forum [BBF] in [BBF TR-069]. Depending on the operator's decision, the HeMS may be located within the operator core network or accessible directly on the public Internet. The latter scenario was included to allow the usage of an existing management infrastructure for home equipment (e.g. for residential gateways and/or DSL routers) to be used also for HeNBs. To cater for the special enrolment and registration needs of a HeNB, which may be bought and connected by the hosting party and not by the operator, the HeMS was logically split into an initial and a serving HeMS. This allows provisioning the HeNB with a factory default configuration containing the address of the initial HeMS only, which is not necessarily an operator-specific address. In addition, the initial HeMS may check and modify the SW and configuration of the HeNB before it connects to the serving network. Thus a location of the initial HeMS in the public Internet may be advantageous, even if the serving HeMS is located in the operator core network. Details are described in section 13.5.

MME and S-GW. The Mobility Management Entity (MME) and the Serving Gateway (S-GW) are the same network elements as specified for EPS in the preceding chapters of this book. Also the interfaces between a HeNB and these network elements are the same S1-MME and S1-U interfaces as defined for macro base stations.

HeNB Gateway (HeNB-GW). The HeNB gateway is specified in [TS36.300] and is an optional element in the EPS architecture. This is a deviation from 3G networks, where the HNB Gateway is a mandatory element, which hides the specific features of HNBs and the Iuh interface

[TS25.467] from other core network elements. It is the task of the HeNB-GW to relieve the MME from keeping track of huge numbers of HeNBs, as the MME was more designed to cater to a limited number of eNBs only. As the HeNB-GW is optional, and has the same S1 interface on both sides, both HeNB and MME are unaware if a HeNB-GW is deployed. This means that a HeNB sees the HeNB-GW as MME, and the MME sees all HeNBs connected to a HeNB-GW as one big eNB. For the realm of security features there is no difference if the HeNB is connected via the S1 interface to a HeNB-GW or an MME, as the secure channel is terminated at the border of the operator core network in the SeGW.

AAA Server. The Authentication, Authorization and Accounting (AAA) server is optional to support. It is used for two optional mechanisms, first for communicating with the HLR/HSS if hosting party authentication is deployed, and second if access authorization for HeNBs is controlled by a AAA server.

OCSP Responder. The Online Certificate Status Protocol (OCSP) server is optionally deployed if the operator uses a certificate revocation infrastructure for SeGW certificates. It may either communicate with the SeGW and the HeMS, if in-band signalling of certificate validity status is used, or directly with the HeNB. In the latter case no communications security is needed despite the fact that this communication goes via the insecure link, as OCSP response messages are protected by a signature. OCSP is specified in [RFC2560].

X2 interface. This interface is intentionally missing from Figure 13.1, as the Release 9 specifications do not foresee a direct interface between HeNBs. This may change in future releases, in particular when HeNB deployment in enterprise scenarios is considered, where handover between HeNBs and procedures for interference mitigation based on principles of self-organizing networks (SONs) may be of greater importance.

Closed Subscriber Group (CSG). The HeNB is intended as a wireless access point operated by a customer, so the possibility to restrict the general access to the HeNB was specified. Three access modes for HeNBs are defined, namely the *closed* mode (giving access to a CSG only), the *open* mode (giving access to all subscribers of an operator and their roaming partners) and the *hybrid* mode as mixture of the other two. The HP of the HeNB can manage the membership of the CSG of his HeNB within certain limits set by the operator. Section 13.6 gives an overview of the security-related features of CSG management.

13.1.2   Threats and Risks

This subsection discusses the reasons for HeNB-specific security measures and gives an overview of the threats and risks discussed during the development of the security specification for HeNBs.

The threat and risk analysis is contained in a technical report [TR33.820] which was started before the normative standardization work began. The normative specification [TS33.320] only contains the requirements deduced from the threat and risk analysis. This is handled in the section 13.1.3.

The list below summarizes the reasons why HeNB-specific security is seen as necessary. The HeNB is a Network Element (NE) under the responsibility of an operator, but is not located in the security domain of the operator as opposed to other NEs. Thus the following new issues arise.

- The link to the core network (e.g. DSL and the Internet) is not secured by operator administrative means.
- The HeNB provides termination of the air-link encryption, thus user and RRC signalling data are available in cleartext in the NE on the customer premises.
- The NE located on the customer premises has direct access to the core network through the secure tunnel, once it is authenticated.
- Experience with, for example, set-top boxes implementing digital rights management shows that the HeNB may be prone to intense offline examination by attackers.
- Once vulnerabilities are discovered, exploits may be easily available from the Internet to a fraudulent HP, and may be applied, for example, to the Ethernet port of the HeNB within the local hosting party Ethernet.

As the HeNB is a consumer-like device, the following features make it more prone to attacks.

- The deployment numbers and thus the distribution are much larger and more widespread than for any current NE.
- The price tag must be much lower than for current commercial operator NEs deployed in smaller numbers, thus not allowing expensive security features.

On the other hand, the mobile operator has the following interests and obligations.

- The HeNB operates in licensed spectrum, contrary to WLAN for instance, so the operator is liable for any violation of regulations (geo-location, transmit power, frequency etc.).
- The operator must prohibit disturbances to their and other networks.
- The operator must ensure integrity, privacy and lawful interception also for UEs connected over HeNBs.

Keeping the above issues in mind, [TR33.820] puts the threats and risks into six main groups, which are given below.

1. *Compromise of HeNB credentials*
   - Credentials may be disclosed by local physical or remote algorithmic attacks, allowing cloning of the credentials for a multitude of devices, or for misuse of the credentials for other purposes.
2. *Physical attacks on a HeNB*
   - The device may be tampered with to compromise its integrity, such as to get access to cleartext data transferred between air link and backhaul link.
   - Faked or cloned credentials may be inserted in the device, leading to otherwise unauthorized devices being admitted to the core network.
   - Fraudulent software and/or false configuration data may be inserted by, for example, physical access to non-volatile memory.
3. *Configuration attacks on a HeNB*
   - Unsuitable or outdated SW versions may be loaded.

- The radio management may be misconfigured.
- Access control lists may be altered, if enforced within the HeNB.

4. *Protocol attacks on a HeNB*
   - A man-in-the-middle attack may be carried out on the backhaul link by manipulating, inserting or dropping messages to the HeNB.
   - Denial-of-service (DoS) attacks on the HeNB may be carried out by sending faked messages to the HeNB.
   - If vulnerabilities of the protocols used on the backhaul link are discovered, these may be exploited for attacks.
   - External time messages and O&M traffic may be disturbed.

5. *Attacks on the core network, including HeNB location-based attacks*
   - A faked HeNB may attach to the core network and attack it subsequently, perhaps by trying DoS attacks or exploits on core NEs.
   - Traffic from other sites may be tunnelled into the core network.
   - A false location may be reported to the core network, giving rise to the network configuring the HeNB with wrong parameters.

6. *User data and identity privacy attacks*
   - As RRC signalling, S1 signalling terminating in the HeNB, and user plane traffic is available in cleartext in the HeNB, eavesdropping on user data and revealing user identities is possible.
   - A faked or manipulated HeNB may masquerade as a valid HeNB to attract other users, such as members of other closed subscriber groups normally not using this HeNB.

13.1.3 Requirements

The technical report [TR33.820] gives a list of 32 single security requirements on the HeNB, derived from the particular threats against HeNBs described in the previous subsection. For better readability the following list combines them under their related main topic:

- *authentication* – mutual authentication for backhaul link and O&M, strong enough cryptographic mechanisms, unique identities for authentication, protected storage for authentication credentials;
- *backhaul link and management traffic* – integrity protection mandatory, confidentiality protection mandatory for management and optional for backhaul link, authorization needed for connection to core network;
- *software integrity, data confidentiality and integrity for the HeNB* – secure boot, authorized software only, hardening of the device, validation of device integrity, secure data storage and secured operations on sensitive data;
- *user privacy* – IMSI hiding in device and over the air, confidentiality of signalling and user plane data;
- *operation and management security* – addressed separately for operator and user data with related access control, ultimate operator control for many data;
- *denial-of-service protection of network* – restriction of the number of connections per HeNB to the network, only allow validated HeNBs into the core network;
- *closed subscriber group management and enforcement* – done by HP under control by operator, access control enforced in core network;

- *location and time* – locking of HeNB to geo-location possible, reliable location information shall be gathered and transferred by the HeNB, time information for the HeNB must be reliable.

Not all of these requirements could be fulfilled during the development of the normative specification [TS33.320] on HNB and HeNB security. To give some examples:

- As there is no single reliable location information available in all possible locations, the specification only recommends using the most adequate combination of methods for each deployment.
- IMSI transfer over the air may not be avoidable, as the resolution of temporary identities of many users passing by the HeNB location and trying to connect may put too high a burden on the core network.

In addition to the above-mentioned requirements coming from HeNB deployment, the general security architecture specification for EPS [TS33.401] sets security requirements for eNBs, which are also valid for HeNBs. These requirements are described in section 6.4.

Clause 4.4 of [TS33.320] gives an extensive list of requirements on the operation and on the different network elements involved. They are not repeated here as all of them are considered in the security architecture and procedures, and thus are handled later in this chapter.

13.1.4   Security Architecture

The security architecture is derived from the requirements and with the intention to deviate as little as possible from existing 3GPP security architectures in Network Domain Security (NDS), covered in section 4.5, and the application of NDS to EPS in [TS33.401], covered in section 8.4. Figure 13.2 repeats the architecture given in Figure 13.1, but with the main area for HeNB-specific security measures highlighted.

Requirements and measures for local security in the network element are given only for the HeNB. Here the device integrity has to be ensured by different measures, to give the basis for the local security features mentioned in section 13.2.

The two bold lines in Figure 13.2 indicate the secure communication paths, the upper one to a HeMS accessible on the public Internet, the lower one to the SeGW, providing secure access to the core network for signalling and user plane traffic, and for management traffic if the HeMS is located within the operator network. Both require mutual authentication to be performed, based on a HeNB device certificate and on a network-side (SeGW or HeMS) certificate, before the communication paths are opened.

The Security Gateway (SeGW) performs HeNB authentication and access control, supported optionally by the AAA server in case of hosting party authentication and of access authorization by the AAA server.

The OCSP responder optionally provides the HeNB with certificate validity information, if the operator configures the HeNB to use such service. Usage of this feature is recommended, and, according to a note in the specification, this validity check may become mandatory in future releases.

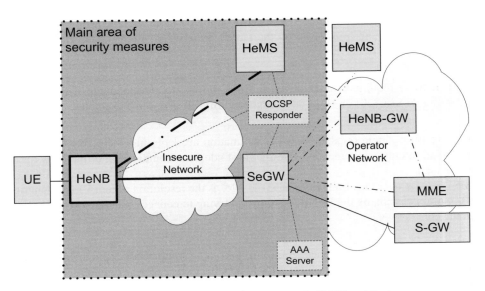

Figure 13.2   Main area for security measures in HeNB architecture

Figure 13.2 does not show the Hosting Party (HP), as it is a role only and not a network element. To distinguish the party having the physical control over the HeNB and managing some features (e.g. the membership of subscribers in the CSG of the HeNB) from other customers/subscribers of the operator, the term 'hosting party' was coined. The HP will also have a contract with the operator about the deployment of the HeNB, except in cases where no separate HP is involved and the operator has direct control over the HeNB.

13.2   Security Features

This section describes the various security features used for securing the HeNB ecosystem. The following sections then describe the procedures to implement these features. Detailed references and technical descriptions of the features are also left to the sections describing the procedures.

13.2.1   Authentication

The device identity of the HeNB is seen as the primary identity that is authenticated by the operator network. Thus this identity necessarily is a globally unique identity. This HeNB unique identity is specified in [TS23.003] on 'Addressing, Numbering and Identification' to allow a general usage of this identity within EPS. The format of this identity is a Fully Qualified Domain Name which facilitates the use of the name in X.509 certificates according to NDS/AF [TS33.310].

As a Public Key Infrastructure (PKI) was chosen as the basis for device authentication, each HeNB device has to be provided with a private/public key pair, and a certificate binding the identity and other properties to the public key. The device certificate will be issued by the operator, manufacturer or vendor of the HeNB, or by another party trusted by the operator. The issuing of the device certificate must in all cases be authorized by the manufacturer or vendor, as the certificate is used to assure the device integrity of the HeNB – see the description of autonomous validation in section 13.3. The choice to use a device certificate not provided by the operator was selected to not oblige the operator to deploy a huge PKI for the expected mass rollout of HeNBs.

Similarly the SeGW has to be provided with a private/public key pair and the associated certificate by the operator. This can be done within the existing NDS/IP and NDS/AF infrastructure available to many operators.

The device authentication comes in two forms: mutual authentication between the HeNB and the SeGW, or between the HeNB and the HeMS. It is explained later under which circumstances which form of device authentication is applied.

Both sides of the authentication procedure may use certificate validity information to check the revocation status of the identity certificates and the certificates in the chain up to and including the root certificate.

Two authentication mechanisms are specified for the device authentication: IKEv2 for establishing an IPsec tunnel to the SeGW, and the TLS handshake for establishing a TLS tunnel to the HeMS.

Certain deployment scenarios require the separate authentication of the Hosting Party (HP). This authentication is optional and is always preceded by a (successful) device authentication. This sequence of authentications is called combined (device and HP) authentication.

The HP authentication uses the AKA mechanism and is, hence, based on a permanent shared secret stored in the USIM and the HLR/HSS. For carrying this authentication within the same protocol as the device authentication, EAP-AKA together with the feature of multiple authentications in IKEv2 is used.

EAP-AKA provides mutual authentication between the operator network and the HP Module (HPM) containing the HP identity and secret. There were two main reasons why the device authentication is mandatory, even when HP authentication is used.

- First, the requirements for device integrity are bound tightly to the Trusted Environment (see later), which also holds the secret (private key) used for certificate-based device authentication. The definition of autonomous validation explicitly uses this fact, so the device authentication is necessary to ascertain the successful device integrity validation to the network.
- Second, the HPM is a removable token and thus not physically bound to the HeNB. On the contrary, the specification explicitly allows transferring HPMs to other HeNB devices, to allow an HP to swap devices for the same HP.

13.2.2   Local Security

The local security comprises secure storage of data and secure execution of software (SW). For a mandatory implementation of the HeNB these features are concentrated in a Trusted Environment (TrE) within the HeNB device. If optional HP authentication (as part of combined authentication) is used, a second trusted environment in the form of the HPM is introduced, realized as UICC. These secure environments are independent of each other.

Trusted environment and secure execution

The TrE is the logical entity within the HeNB that is responsible for securing a root of trust used for the secure boot of the HeNB. The term 'logical entity' implies that its implementation need not be physically separated from the rest of the HeNB, but still the implementations of all functions in this logical entity must be physically bound to the HeNB device. The TrE will first perform a self-check based on a root of trust, and then verify further SW modules. Once all SW modules necessary for the trusted operation of the HeNB have been successfully started, the HeNB has successfully passed the local device integrity check and is enabled for further operation. See section 13.3.1 for a more detailed description.

A second task of the TrE is the secure storage of sensitive parameters used during operation of the HeNB. In addition, all sensitive functions used for the device authentication described in the previous subsection must be executed within the TrE. This refers mainly to all operations involving the private key of the HeNB, as this secret will never leave the TrE.

The network, represented by SeGW or HeMS, will be assured that the above-mentioned secure boot has happened, and that in consequence the HeNB has passed the local device integrity check. This verification result can be communicated to the network either explicitly or implicitly. The combination of the above-mentioned properties of the TrE, namely performing the integrity check and providing the sensitive functions for the device authentication, leads to an elegant implicit form of communicating the successful device verification. The TrE is in control of the private key used for authentication, so it can also enforce that authentication may happen only under certain conditions. Therefore it is specified that the TrE will perform the functions necessary for authentication and involving the private key only after a successful device integrity check. Then the network is assured after successful authentication that only an integrity-checked device could have performed this authentication successfully. This feature is called 'autonomous validation', as all action for the validation is performed autonomously within the HeNB, and the network can implicitly validate the integrity status of the HeNB. For this reason, no explicit communication about the verification result was specified.

Hosting Party Module

The Hosting Party Module (HPM) is specified to be a Universal Integrated Circuit Card (UICC) [ETSI TS 102 221]. It thus provides secure storage for the shared secret and a secure environment for the execution of the sensitive functions using the shared secret for EAP-AKA authentication.

The HPM is bound to the HP by organizational measures of the operator.

Physical Security

Physical security is required to avoid easy local access to stored secrets, sensitive configuration parameters and SW. In particular, the root of trust within the TrE must be physically secured as otherwise the whole local security of the HeNB device cannot be guaranteed. For the HeNB device the design and implementation of physical security features is left to the manufacturer. It is up to the manufacturer to assure the operator of a secure design of the HeNB. Evaluation according to some externally specified standards was not seen as adequate. The same arguments as given in section 6.4 for macro base stations apply, with the additional restriction that the

HeNB is a consumer device with a much lower price tag than the commercial macro base station.

For the HPM, the physical security is given by the fact that it is a UICC.

13.2.3   Communications Security

For communications security two mechanisms are specified:

- IPsec with Encapsulating Security Payload (ESP) in tunnel mode for the backhaul link to the SeGW;
- TLS for the management traffic to an HeMS accessible on the public Internet.

13.2.4   Location Verification and Time Synchronization

The geo-location of the HeNB is checked by the HeMS before the HeNB is allowed to radiate. This avoids operation of a HeNB in an area where the operator is not allowed to operate base stations, or where the operator does not allow the hosting party to operate the HeNB. See section 13.5.8 for details of geo-location checking.

The availability of the correct time in the HeNB is important for checking certificate expiry time. For this purpose, time synchronization messages are sent from a time server. The transfer of these messages has to be protected. Furthermore, the time server must provide a reliable time signal. Support for protecting time synchronization messages by sending them via the secure backhaul link is mandatory, but optionally also other communication paths may be used provided that time server and transmission are secured. To allow the HeNB the validation of certificate expiry also with missing or faulty local clock, the HeNB must save the time when it is powered down in non-volatile secured memory, and use this time at power-up, if no continuous time is available.

13.3   Security Procedures Internal to the Home Base Station

This section deals with the security related procedures that are executed locally within the HeNB. Security procedures involving the network elements SeGW and HeMS are described in the subsequent sections.

13.3.1   Secure Boot and Device Integrity Check

As described in section 13.2, the HeNB contains a Trusted Environment (TrE) with a built-in root of trust. On power-up of the HeNB, first the TrE itself is checked for integrity using the root of trust. This secure boot process for the TrE itself ensures that only successfully verified software (SW) components are loaded or started. Once the TrE has been started successfully, it proceeds to verify other SW components of the HeNB (e.g. the operating system and further programs) that are necessary for the trusted operation of the HeNB.

The above-mentioned verification process consists of the comparison of measurement values (e.g. hash values) over the SW component to be loaded with the associated trusted reference values stored in secured memory. If the values match, then the verification was successful. Normally these reference values will be hashes over the SW components to be loaded, but also hashes over data (e.g. configuration parameters) included in the downloaded SW package are possible.

The device integrity check has been performed successfully once all SW components necessary for the trusted operation of the HeNB have been verified and started.

To be able to perform the verification of the SW components, the downloaded SW package must contain the associated trusted reference values. These have to be stored in secured memory after the downloaded SW package was verified according to the procedures described in section 13.5.7 on SW download. This secured memory must be protected against unauthorized modifications as the validity of the device integrity check completely depends on the trustworthiness of the reference values.

13.3.2 Removal of Hosting Party Module

The Hosting Party Module (HPM) provides secure storage of the credentials used for hosting party authentication. Critical security functions for support of the EAP-AKA authentication are performed in the HPM. Thus it is ensured that the HPM is available to the HeNB at the time of a combined device-hosting party authentication (see section 13.4.5).

To avoid misuse of the hosting party credentials to be used during a second authentication within some other device while the current HeNB is still operating, the HeNB must monitor the availability of the HPM during subsequent operation. If the HeNB discovers the removal of the HPM, the HeNB must shut down its air interface and disconnect from the operator's core network according to clause 4.4.2 of [TS33.320].

To bring the HeNB back into operation, the HeNB must establish a new connection to the SeGW. If hosting party authentication has to be performed, this requires the insertion of the same or another HPM into the HeNB.

13.3.3 Loss of Backhaul Link

To prevent uncontrolled transmission of the HeNB in case of loss of the connection to the core network, the HeNB must implement a mechanism to shut down the air interface within a certain time period after being disconnected. The usage of this mechanism and the configuration of the time period is up to operator policy.

13.3.4 Secure Time Base

On establishment of the secure backhaul link to the SeGW, the HeNB has to validate the certificate of the SeGW. This includes the check of the expiry times of the SeGW certificate, which must be based on the current time. Thus a time source must be available. Similarly, when a TLS tunnel to the HeMS is established the validity of the HeMS certificate has to be checked.

The only mandatorily supported communication with the time server goes via the backhaul link (see section 13.4.8). Therefore a secure external time may not yet be available to the HeNB when the secure backhaul link is being established. Thus the HeNB needs an internal time source at start-up. Even if it can be expected that many HeNBs will be equipped with a continuously running local clock, the specification does not mandate such a clock. The reason is that, for low-cost devices, such a clock could be left out, and that even with a continuous clock some clock fault (e.g. a flat battery) should not prevent the HeNB from connecting to the operator network. Thus the specification provides a solution requiring only a secure non-volatile storage within the HeNB. It is required for every HeNB to save the current time in the TrE when it is powered down. On subsequent power-up, the HeNB may use this last saved time directly, if it has no continuous clock.[1]

If there is a continuous clock available, then the HeNB will compare the last saved time and the continuous time, and may continue counting with the time of the continuous clock if it is later than the last saved time. This comparison was introduced because, after a fault, local clocks often start on power-up at some fixed point in time, such as the UNIX epoch at 1970-01-01 as specified in section 4.15 of [IEEE Std 1003.1], and a time so far in the past will probably exceed the validity period of all certificates and prevent a successful connection to the operator network.

After establishment of the backhaul connection, a re-synchronization of the local clock is mandatory. This is necessary not only for the HeNBs without continuous clocks, but also caters for a possible drift in time for continuous clocks. The procedure for time synchronization is described in section 13.4.8.

13.3.5 Handling of Internal Transient Data

The HeNB terminates the air-link security and the backhaul-link security to the core network. As only NAS messages are end-to-end protected between UE and MME, all radio level signalling and all user plane traffic is available in cleartext during transfer inside the HeNB. According to a requirement on the HeNB in clause 4.4.2 of [TS33.320], this traffic has to be protected against unauthorized access. This means that the modules handling the security of both air and backhaul link must be located in a protected area of the HeNB and that the transfer of data between these endpoints has to be performed securely. This may be achieved either by putting both endpoints into the same secure area, or by deploying a protected (e.g. encrypted) link between both endpoints even when both are inside the HeNB device.

13.4 Security Procedures between Home Base Station and Security Gateway

13.4.1 Device Integrity Validation

A prerequisite for any connection establishment between HeNB and operator network is the successful device integrity validation of the HeNB by the network. This validation is based on the secure boot of the HeNB and a device integrity check, which are both described in section 13.3. The validation itself is done implicitly by the network, as successful authentication can be performed only if secure boot and device integrity check succeeded (see below). As no active collaboration of the network is needed, this validation is termed 'autonomous validation'.

---

[1] This feature may require the operator to also take care of the start time ('not valid before') of the validity period of the network side certificate. This must allow HeNBs to connect to the network even if they have, for example, a 'last saved time' from manufacturing only, and were not connected for quite some time.

The dependency of authentication on device validation is enforced by the Trusted Environment (TrE) of the HeNB. The access to the private key used for device authentication is only given based on a positive device integrity result. As device authentication is also part of the combined authentication described later in section 13.4.4, this arrangement ensures the correct behaviour of the HeNB for both device authentication and combined authentication. In addition, there is the same dependency between authentication and device validation for any separate secure connection to the management system described in section 13.5, as the client authentication in TLS also needs to make use of the private key secured by the TrE.

13.4.2   Device Authentication

The mutual authentication of the HeNB device and the Security Gateway (SeGW) is mandatory according to clause 4.4 of [TS33.320]. The use of digital signature-based authentication with certificates according to [RFC4306] for this purpose is specified in clause 7.2 of [TS33.320].

The HeNB will authenticate itself to the SeGW with its permanent and unique identity that is described in section 13.2. The identity of the SeGW is not specified by 3GPP, but it is under control of the operator. The identity needs to be contained in the subjectAltName field of the SeGW certificate, which is signed by a Certificate Authority (CA) trusted by the operator. The format of the SeGW identity is a Fully Qualified Domain Name (FQDN) [RFC1912] if DNS is available, otherwise it is simply an IP address.

The authentication procedure is based on a private key and a certificate, in both HeNB and SeGW. As the private keys have to be kept confidential, they must be securely provided to and stored in both elements. In addition, both sides must have access to a root certificate against which the element certificate of the other side is to be validated. These root certificates are public and thus not subject to any confidentiality requirements; but as they constitute the trust anchors for certificate validation, unauthorized exchange has to be prevented.

The provisioning of the HeNB with the required data is described in section 13.5.

The provisioning methods and security requirements for the SeGW are not specified in 3GPP. As the SeGW is located at the border of the core network and seen as part of the core network, it is left to the operator security policy how the private key for the SeGW authentication and the root certificate for validation of HeNB certificates are provided and stored.

The details of the mutual device authentication are specified in clause 7.2 of [TS33.320].

Figure 13.3, which is adapted from Figure A.1 of [TS33.320], gives an example flow diagram for the certificate-based authentication. Details of the description of the payloads in the messages are taken from [RFC4306]. This diagram takes into account that both sides request certificates from the other side, and assumes that the HeNB requests configuration data from the network side. The following text describes the single steps in more detail.

1. The HeNB is securely booted with the help of the TrE. The following steps are executed only if the device integrity validation succeeds.
2. To initiate the IKEv2-based authentication, the HeNB sends an IKE_SA_INIT Request to the SeGW. The connection is set up to the SeGW identity which was provisioned to the HeNB either by initial vendor provisioning or by management (e.g. from the initial HeMS – see section 13.5). HDR is the IKE header. The SAi1 payload states the cryptographic algorithms the initiator supports for the IKE_SA. The KEi payload sends the initiators Diffie–Hellman value. Ni is the initiators nonce.

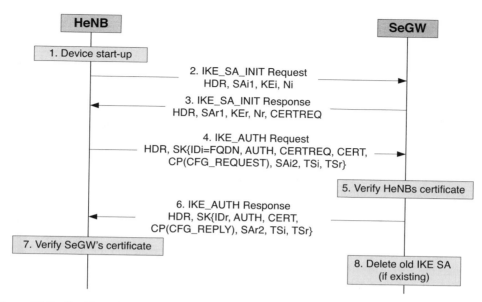

Figure 13.3 Certificate-based authentication with device integrity validation. Adapted with permission from © 2010, 3GPP™

3. The SeGW sends an IKE_SA_INIT Response, requesting a certificate from the HeNB. The responder chooses a cryptographic suite from the initiators offered choices and expresses that choice in the SAr1 payload, completes the Diffie–Hellman exchange with the KEr payload, and sends its nonce in the Nr payload. In addition, it requests a certificate from the HeNB.

4. The HeNB sends its identity in the IDi payload in this first message of the IKE_AUTH phase. This identity is identical to the one provided in the HeNB certificate. The HeNB sends the AUTH payload and its own certificate, and also requests a certificate from the SeGW. As all sensitive functions used for device authentication are to be performed within the TrE (as specified in clauses 5.1.2 and 7.2.2 of [TS33.320]), the computation of the AUTH parameter authenticating the first IKE_SA_INIT message is performed within the HeNBs TrE. If the HeNB is configured to check the validity of the SeGW certificate (see section 13.4.4), it may add an OCSP request to the IKE message. Alternatively the HeNB may retrieve the SeGW certificate status information from the OCSP responder later (in step 7). A configuration payload CP(CFG_REQUEST) is carried in this message if the H(e)NB's remote IP address should be configured dynamically. The Security Association Payload SAi2 is used to negotiate attributes of the security association established by the messages in steps 4 and 6 – for the tricky details see [RFC4306]. The TSi and TSr payloads contain the proposed traffic selectors. The notation SK {. . .} indicates that these payloads are encrypted and integrity protected.

5. On receipt of this message the SeGW may optionally select a user profile based on the HeNBs identity presented in the IDi payload. The SeGW checks the correctness of the AUTH received from the HeNB and calculates the AUTH parameter which authenticates the second IKE_SA_INIT message. The SeGW verifies the certificate received from the

HeNB against the vendor root certificate stored within SeGW. The SeGW may check the validity of the certificates using CRL or OCSP if configured to do so.

6. The SeGW sends its identity in the IDr payload, the AUTH parameter and its certificate to the HeNB together with the rest of the IKEv2 parameters, and the IKEv2 negotiation terminates. If the request received in step 4 contained an OCSP request, or if the SeGW is configured to provide its certificate revocation status to the HeNB in an IKEv2 message, the SeGW retrieves SeGW certificate status information from the OCSP server, or uses a valid cached response if one is available. The Remote IP address is assigned in the configuration payload (CFG_REPLY), if the HeNB requested it by sending CFG_REQUEST in step 4. The traffic selectors for traffic to be sent on that SA are specified in the TSi and TSr payloads, which may be a subset of what the initiator proposed in the message in step 4. The payload SAr2 contains the offer accepted by the responder.

7. The HeNB verifies the SeGW certificate using its stored operator root certificate. This root certificate must be secured against unauthorized exchange, so it has to be stored within the TrE of the HeNB. Also the signature verification process has to be performed within the TrE. The HeNB checks that the SeGW identity as contained in the SeGW certificate equals the SeGW identity as used for connection establishment in step 2. The HeNB checks the validity of the SeGW certificates using the OCSP response if configured to do so. The HeNB checks the correctness of the AUTH parameter received from the SeGW.

8. If the SeGW detects that an old IKE SA for that HeNB already exists, it will delete the IKE SA and start with the HeNB an INFORMATIONAL exchange with a Delete payload in order to delete the old IKE SA in HeNB (not shown in the figure).

After successful completion of this procedure the IKE security association (IKE_SA) and a first CHILD_SA are established, and further CHILD_SAs for IPsec tunnels may be created. This is described in section 13.4.7.

In case any of the steps given here failed, a working backhaul link between HeNB and operator network cannot be initiated. The current specification for Release 9 does not give any standardized remediation for this case. Thus it is implementation and configuration dependent, for example if a separate connection (not through the SeGW) to the HeMS is established to try remediation of the device by management means, or if a visible indication is given to the customer on the HeNB device to prompt him to contact customer care.

13.4.3   IKEv2 and Certificate Profiling

The profiles for IKEv2 and the related certificates follow as closely as possible the 3GPP specifications on Network Domain Security (NDS). These specifications on NDS/IP [TS33.210] and NDS/AF (Authentication Framework) [TS33.310] are handled in sections 4.5 and 8.4. Additional profiling is, however, necessary to adapt the profiles to the particular environment of HeNBs. One big difference is that IKEv1 is not allowed for HeNB connections, so only the IKEv2-related parts of the NDS specifications are valid. The reason for this difference is that HeNB and SeWG were defined for 3GPP Release 9 and, hence, no legacy systems supporting only IKEv1 had to be taken into account.

The IKEv2 profile is given in clause 5.4.2 of [TS33.210]. It states the mandatory algorithms to be supported for IKE_SA_INIT exchange and IKE_AUTH exchange. In addition [TS33.320] explicitly excludes the usage of pre-shared keys for the IKE_AUTH exchange. For the certificate-based authentication the following additional rules are given.

- RSA signatures for authentication will be supported.
- The HeNB will include its identity into the IDi payload of the IKE_AUTH Request, to allow the SeGW policy checks based on HeNB identity before any certificate handling is done in SeGW. Usage of this unauthenticated identity does not constitute a security risk, as [RFC4306] anyway requires the cross-check of IDi with the identity carried in the certificate after certificate validation.
- Certificate requests and certificates (including any certificate chains up to the root certificate) must be sent by both sides within the IKEv2 exchange. These certificates shall all be of type 'X.509 Certificate – Signature' (type 4 as specified in [RFC4306]).

For certificate profiles of the X.509 certificates, the HeNB security specification references the relevant clauses of [TS33.310]. For all certificates that are not root certificates (i.e. HeNB and SeGW certificates), and for all certificates up in the chain to the root certificate, clause 6.1.3 of this specification about Security Gateway (SEG) certificates is referenced. This clause in turn references clause 6.1.1, which is valid for all certificates used in [TS33.310].

For the HeNB certificates the following additional rules are given.

- The certificate must be signed by a CA that is authorized by the operator, such as a CA of the manufacturer or the vendor. This is a deviation from the rules in [TS33.310], where all certificates are signed directly by the operator CA. This deviation is necessary as the HeNB device certificate is installed into the HeNB by the manufacturer, and the specification should not impose any need to exchange this certificate during the whole lifetime of the HeNB. In addition, the operator is not the only party taking responsibility for the HeNB, as the manufacturer stands for the device integrity of the HeNB when signing and providing the certificate. Thus in many cases the manufacturer will sign the HeNB certificates using their own CA, but also a (e.g. commercial) third-party CA can be involved here if it is trusted by both the manufacturer and the operator.
- The HeNB certificate must carry an identity in FQDN format in the subjectAltName field of the X.509 certificate [ITU X.509]. This clarification is necessary as TR-069 [BBF TR-069] allows other formats for the HeNB identity.
- The HeNB certificate may carry information about the location of a revocation information server of the manufacturer or vendor of the device. When the revocation information is provided in the form of a Certificate Revocation List (CRL), the CRL distribution point as specified in [TS33.310] shall be given in the certificate. If an OCSP server is deployed for online retrieval of certificate status information, the OCSP server information (AIA extension) as specified in [RFC5280] and [RFC2560] shall be given.
- One requirement on the HeNB certificate is not stated explicitly, but can be deduced from the usage scenario of the HeNB certificates. Basically the certificate must be unalterable once stored within the TrE. As there is no certificate renewal procedure specified, normally the certificate must be valid for the complete expected lifetime of the HeNB and the expiration time ('not valid after') has to be set accordingly. If the manufacturer provides a proprietary method for renewal of the HeNB certificate, an authorization mechanism must be included, as only authorized parties shall be able to perform this renewal.

For the SeGW certificate, only one rule is given in the specification in addition to the provisions of clause 6.1.3 of [TS33.310]:

- The operator may populate the certificate with an OCSP server information as specified in [RFC2560]. If configured to do so, the HeNB may use this information to request certificate validity status information from the OCSP server.

For the CA certificates used for validation of the HeNB and SeGW certificates, the requirements in clause 6.1.4b of [TS33.310] for CAs issuing certificates for network elements are referenced, and not the requirements for SEG CAs where SEG is defined as for NDS/IP – see section 4.5. This was done as connections between different security domains and thus interconnection CAs are not required.

The validity periods of the root certificates are not specified and are a matter of policies of the bodies operating the CAs. For the CAs used in the certificate chain of SeGW certificates, most probably any existing PKI policies of the operator will be deployed. Replacement of certificates in network elements is common practice in operator networks. A secure replacement procedure for the operator root certificate in the HeNB is not specified in [TS33.320], but may be implemented by management procedures. For the certificate chain used for validating HeNB certificates, the situation is different because the HeNB device certificates are lifetime certificates. No example policies for the PKI infrastructure on the manufacturer side are mentioned in the specifications, but two different approaches are given in the following.

- If the CAs have a conservative approach, meaning that they are not allowed to sign certificates with a later expiration time than that contained in their own current certificate, then also the root certificate and all intermediate certificates must have an expiration time which exceeds or at least equals the expected lifetime of the HeNBs.
- A more open approach is possible, if the CA signing the HeNB certificates is allowed to issue certificates with longer lifetime than contained in its own certificate or, if it is not the root CA, in any of the certificates in the chain to the root certificate. In this case there is no strong requirement on the expiry time of these certificates. Given normal validation policies, the earliest expiry time in the chain of certificates will anyway limit the HeNB certificate lifetime. But if at a later point in time new certificates are issued with longer lifetime, then also the actual validity of the HeNB certificate is extended at most up to the expiry time of the HeNB certificate itself. This procedure implies two hard conditions, namely that any new root certificate must be provided to the entity checking the HeNB certificate, and that any new certificate for the CA having signed the HeNB certificate must certify the same subject name and private/public key pair as with the old certificate.

Based on the described certificate profiles, it is clear that the CA infrastructure for NDS and the one for HeNB backhaul link differ from each other. While NDS assumes a common CA for both sides of the authentication, or at least specific interconnection CAs when bridging two security domains, for HeNB purposes the approach of two entirely separated PKIs was chosen. Reasons for this decision are given in section 13.1. Thus the following CA infrastructure is envisaged.

- For the HeNBs the root CA logically lies with the manufacturer of the device. It is a decision of the manufacturer either to deploy their own root CA, and to sign all HeNB certificates locally, or to rely on the service of a trusted third party. The latter approach may use a third-party CA either as root CA only while the manufacturer deploys their own signing

CA, or for signing for all single HeNB certificates. The root CA (and all intermediate CAs that may be there) must be trusted by all operators who allow HeNBs of this manufacturer to have access to their core networks. This also implies that the HeNB manufacturer must provide any operator deploying their HeNBs with the trust anchor used for signing the HeNB certificates.

- For the SeGW root CA the PKI deployment resembles more the ordinary NDS usage. The operator may issue SeGW certificates just like certificates for any other network element within their own network. The root certificate of the operator must be provisioned to all HeNBs. This provisioning is described in section 13.5.

13.4.4   Certificate Processing

Certificate processing during IKEv2 authentication is profiled for HeNB and SeGW in the following way.

- Certificate validating entities are required to only have the related root certificate available locally. All other certificates in the certificate chain including the end entity certificates must be provided by the end entities whose certificates are to be validated.
- Certificate chains sent to the other entity shall have a maximum path length of four certificates, to limit the processing requirements and the amount of data transferred.
- Certificate validity times ('not valid before' and 'not valid after') must be checked, and invalid certificates must be rejected.
- Certificate revocation checking is performed based on local policy. The mechanisms used and the requirements on support and usage are different for HeNB and SeGW, and are described below.

Revocation status check in HeNB

To ease the implementation of the mass-deployed HeNB, only the Online Certificate Status Protocol (OCSP) mechanism [RFC2560] is specified for the HeNB. Usage of OCSP is optional, while support for the protocol in HeNBs is strongly recommended. This should ease the migration path for operators intending to introduce mandatory certificate validation at a later time even with a potentially huge number of already deployed HeNBs. To further reduce the implementation requirements on the HeNB, usage of in-band signalling of certificate revocation status in IKEv2 according to [RFC4806] is made optional. This in-band signalling avoids a separate connection from the HeNB to an OCSP server, and the operator does not have to deploy an OCSP server in the public Internet.

Revocation status check in SeGW

For validation of HeNB certificate status, the SeGW may use two different mechanisms: one is OCSP as used by the HeNB, and the other is the revocation by Certificate Revocation Lists (CRLs), the latter being the standardized way in NDS/AF. Implementation for at least one of the two mechanisms is mandatory in SeGW, while usage is at the discretion of the operator.

Input for certificate revocation may come from both the operator and the manufacturer, as both may have reasons to revoke a certificate. These reasons may be, for example, that the manufacturer finds out that some series of HeNBs have newly discovered flaws in the integrity protection, and thus the certificate is no longer valid for usage with autonomous validation, or the operator may deem only certain series of HeNBs as valid to access their core network.

Based on this dual input for revocation, either both parties operate their own revocation server, which are both queried by the SeGW, or the manufacturer provides their revocation lists to the operator, who combines them with their own revocation lists. [TS33.320] explicitly states that the manufacturer is obliged to provide such revocation data, if the operator uses certificate status checking at all.

13.4.5   Combined Device-Hosting Party Authentication

The mutual authentication of the HeNB device and the Security Gateway (SeGW) is mandatory according to clause 4.4 of [TS33.320]. The same clause gives the option to also authenticate a Hosting Party (HP) to the network. This HP authentication is performed in addition to and following the mutual device authentication, described in section 13.4.2. It uses the feature of IKEv2 to provide authentication of the initiator by means of an EAP method. The EAP method used for this purpose is EAP-AKA, which is described in Chapter 5. Section 5.2.3 describes for the case of 3G-WLAN interworking how EAP-AKA is used in the context of IKEv2. Both authentications, EAP-AKA and certificate-based authentication, are embedded into the multiple authentication procedure specified for IKEv2 in [RFC4739]. EAP-AKA' would provide no security advantage over EAP-AKA for this purpose because IKEv2 mandates the use of certificates for responder (SeGW) authentication – see the similar case for untrusted access network discussed in section 11.2.1.

The reasons why HP authentication is not used as a standalone solution, but always combined with device authentication, are given in section 13.2.

The AKA functions used to support the EAP-AKA authentication must be provided by the Hosting Party Module (HPM), which is described in section 13.3. The storage of the secrets used for this authentication, and the calculation of the authentication parameters, must take place within the HPM.

As can be seen from Figure 13.4, the HP authentication requires a AAA server as an additional network element as compared to device authentication according to section 13.4.2, and the involvement of the HLR/HSS. The setting is similar to the architecture needed for 3GPP IP access in 3G-WLAN interworking specified in [TS33.234] and described in section 5.2.3, and, if already deployed, could be reused. A difference to the case of 3G–WLAN interworking is that the HP identity is not an ordinary subscriber identity used, for example, in a UE, but an identity used for a network element. On the other hand, if the HLR/HSS infrastructure is to be reused, the HP must have an entry there like any other subscriber. But to avoid misuse of the HP identity, this entry should be clearly differentiated in the subscriber profile from ordinary UE subscriptions.

Figure 13.4, adapted from Figure A.2 in [TS33.320], gives a message flow diagram for the combined (device and HP) mutual authentication, which is explained in more detail below. As with Figure 13.3, the diagram takes into account that both sides request certificates from the other side.

1. This and steps 2 to 7 resemble the execution of the mutual device authentication as described in section 13.4.2 and Figure 13.3. In steps 3 and 4, both sides have to state

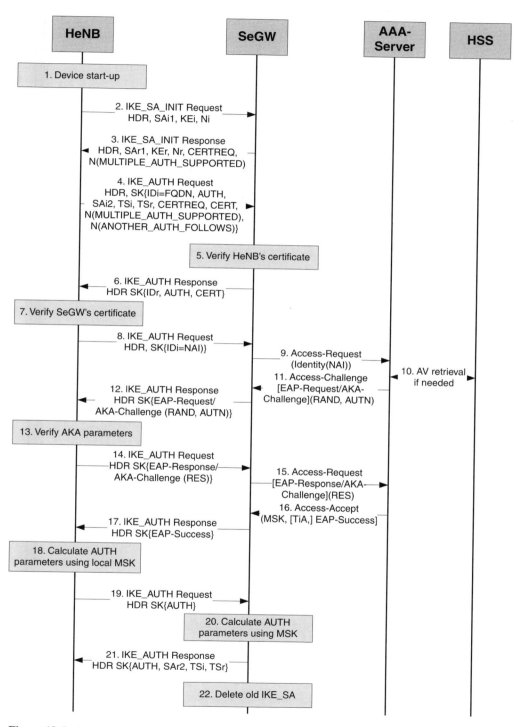

Figure 13.4 Device and hosting party identity-based authentication. Adapted with permission from © 2010, 3GPP™

their support for multiple authentications and the HeNB indicates in step 4 that a second authentication will follow after the device authentication. The indication of 'multiple authentications supported' in step 3 could be interpreted by the HeNB that HP authentication is wanted by the SeGW. But such interpretations are not specified in any way, and depend solely on the policies given to SeGW and HeNB. The exchange of some SA-related parameters sent in step 6 of section 13.4.2 is postponed to step 21.

8. In this and steps 9 and 10, the HeNB starts the second authentication by sending its HP identity to the SeGW in the IDi payload. The SeGW forwards this identity to the AAA Server in an Access Request message. The AAA Server fetches authentication vectors for this identity from HSS/HLR together with subscription data if needed.

11. In this and steps 12 to 15, the AAA Server performs the EAP-AKA authentication with the HeNB via the SeGW. The SeGW acting as IKEv2 responder inserts the EAP messages into the IKEv2 messages. Within the HeNB the AKA functions are performed within the HPM holding the shared secret key. It is expected that an ordinary UICC with USIM similar to a subscriber UICC can be used as HPM.

16. When all checks are successful, the AAA Server sends the Authentication Answer including an EAP Success message and the MSK generated from EAP-AKA procedure to the SeGW.

17. The EAP Success message is forwarded to the HeNB over IKEv2.

18. In this and step 19, the HeNB calculates the AUTH parameters for authentication of the IKE_SA_INIT phase messages using the locally generated MSK. The AUTH parameter sent by SeGW in step 17 is checked.

20. In this and step 21, the SeGW calculates the AUTH parameters for authentication of the IKE_SA_INIT phase messages using the MSK received in step 16. The SeGW checks the correctness of the AUTH parameter received from the HeNB in step 19. The message sent to HeNB terminates the IKEv2 negotiation and contains all remaining parameters from step 6 of the device mutual authentication, which were not sent in step 6 of the current procedure.

22. If the SeGW detects that an old IKE SA for that HeNB already exists, it deletes the IKE SA and starts with the HeNB an INFORMATIONAL exchange with a Delete payload in order to delete the old IKE SA in HeNB (not shown in the figure).

13.4.6 Authorization and Access Control

The simplest deployment model of HeNBs does not need any explicit authorization and access control for access to the operator core network. To allow successful execution of the certificate-based device authentication, the operator has to provide all SeGWs with the root CA certificate(s) of the HeNBs. This provisioning of the root certificate(s) by the operator is the prerequisite of implicit access control. This form of access control admits all devices to the operator network that successfully pass device authentication; and only HeNBs with certificates rooted in the provisioned root certificate(s) can do that. Implicit access control ascertains that each HeNB having access to the network comes from a manufacturer accredited by the operator, and that the HeNB conforms to the integrity rules set up by the manufacturer and operator.

The authorization scheme described above requires that the root CAs only sign HeNB device certificates. If also other certificates besides HeNB device certificates may be issued under the same root CA, then the mere validation against the root certificate is not sufficient. One solution for this problem is to restrict access to HeNB identities that are verified by certificates signed by an intermediate CA exclusively used for device certificates. Other solutions that take information elements of the certificates (e.g. subject names) also into account are described below.

If there is no certificate revocation handling, this access control scheme is static. This means that the access rights last until the expiry of the HeNB certificate or until any of the certificates in the chain to the root certificate expires, whichever happens first. To allow also the barring of single devices or complete series of HeNBs, perhaps if some devices were found vulnerable or compromised, a revocation infrastructure may be deployed. Alternatively, blacklists or whitelists may be used (see below). Details of an envisaged architecture of an optional revocation infrastructure are given in section 13.4.4.

A more fine-grained access control may be desirable for the following reasons.

1. The operator does not want to admit all products of a certain manufacturer to their network, but wants to restrict the access to accredited types or series only. The criteria used may be, for example, the feature set of the HeNB or the management capabilities in their network.
2. Some manufacturers may share a common third party root CA for HeNB device certificates, but the operator may want to allow access for HeNBs of certain manufacturers only. The differentiating criterion here may be the manufacturer ID that is part of the device identity.
3. The operator may have commercial or business reasons to admit only certain HeNBs to their network. For example, only HeNB devices explicitly registered with or provided by the operator are admitted. In addition, a HeNB device may be bound to a specific HeNB subscription, and only devices with such subscriptions are allowed. This access control may be extended if the operator provides separate Hosting Party (HP) subscriptions, and wants to bind single HeNB device identities to specific HP identities.
4. The operator wants a separate control over allowed devices, based for example on data gathered by the management system about possible irregularities, or for commercial reasons, such as to block certain hosting parties out temporarily or indefinitely.

Such fine-grained access control is mentioned in clause 7.5 of [TS33.320], but no specific method (e.g. blacklisting or whitelisting) is specified. Also the location of the involved logical entities is not given. Naturally the SeGW is the access enforcement point, but the location of the access decision point may be within the SeGW, in a separate AAA server, or in the management system.

There may be concerns about the usage of an external AAA server, as the interfaces between SeGWs and such servers are not well specified, and vary considerably. Clause 7.5 of [TS33.320] mentions the possibility to use the standardized OCSP protocol [RFC2560] for this purpose. As in OCSP a separate request is sent to the OCSP server for each certificate to be validated, the OCSP server may be enhanced with AAA functionality to additionally look up a blacklist or a whitelist. In cases of denied access the OCSP server may respond with a 'certificate invalid' message. This is not the intended purpose of the OCSP protocol, which should only report the revocation status of certificates. Therefore, this solution may only be deployed in a proprietary manner. Also, OCSP servers with this extended functionality may not be available.

13.4.7   IPsec Tunnel Establishment

As a result of the authentication by means of IKEv2 as described in sections 13.4.2 and 13.4.5, at least one IPsec tunnel is established between HeNB and SeGW. The profiling of this IPsec tunnel is done in accordance with clause 5.3 of the NDS/IP specification [TS33.210]. The mandated security protocol is the Encapsulating Security Payload (ESP) [RFC4303] in the tunnel mode.

All traffic between HeNB and SeGW is carried through this tunnel – signalling, user plane and management plane traffic. For management traffic a second option for a secure tunnel exists, which is described in section 13.5.

The NDS/IP specification is referenced also for the list of supported ESP authentication transforms and ESP encryption transforms. This means according to clause 5.3.3 of NDS/IP that for encryption the algorithms ESP_NULL, TripleDES-CBC [RFC2451] and AES-CBC with 128-bit keys [RFC3602] must be supported. For integrity protection, clause 5.3.4 of NDS/IP states that the ESP_NULL, ESP_HMAC_MD5 and ESP_HMAC_SHA-1 transforms must be supported.

The HeNB is normally located in the home network of the HP, with access to the Internet via a broadband connection. Thus most probably there will be Network Address/Port Translators (NATs/NAPTs) and firewalls in the home network and the access network. Support for such environments is mandated in clause 7.2.4 of [TS33.320] by requiring the support of mechanisms for detection of NAT [RFC3947], UDP encapsulation for NAT Traversal and HeNB-initiated NAT keep-alive [RFC3948], and Dead Peer Detection.

13.4.8   Time Synchronization

In the context of HeNB security, reliable time information is required for the checking of certificate validity times. Every certificate has a 'not valid before' and 'not valid after' entry, which have to be checked on every usage of the certificate. Therefore the HeNB requires a secure time base, which is described in section 13.3.4.

This local clock must be synchronized with a reliable external clock periodically. The specification requires a maximum time interval of 48 hours between synchronizations, when the HeNB is connected to the network. Local reception of time signals, such as from a Global Navigation Satellite System (GNSS), was not seen as sufficient for standardization purposes as there may be locations for HeNBs where such reception is not possible. Also such signals may be disturbed or faked quite easily, given that GNSS test kits are commercially available at reasonable prices.

Based on these considerations, 3GPP decided that the provisioning of the HeNB with time synchronization messages over the secure backhaul link is mandatory to implement. The Network Time Protocol (NTP) [RFC1305] would be a candidate, but the exact protocol is not specified in the HeNB security specification.

The selected approach has a twofold advantage:

- Regardless of which time protocol is used, the operator has full control over the time messages. This applies to both cases when either the operator deploys a time server, or if they have access to a reliable time source from elsewhere, and feed the messages into the backhaul links to the HeNBs under their control.

• There is no need to specify the usage of specific security mechanisms for the time protocol itself.

Disadvantages of the approach were not seen, as the latency of the time synchronization messages was not seen to be increased significantly by the handling within the IPsec stack, as compared to the latencies introduced by transmission over the subscriber access line, even if QoS mechanisms are used. Also the increase in required bandwidth on the subscriber line and the additional processing capacity required in the SeGW for handling the IPsec packet overhead were not seen as significant.

In addition to the reception of time synchronization signals over the backhaul link, the HeNB may also receive time synchronization signals from other time servers. [TS33.320] gives here the general requirement that the time server and the communication must be secured.

One special error case for clock synchronization is also mentioned in this specification. It may happen that by mistake the HeNB receives time messages with a time value far in the future. If this error is not corrected before power-down of the HeNB, this time will be stored and used on next power-up. If this time is beyond the validity of the SeGW certificate, connecting to the SeGW is not possible. No standardized solution is given for this error case in Release 9 of 3GPP, but it is expected that it will be handled together with extended management features in future releases.

13.5   Security Aspects of Home Base Station Management

13.5.1   Management Architecture

As described in section 13.1, HeNBs (and HNBs as their UMTS counterparts) are the first mobile network elements deployed in customer premises. Thus for the first time in 3GPP specifications the security for management of a network element is specified in such detail. Driven by the expected mass deployment of HeNBs, the management interface should be completely specified to allow unlimited interworking between HeNBs and HeNB management systems (HeMS) of different vendors.

The security architecture for HeNB management builds on the 3GPP specifications on HeNB management. The requirements are given in [TS32.591], and the architecture and procedures in [TS32.593]. These specifications describe the 'Type 1 Interface' which is defined as the interface between network element management operations systems and the network element. The basic management procedures are taken from the Broadband Forum [BBF] specification TR-069 [BBF TR-069]. This protocol allows the online communication between HeMS and HeNB, and specifies commands and data formats to be used. In addition, the usage of file transfer mechanisms is specified for the download of software and bulk configuration data and the upload of, for example, performance measurement data. The data model used for the managed information elements is specified in [TS32.592]. The content of this specification is based on the comparable specification for HNBs [TS32.582], amended by EPS-specific elements. These data models are heavily based on BBF specifications with the general data model in [BBF TR-098] and the HNB-specific data model in [BBF TR-196]. The 3GPP

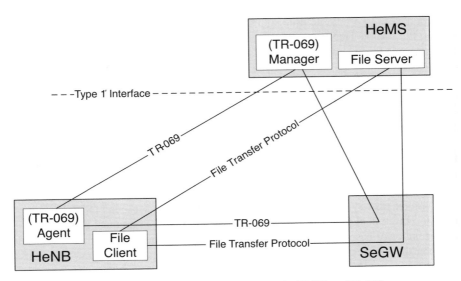

Figure 13.5   Management architecture for HeNB and HeMS

working group TSG SA WG5 on Telecom Management is in charge of the communication between 3GPP and BBF on these data models.

Figure 13.5 shows the basic management architecture for HeNBs. The HeMS may be located within the operator network or in the public Internet. If the HeMS is located in the operator domain, the management traffic is routed through the SeGW, as traffic from the Internet shall never access the operator security domain directly. If the HeMS is located in the public Internet then a direct connection between HeNB and HeMS is foreseen.

The security for the management traffic is specified in clauses 8.3 and 8.4 of [TS33.320]. Depending on the location of the HeMS, different security mechanisms are required. In addition, it has to be considered that the HeMS may be distributed; for example, the TR-069 manager (called TR-069 Auto-Configuration Server – ACS – by BBF) and the file server might be physically separated. This may be the case if an existing file server infrastructure that is accessible on the public Internet and is used for support of existing home gateways or DSL routers is to be reused for HeNBs.

HeMS in operator domain. When the HeMS is located within the operator security domain, the management traffic is tunnelled through the same IPsec tunnel that is used for signalling and user plane traffic between HeNB and core network. This is described in the preceding section 13.4. In addition, the operator may optionally deploy the security mechanisms specified for access to an HeMS accessible on the public Internet, if end-to-end security is required between HeNB and HeMS.

HeMS accessible on public Internet. When the HeMS is accessible on the public Internet, the HeNB has to establish a secure tunnel to this HeMS for the management traffic. Such secure

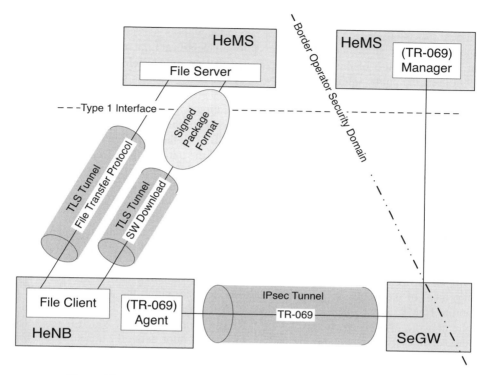

Figure 13.6   Security mechanisms used depending on location of HeMS

tunnel using TLS is specified as optional in TR-069, but [TS33.320] requires the usage of TLS (e.g. HTTPS [RFC2818] or FTPS [RFC4217]) as mandatory.

Figure 13.6 shows an example management architecture with distributed HeMS and the mandatory security mechanisms for this deployment. It shows the basic types of connections that can be used in this configuration. The management traffic between TR-069 agent in HeNB and TR-069 manager in HeMS is secured by the IPsec tunnel between HeNB and SeGW. According to operator policies, the interface between SeGW and HeMS and other network internal interfaces may be secured using the Zb interface for elements in the same security domain, and using a sequence of Zb and Za interfaces for elements in different security domains (see section 4.5 on Network Domain Security). But this is not required for HeNB security. For SW Download or any other file transfer, the HeNB has to establish a TLS tunnel with the file manager in HeMS before any data may be downloaded or uploaded.

The management architecture for HeNBs defined by 3GPP comprises two different kinds of management systems. They are described in detail in the following.

Initial HeMS. The initial HeMS is specified as the first management contact point for the HeNB after first power-up, or after a reset of the HeNB to factory default values. The access URL of this initial HeMS may be hard-coded into the HeNB, or provisioned at the factory. The 3GPP specifications do not specify if the initial HeMS is owned or operated by the operator, by the HeNB manufacturer or vendor or by a third party. This is to allow for a flexible enrolment procedure of HeNBs to operator networks, without necessitating the provisioning of operator-specific parameters into the HeNB for all HeNBs at time of production or delivery from factory. Because of this it is also expected that an initial HeMS is more likely to be accessible on the public Internet, as otherwise also a SeGW address must be pre-provided in the HeNB.

The initial HeMS provides the HeNB with operational addresses and parameters for later operation in a specific operator network. The selection of the addresses and parameters may be based on the geo-location as reported by the HeNB or on the globally unique identity of the HeNB. Also a first software download may be done if the initial HeMS detects an outdated or inappropriate version on the HeNB. With respect to security mechanisms, the general security requirements for HeMSs apply also to the initial HeMS. If the initial HeMS is located in the operator network behind a SeGW, this SeGW is called 'initial SeGW', and the address of this SeGW must also be pre-provisioned to the HeNB. The term 'initial SeGW' is a logical name used in management, and does not require a SeGW that is physically separated from the SeGW used for (for example) S1 interface data or for the connections to the serving HeMS.

Serving HeMS. The serving HeMS is the management system that takes care of the everyday management of the HeNB. It is more likely to be located within the operator network than the initial HeMS, as the management tasks are closely related to the actual operation of the mobile network. This is in contrast to the tasks of the initial HeMS, which is restricted to the task of initially provisioning the HeNB for one operator network. Based on the HeNB management specification [TS32.593], the HeNB has to register with the serving HeMS when first connecting to the network. Later on, the serving HeMS performs configuration management and SW updates and is the receiver of performance measurement data collected by the HeNB. If the serving HeMS is located in the operator network, then the SeGW used for the management traffic is called 'serving SeGW'. This requires neither a physically separated SeGW nor a separate IPsec tunnel for management traffic.

The distinction of initial and serving HeMS is primarily logical, thus there is no need to deploy two HeMSs if the particular deployment scenario of an operator does not require separate entities.

Figure 13.7 gives a possible architecture for HeNB deployment with the initial HeMS accessible on the public Internet and the serving HeMS on the operator network. On first power-up the HeNB connects to the initial HeMS and is configured with the FQDN of the operator SeGW and the (inner) FQDN of the serving HeMS. The FQDNs may be replaced by IP addresses, if no DNS is available for resolving domain names. All this happens via a TLS tunnel. Then the HeNB disconnects from the initial HeMS, establishes the secure backhaul tunnel as described in section 13.4, and then connects to the serving HeMS in the operator network.

The paths for the S1 interface are also given in this figure, to show that the management data is not handled differently from user and signalling data, when seen from the point of view of the backhaul link. Naturally a separation according to the type of traffic may be done on the backhaul link, if for example QoS mechanisms are applied there. If separate Security Associations (SAs) are deployed for traffic separation, all these SAs are derived as child SAs from the same IKE SA.

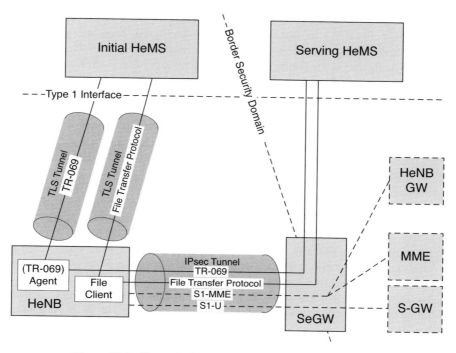

Figure 13.7   Example deployment of initial and serving HeMS

13.5.2   *Management and Provisioning during Manufacturing*

For the whole concept of HeNB security to work the manufacturer has to pre-provision some data into the HeNB. The data described here are independent from the target operator network, where the HeNB will later be connected to.

As the authentication of HeNBs to both the operator network and the HeMS accessible on the public Internet is based on manufacturer-provided certificates, the manufacturer has to provide a private/public key pair and a related certificate to the HeNB device.

The private key has to be provided to and stored securely within the Trusted Environment (TrE) of the HeNB. The best security level is typically achieved if the private key is generated within the element itself inside the TrE, and the private key never leaves the TrE. The device certificate is public and thus not subject to specific security requirements. If somebody tampers with the certificate, then it can no longer be verified. For HeNBs the exact method of key generation is not specified in [TS33.320]. It is left to the manufacturer to either generate the private key internally in the TrE or to generate it externally and provide it to the TrE later on. If external generation is used, the process of key-pair generation and provisioning of the private key must be performed in a secure environment. As the private key is only used for authentication and tunnel establishment, and without any need for possible later key recovery, there is no need to keep a copy of the private key with the manufacturer.

For generation of the HeNB certificate, the manufacturer either has to deploy their own Certification Authority (CA), or has to send the certificate signing request to a third party CA which is trusted by the manufacturer and all possible customers of the manufacturer. This signing request normally carries the public key, a list of intended certificate attributes including the device identity and validity period, and some proof-of-possession of the associated private key (e.g. a signature with the private key over some data). If the manufacturer provides certificate revocation information online by CRL or OCSP server, the associated server information has to be included into the request. The validity time should include the complete expected lifetime of the HeNB, as a renewal of this manufacturer-provided device certificate is not envisaged in the specification. The resulting certificate has to be stored in the HeNB. For the certificate profile, see section 13.5.6.

For authentication to the initial HeMS, a root certificate for validation of the certificate used by the network side has to be provided. This is either the root certificate for the TLS server certificate if an initial HeMS accessible on the public Internet is contacted via a mutually authenticated TLS tunnel, or a root certificate for the validation of the SeGW certificate in front of the initial HeMS. This root certificate is not confidential, but it must be secured against unauthorized replacement as it is the root of trust for authentication of the network side.

Section 13.5.7 describes the secure software download specified for SW updates of the HeNB. To allow the validation of such SW downloads within the HeNB, a root certificate for the validation of the signed data object must be provisioned to the HeNB. If the root of trust for the SW download lies with the manufacturer of the HeNB, a certificate of a CA acting as a root for SW signing must be provisioned to the HeNB. This certificate must be stored securely, as any modification can only be possible by authorized access to the HeNB.

13.5.3   Preparation for Operator-specific Deployment

The following data have to be provided to the HeNB before registration at the serving HeMS and ordinary operation in the operator network. The provisioning of the data requires authorized access to the HeNB, as the root certificates provided are the root of trust for authentication of the operator network, and thus must be secured against unauthorized modification.

The minimum set of operator specific data needed for registration with the operator network depends on the architecture used by the operator.

- When the serving HeMS is accessible on the public Internet, only the FQDN of the HeMS and the root certificate for validation of the HeMS certificate are necessary.
- For the deployment scenario where all communication of the HeNB with the operator network is routed through the SeGW, the FQDN of the serving SeGW and the root certificate for validation of the SeGW certificate are necessary. In addition, the operator network internal FQDN for connecting to the HeMS must be available. A root certificate for validating the HeMS identity is only necessary if the operator uses the optional end-to-end TLS tunnel within the IPsec tunnel.

The operator may decide to base the secure SW download on an operator-provided signature in the signed data object as part of the SW download package (see section 13.5.7). In this case the HeNB must be provisioned with the root certificate of the operator which is to be used to validate the signed data object.

When a HeNB is branded for a specific operator, all data mentioned above in this subsection may be provided at the operator premises before delivery to the HP or by the manufacturer together with the data described in the previous subchapter. With an unbranded HeNB, the most viable method is the use of an operator-independent initial HeMS. The FQDN of this HeMS may be pre-provisioned according to the previous subsection, and the operator-specific data are configured by the initial HeMS, based for example on the globally unique identity of the HeNB authenticated by the HeMS and/or based on the geo-location of the HeNB.

Provisioning at customer care points of the operator is also possible if a local management interface is supported by the HeNB. Furthermore, a distribution of these data by removable storage media is possible, such as by a UICC used for Hosting Party authentication. Such local management is not part of the 3GPP specifications and the security measures for such vendor-proprietary procedures are not handled by 3GPP. In particular, the provisioning of these data by removable media would require additional security measures as root certificates are required to be inserted into the HeNB by authorized access only and must be securely stored within the TrE of the HeNB.

13.5.4   Relationships between HeNB Manufacturer and Operator

To enable authentication of the HeNB to the operator network, some interactions between the HeNB manufacturer and the mobile operator are necessary. The decision to base the HeNB device authentication on a manufacturer-provided certificate has the consequence that the manufacturer has a responsibility for the HeNB for its complete deployment time. This relates to the integrity and validity of the HeNB certificate, but also to the integrity protection and validation of the HeNB device itself, as the device integrity validation is closely bound to authentication in the concept of autonomous validation as described in section 13.3.

The first relationship between the manufacturer and the operator is of organizational nature, and refers to the trust of the latter into the former. If any CA involved in the signing of the HeNB certificate is not operated by the manufacturer, the operator must also trust the CA which signed the HeNB device certificate and the CA chain up to the root CA.

With respect to the root certificate used for the validation of the HeNB device certificates, it is required that the currently valid root certificate has to be handed to each operator who allows HeNBs of this particular manufacturer to authenticate to their network.

The expiry time of the root certificate must be sufficiently far in the future to allow the validation of all certificates issued during the expected deployment time of the HeNB device. However, if the root certificate is about to expire, a renewed certificate is to be distributed. Such renewal must be done in such a manner that the public key and the subject name of the CA issuing the device certificate stays the same, as otherwise older device certificates could no longer be validated.

With respect to the certificates of individual devices, the manufacturer is obliged to provide the operator with revocation information, in case the operator wants to establish revocation lists. Such manufacturer-generated revocation information may be necessary, for example, if some HeNB devices or types are prone to be compromised or even already found to be compromised. This compromise refers to both disclosure of the private key, which would allow cloning of the HeNB identity, and to weaknesses in device integrity protection, which would lead to failures of the autonomous validation mechanism.

13.5.5   Security Management in Operator Network

For the deployment of the security mechanisms described in this chapter, some management operations in the operator network are necessary.

As the network must authenticate the identity of the HeNBs, the network elements performing this authentication must be provided with the root certificate(s) for validation of the HeNB certificates. This applies to the SeGW, and also to the HeMS in case the HeMS uses TLS. These root certificate(s) are received from the HeNB manufacturer(s), as described in the previous subsection.

If the operator deploys a certificate revocation infrastructure for SeGW certificates, they have to operate an OCSP server. This OCSP server must be provided with a certificate signed by the operator root CA. This should be the same root CA as used for validating the SeGW certificates, as otherwise the HeNB would have to be provided with two different root certificates of the same operator.

If the operator uses authorization and access control (see section 13.4.6), then the related access control lists (e.g. blacklists or whitelists) have to be managed. Procedures for the management of such lists are out of scope of [TS33.320].

13.5.6   Protection of Management Traffic

The security for all connections carrying management traffic between HeNB and HeMS is specified in clause 8.3 of [TS33.320], based on the requirements given in clause 4.4.6. This clause reads that the HeMS link shall provide integrity, confidentiality and replay protection of the transmitted data. The required security mechanisms are not different between initial and serving HeMS and initial and serving SeGWs, so all the text in this subsection applies to all scenarios. The requirements also equally apply to the transfer of the TR-069 management protocol data and any file transfers.

It is a general requirement for the communication between HeNB and HeMS that both entities must be mutually authenticated and that a secure communication channel be established between them. For management traffic through the SeGW (see below), the SeGW is the authentication partner on networks side instead of the HeMS.

Management traffic scenarios

Clause 8.3 of [TS33.320] gives the security requirements for the two different connection scenarios, i.e. for the HeMS accessible on the operator network and on the public Internet. The main features for both scenarios are given below.

Management traffic via SeGW
If the HeMS is located within the operator network, all management traffic will be sent through an IPsec tunnel that terminates in the SeGW at the border of the operator network. The most

common deployment will be that this tunnel is the same tunnel that also carries S1 signalling and user plane traffic (see the example in Figure 13.7). But a tunnel to a separate SeGW for management is also allowed by the specification. In all cases, the description of this tunnel and its establishment using mutual authentication follow the generic procedures given in section 13.4 for the secure connection to SeGW. This deployment scenario is described in section 13.5.1 on management architecture.

Management traffic between HeNB and HeMS accessible on public Internet

If the HeMS is accessible on the public Internet, then the IPsec tunnel to the SeGW cannot be used to secure the management traffic. Instead security mechanisms as given by [BBF TR-069] are used. It is required to establish a TLS connection between HeNB and HeMS based on mutual authentication using entity certificates. All procedures for this establishment are specified to be as close as possible to the NDS/AF specification [TS33.310] and the procedures for tunnel establishment to the SeGW described in section 13.4. We have also here the precondition of a successful device validation as described in section 13.4.1. Furthermore, all sensitive functions needed for the TLS handshake (e.g. cryptographic calculations using the private key) have to be executed within the TrE. The rules for certificate processing and validation are equivalent to the rules given for IKEv2 in section 13.4.4. In-band transport for certificate revocation status information may be optionally deployed for TLS according to [RFC4366] in a similar fashion as for IKEv2.

It should be noted that [TS33.320] in Release 9 does not specify a method to use HP authentication in conjunction with TLS tunnel establishment. Thus even if the operator requires the combined authentication as described in section 13.4.5 for connection to the SeGW, for access to an HeMS in the public Internet it is only possible to authenticate the HeNB device identity to the HeMS. As the deployment of HP authentication normally implies that also the HeMS should authenticate the HP, in this case only an HeMS accessible on the operator network via the SeGW should be deployed.

TLS certificate profiles

The TLS entity certificates are specified for HeNBs according to [TS33.310] and the additional profiling as given for IKEv2 in section 13.4.3.

In particular, the profile for the HeNB TLS certificate was chosen to allow the reuse of the X.509 certificate used for device authentication. The only extension is that the globally unique identity (FQDN) of the HeNB shall also be contained in the common name field of the certificate. The reason is that many HTTPS implementations use this field for entity name validation even if this violates the recommendation to use the subjectAltName field in [RFC2818]. This field is in addition to the subjectAltName field as needed for IKEv2 and does not prevent the usage of the same certificate in IKEv2.

The profile for the TLS CA certificate deviates from the specification in [TS33.310] only in the requirement that the issuer does not have to be an interconnection CA. This stems from the different deployment scenarios, where [TS33.310] assumes an inter-security-domain scenario, while the usage of TLS for HeNB management traffic is meant to connect two 'outlying' entities which are under control of the same operator.

TR-069 profiling

In the context of [TS33.320], a profiling of the security requirements given in [BBF TR-069] was necessary. The two main reasons were that some requirements referred to outdated security specifications (e.g. SSLv3), and that the usage of the HeNB as a radio device operating in licensed spectrum is under regulatory control and requires higher security than ordinary consumer devices. The details are given in the following.

- Owing to the increased security requirements of HeNBs as opposed to common consumer devices within customer premises, the use of TLS to transport management traffic is mandatory when the HeMS is accessible on the public Internet. This is more stringent than the optional usage of TLS in TR-069.
- The above requirement also rules out the ACS connection request carried over HTTP as specified in section 3.2.2 of TR-069 and the connection request via NAT gateway as specified in annex G of TR-069, when the HeMS would like to send this from the public Internet, as TR-069 forbids HTTPS for this request. This is not a severe restriction compared to TR-069, as the support of this feature in the HeMS is not mandated. In addition, if the ACS connection request is needed for a particular deployment, the network configuration with the HeMS accessible on the operator network can be deployed.
- SSL 3.0 [draft-freier-ssl-version3-02] and TLS 1.0 [RFC2246] must not be used as they are outdated.
- At least TLS 1.1 [RFC4346] must be supported and TLS 1.2 [RFC5246] should be supported. Ideally only TLS 1.2 would have been specified, as it contains the most up-to-date list of algorithms, but TLS 1.1 was allowed also, as implementations of TLS 1.2 are not yet widely deployed. If possible by any means, TLS 1.2 should be implemented. Even with TLS 1.1 allowed, the list of allowed and mandated cipher suites is taken from TLS 1.2. Thus TLS_RSA_WITH_AES_128_CBC_SHA is mandatory to be supported, and the support of RSA_WITH_RC4_128_SHA is not mandatory. Also the usage of RC4-based cipher suites is discouraged.
- As the design decision in 3GPP was to use PKI-based authentication for HeNB devices, and not to mandate any shared key infrastructure, shared-secret-based authentication between HeNB and HeMS is disallowed. Only certificate-based authentication is allowed. This is intentionally specified contrary to the current version of TR-069 Issue 1 Amendment 2 [BBF TR-069]. This profiling also means that the digest authentication for the ACS connection request carried over HTTP as specified in TR-069 and the password-based signature verification of the connection request via NAT gateway as specified in annex G of TR-069 is not used in the context of [TS33.320]. As mutual authentication is required, the above decision includes that TLS must support client-side authentication with certificates in addition to the server-side authentication commonly used with TLS.
- The HeNB may not have accurate absolute time when being powered up, as it may use the 'last-saved time' stored at the point in time of the last power-down (see section 13.4.8). On the other hand, the current time is needed for the validation of certificates during TLS tunnel establishment. The specification requires the use of such 'inaccurate local time' for certificate validation, which is (intentionally) contrary to section 3.3 of the current version of [BBF TR-069].

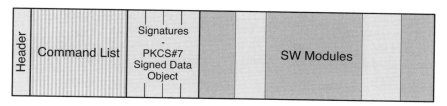

Figure 13.8   Signed package format

13.5.7   Software Download

The general security requirements for eNBs given in clause 5.3.2 of [TS33.401] require that SW download be integrity- and confidentiality-protected, and that the SW be authorized. Similar requirements specifically for HeNBs are given in clause 4.4.2 of [TS33.320]. In addition, the requirements on communication security given in clause 4.4.6 of [TS33.320] apply, which were described in the previous subsection.

The communications security with integrity and confidentiality protection is provided by the secure channel for file transfer described in the previous subsection. In addition, clause 8.4 of [TS33.320] gives measures for integrity protection and authorization of the downloaded SW package itself.

For integrity protection and authorization control the downloaded SW must be signed. The HeNB must verify the signature after download and install the downloaded SW on success only.

Annex E of [BBF TR-069] specifies a signed package format, which combines SW modules to be loaded and the related installation commands into one package, and adds one or more signatures to the package for proof of origin and integrity protection. Figure 13.8 shows the outline of this signed package.

The signature part contains one PKCS#7 signed data object according to the Cryptographic Message Syntax (CMS) specified in [RFC2315]. The only hash algorithm specified is SHA-1 [RFC3174]. The signature algorithm is specified to be RSA.

The PKCS#7 signed data object contains external signatures, which means that the signed data object does not contain the signed data itself. The signature is taken over the header and command list, while the SW modules are not included. To still get an integrity protection of the complete package, certain commands acting on the SW modules, such as extract and add commands, contain the hash value of the SW module they act upon. This indirectly protects all SW modules that are handled by commands in the command list, and requires that on execution of a particular command the hash of the related SW module has to be verified.

The standard allows multiple signatures so that the data can be signed by different parties. It is specified that one valid signature is sufficient to validate the package, even if multiple signatures are contained in the signed data object. This allows basing different signatures on different root certificates, for example one root certificate of the manufacturer and one of the operator. One operator may decide to leave the manufacturer root certificate in the HeNB, and distribute the new SW versions as signed by the manufacturer. Another operator may prefer to validate the SW against their own root certificate. Then such an operator first checks the manufacturer signature and then signs the package a second time with their own signing authority based on their own root certificate. This would even allow the operator to change the SW and any parameters loaded with the SW specifically for their own usage.

For usage of the signed package format in the context of HeNB deployment, the following requirements in addition to TR-069 are given in clause 8.4 of [TS33.320].

- *Root for signatures*. At least one of the signatures in the signature part of the signed package must be from a signing authority with a certificate issued by an operator-trusted CA. This trust is automatically given if an operator root certificate is used. With, for example, a manufacturer root certificate the operator must trust the manufacturer and their root CA. But such a trust relation would not really extend the trust relation of the operator to the manufacturer, as the operator has to trust the manufacturer's root CA anyway for device authentication of the HeNB.
- *Validation in the Trusted Environment*. The TrE of the HeNB must perform the signature and certificate validation based on an operator-trusted root certificate, which shall be securely stored within the TrE.
- *SW reference values in signed package*. The reference values used by the TrE to secure the boot procedure will also be contained in the signed package. The usage of these values is described in section 13.3 on security procedures internal to the home base station.

13.5.8   Location Verification

A prerequisite for the operation of a HeNB is the provisioning of the HeNB with correct operational parameters, such as frequency range and allowed transmit power levels. Some of these parameters are necessary to ensure correct operation (e.g. to avoid interference with macro-cells or other HeNBs), while others may depend on regulatory requirements (e.g. restriction of an operator coverage to certain areas or countries). As these parameters depend on the geo-location of the HeNB and the macro-cell coverage, the operators require assurance of the HeNB location.

As the customer may move the HeNB and may connect it to the Internet (and thus to the SeGW or HeMS) at any location he wants to, the operator cannot rely solely on administrative data for location determination, such as the intended location as given in a HeNB hosting party contract. Thus an online location verification has to take place, both to supply the HeNB with the correct intended configuration parameters, and to allow the radio transmission to be turned on.

Clause 8.1 of [TS33.320] requires that the location verification be performed within the HeMS, and introduces the term 'verifying node' for this function. The S1 signalling protocol does not contain any location information data elements, and only the TR-069 management protocol is able to transfer location-related data. Thus only the HeMS is able to verify the geo-location of the HeNB. Therefore the following text refers to the HeMS as verifying node.

[TS33.320] specifies four methods for determining the HeNB location, which are described in the following. All of them rely on messages from the HeNB itself, and not on measurements taken by other elements.

Public IP address

This method uses the public (Internet) IP address of the HeNB to determine the location of the HeNB based on the geographical assignment of IP addresses. The HeNB will determine its IP address and send it as location information to the verifying node. This is no problem if the HeNB is directly connected to the Internet, but normally the HeNB may have one or even more NAT devices within the connection to the Internet, and thus only knows the private address assigned to it. The specification [TS33.320] mentions the public IP address of the broadband access device, which may be the IP address of the HeNB itself if the HeNB is integrated with the broadband access device. But even the IP address of the broadband access device may not be the public IP address, if there is another NAT device at the border of the broadband access provider network.

Based on the above considerations, this method has the following restrictions.

- It has the precondition that the HeNB can determine its public IP address if located behind a NAT, for example by using a STUN server [RFC3489] reporting the public IP address back to the HeNB.
- The public IP address may allow only a very rough estimate of the actual location. For example, the public IP address may be assigned to an Internet service provider with country-wide coverage.
- The hosting party may actively try to simulate a location by, for instance, proxying the connection from a remotely connected HeNB via his own home based network (i.e. the network at the customer premises) to the operator network.

IP address or location identifier provided by broadband access provider

In some access networks the broadband access provider may provide the location of a certain subscriber to other entities. Annex B of [TS33.320] gives the example of an access network according to the TISPAN Network Attachment Sub-System (NASS) specifications [ETSI ES 282 004]. For this network architecture a Connectivity Session Location and Repository Function (CLF) is defined, which can be queried using the e2 interface [ETSI ES 283 035] for access line identifier and geo-location of a subscriber based on the IP address used within the NASS. The verifying node may then verify the data in the e2 response against the contracted data, i.e. the line identifier and/or the geo-location of the HP.

This method has the following prerequisites and limitations.

- The HeNB can determine its IP address within the access network. If the HeNB is connected via a NAT device to the access network, the HeNB may need similar mechanisms for the determination of the IP address as for the method with public IP address described above.
- The access network must be able to provide location information to other entities.
- The mobile operator has a contract with the broadband access provider, if the access network is not under control of the mobile operator themselves.
- The mobile operator must have online access to some repository in the access network, such as to the CLF via e2 interface in case of NASS.
- If the hosting party provides the HeNB with a faked IP address, perhaps by spoofing STUN responses, the HeNB may report this faked address.

- The proxying attack described above for the method with public IP address may be applied also in this case.

Measurement of surrounding macro-cells

The HeNB measures the coverage of surrounding macro-cells and sends this information to the verifying node. Based on the knowledge of the macro-cell locations, the verifying node determines the location of the HeNB.

This method has quite a high probability to yield correct location information, as it may be hard for an attacker to simulate an environment of certain macro-cells as expected at the contracted location. The main disadvantage of this method is that many HeNBs will be deployed in places where no macro-cell coverage is given, for example within buildings or in remote areas.

Geo-coordinates from a GNSS

Global Navigation Satellite Systems (GNSS) may be used to determine the geo-coordinates of the HeNB location. By, for example, using an embedded Global Positioning System (GPS) receiver, the HeNB is able to measure its geographical longitude and latitude. This is sent to the verifying node for comparison with the contracted location.

Theoretically this method yields exact location data with an accuracy much better than needed by the verifying node. But still this method has limitations.

- Reception of GNSS signals is often disturbed or impossible within buildings or in underground locations.
- Commercially available GNSS test kits which are available at reasonable prices may simulate any geo-location by overriding the signal received from the satellites.
- Even if GNSS reception is possible, it may take many minutes until the GNSS receiver is synchronized to the satellites and provides a result. This may be too long a waiting time for the customer after power-up.

Location measurement without cooperation of the HeNB

The following two external methods were discussed, where the HeNB is not actively involved in the measurement.

- Measurement of the location of the HeNB by adjacent base stations would work only if the HeNB is near enough to a macro-cell base station to be receivable. In addition, such measurement would be possible only after the HeNB starts radiating, while the specification requires that the location may be determined before the HeNB may start radiating.
- The public IP address of the HeNB is determined from the source IP address of the management connection by an operator network element, such as the SeGW or the HeMS. This method fails for any connection via the SeGW, as the SeGW has no interface to the HeMS to report on identity and source IP address of the HeNB, and the HeMS only knows the inner

IP address, by which the HeNB communicates with the HeMS through the IPsec tunnel. In addition this method has the same pitfalls as the method with public IP address reported by the HeNB described above as first method.

From the reasons given above, neither method was seen as suitable to be included into the specification.

Requirements on location verification

Different deployment scenarios and HeNB configurations will influence the availability, accuracy and reliability of these types of location information. Thus no single method was specified as mandatory, but at least one of the methods must be deployed. The selection of the method is left to operator policy. In addition, the following requirements on location verification are given in clause 8.1.6 of [TS33.320]. Many of them stem from the fact that none of the methods is really reliable, and that the selection and combination of the methods will depend a lot on actual location and use case of the HeNB and on the contract with the hosting party.

- The verifying node must be able to request one or more of the four types of location information listed above from the HeNB. This allows the operator to selectively define the method for the particular HeNB. In addition the HeNB may provide such information automatically.
- The verifying node must be configurable with policies for location verification, including type and frequency of the requests.
- The verifying node must be able to use ancillary information such as geo-coordinates of surrounding macro-cells, postal address of HeNB as claimed by the hosting party, IP address location information, and so on.
- Location verification by the verifying node must be possible both before and after switching on the HeNB radio.
- Possible actions of the verifying node shall be to raise an alarm, to permit the HeNB to radiate or to prevent the HeNB from radiating.
- A HeNB must be configurable for how it reacts when ordered to cease radiating. Either it stops radiating immediately, or it waits until any calls in progress have been completed. In no circumstances are new calls allowed to be established after such an order.

13.6 Closed Subscriber Groups and Emergency Call Handling

HeNBs may operate in one of the three access modes: closed, hybrid or open (see section 13.1.1). A HeNB operating in the closed or hybrid access mode broadcasts on the radio interface a specific Closed Subscriber Group Identification (CSG-ID) assigned to the Hosting Party (HP). Only members of this specific CSG may camp on a HeNB operating in closed mode. In hybrid mode, UEs who are not members of the broadcasted CSG are allowed to camp on the HeNB, but the members of the particular CSG may have privileges regarding QoS, the number of allowed connections, and so on.

This section gives an overview of the security-related features of CSG handling, and then covers the complications for emergency call handling caused by HeNBs operating in closed mode.

13.6.1   UE Access Control to HeNBs

Many HeNBs may broadcast the same CSG-ID, but for each HeNB only one CSG-ID is possible. [TS22.220] specifies the general service requirements applicable to UE access control by means of CSGs. The detailed requirements on CSGs are given in clause 4.3.1 of [TR23.830], while the related normative consequences are included in the EPS specifications where applicable. The membership of a specific CSG is managed by the HP of the HeNB together with the mobile network operator. This happens under ultimate control of the operator, as the membership status needs to be reflected in various core network nodes and also updated on the UICC of the subscriber.

Any EPS-capable UE is aware of the feature of CSGs. This means that any UE compliant to Release 8 and higher at least understands the indication of the HeNB that it is operated in closed mode. If the UE is capable of CSG handling, it will interpret the CSG-ID and may try to camp on this HeNB if allowed to. A UE not capable of CSG handling will not at all try to camp on a HeNB allowing access to CSG members only, except for emergency calls (see section 13.6.2).

The lists of CSG-IDs allowed for a certain subscriber are held within UE and HLR/HSS as part of the subscription profile. The ME or the USIM within UE holds a list of CSG-IDs and human-readable CSG names where the UE may camp on. If the UE receives one of the allowed CSG-IDs, the UE may try to camp on the related HeNB. The enforcement of access control to a HeNB is performed in the MME during UE attach (clause 5.3.2 of [TS23.401]).

This handling of CSGs for HeNBs differs from the handling for HNBs, as in 3G non-CSG aware UEs may also camp on a HNB. This case requires different procedures which are not necessary in EPS.

With respect to CSG access control, the H(e)NB security specification [TS33.320] only quotes the other specifications that the MME shall perform the CSG access control and in addition references the stage 3 specification [TS24.301] which defines the NAS protocol.

13.6.2   Emergency Calls

For home base stations in general, the same requirements and procedures exist for emergency access to the EPS as for macro base stations given in [TS33.401]. These are described in section 8.6.

Contrary to macro-cells, HeNBs may operate in closed mode, so allowing only members of a CSG to camp on this HeNB. Still UEs not belonging to the particular CSG of the HeNB must be able to make emergency calls or IMS emergency sessions. Any 'normal' call of such UE will be blocked by the network.

14

Future Challenges

So far in this book we have described LTE security as it has been defined by 3GPP up to March 2010, for 3GPP Releases 8 and 9. In this chapter we present our views on likely future developments in LTE security and beyond. Section 14.1 describes activities already under discussion in 3GPP standardization that may bear fruit in the near term, in 3GPP Releases 10 or 11. Section 14.2 covers studies and research activities that may have an impact on the security of LTE and potential successor systems in the longer run.

14.1 Near-term Outlook

At the time of writing this book, 3GPP is working on enhancements for LTE, especially regarding the radio aspects. This activity is known by the name 'LTE-Advanced' (LTE-A). The key features of LTE-A include, in particular, enhanced peak data rates to support advanced services and applications (100 Mbps for high mobility and 1 Gbps for low mobility). More information on LTE-A can be found from the 3GPP website [3GPP]. Probably the most important feature in LTE-A that will have an impact on LTE security is the work on relay node architectures. This work aims to introduce a relay node between the user equipment and the base station. Apart from LTE-A, other important developments with potential security impact lie in the areas of interworking of 3GPP networks and fixed broadband networks, voice over LTE, machine-type communication, and home base stations. We will address all these developments in turn. We would like to caution, though, that all this is work in progress and subject to change.

Security for relay node architectures

Relaying is seen by 3GPP [TR36.912] as:

> ... a tool to improve e.g. the coverage of high data rates, group mobility, temporary network deployment, the cell-edge throughput and/or to provide coverage in new areas. The relay node (RN) is wirelessly connected to a Donor cell of a Donor eNB via the Un interface, and UEs connect to the RN via the Uu interface.[1]

---

[1] Extract reproduced with permission from © 2010, 3GPP™.

LTE Security   Dan Forsberg, Günther Horn, Wolf-Dietrich Moeller, and Valtteri Niemi
© 2010 John Wiley & Sons, Ltd

Figure 14.1   Relay nodes in LTE

Figure 14.1, which has been adapted from Figure 9.1-1 in [TR36.912], shows the position of the new functional element, the relay node, in relation to the User Equipment (UE), the Donor base station (eNB) and the Evolved Packet Core (EPC) already extensively dealt with in the preceding chapters of this book.

3GPP has completed a study on 'Relay architectures for E-UTRA' and selected one architecture as the basis for future work [TS36.806]. One goal of this work is to maximize the commonality between the Uu interface and the Un interface. The following properties of the selected architecture, which exhibit the dual role played by an RN, are worth noting here as they may be security-relevant.

Relay node in the role of a base station
An RN appears to a UE as an eNB as defined in 3GPP Release 8. The UE does not experience any differences between an architecture with relay nodes and one without relay nodes. When the UE attaches to the network the UE communicates with an MME serving the UE; the S1 signalling traffic between the RN and the MME serving the UE is carried over the Un interface via the Donor eNB. It is particularly worth noting that the S1 signalling traffic is carried over Un in a Data Radio Bearer (DRB) as, from a radio interface point of view, it is user data.

This creates the following security problem. It is clear that S1 signalling traffic needs to be integrity-protected as explained in section 8.4. But, for the Uu interface in EPS according to 3GPP Release 8, user data – and hence the traffic carried in DRBs – is only confidentiality-protected, not integrity protected. This means that the protection mechanisms available on the Uu interface cannot be copied one-to-one to the Un interface as then the S1 signalling traffic would not be integrity-protected. One option would be changing the protection mechanisms at the PDCP layer on the Un interface with respect to the Uu interface so as to also provide integrity protection for (at least some) DRBs; but this would go against the goal stated above of maximizing the commonality between the two interfaces. Another option would be protecting the IP traffic sent over the Un interface at the IP layer, by means of IPsec. Remember that IPsec is also used to protect S1 signalling traffic in the Release 8 architecture. If IPsec was adopted it would then have to be decided further how IP layer protection would work together with PDCP layer protection.

Relay node in the role of a user equipment
When the RN is started up, a connection over the Un interface is to be established in a way similar to how a connection over the Uu interface between the UE and the base station is set up in Release 8. This means that the RN acts in the role of a user equipment, and it implies that the RN contains a USIM (or at least equivalent functionality). In the RN start-up procedure, the RN communicates with an MME serving the RN (that may be different from the MME

serving the UE); in this case, the S1 signalling traffic is exchanged between the Donor eNB and the MME serving the RN, which means it is not carried over the Un interface.

For the RN start-up procedure, the protection mechanisms afforded by the Uu interface seem adequate for the Un interface as well, but the role of the USIM may require additional considerations as the assumptions on the environment, in which a USIM is used, are quite different from those on the use of a USIM in a user equipment. After all, a USIM in an RN is much more easily accessible by unauthorized persons than a USIM in a UE controlled by a human user; it could, hence, possibly be removed by an attacker and inserted into another RN, or even a UE. Potential related threats and countermeasures therefore need careful study.

Platform security

The degree of exposure of a relay node to physical attacks is likely to be somewhere in between that of a home base station and a macro base station. This suggests that the degree of platform security that the relay node needs to provide could also lie somewhere in between that of the other two types of base station, but details are for further study.

Multi-hop relay node architectures

A multi-hop relay node architecture is one where traffic between a UE and a Donor eNB is forwarded across several relay nodes in a multi-hop fashion. Such architectures currently seem not to be high on 3GPP's priority list, but they would certainly pose challenging security problems, notably in the area of key management.

Security for interworking of 3GPP networks and fixed broadband networks

In February 2010, a workshop on Fixed Mobile Convergence (FMC) was held that was jointly organized by 3GPP and the Broadband Forum [BBF]. The workshop discussed how the interworking between mobile networks, defined by 3GPP, and fixed broadband networks, defined by the BBF, could be improved. The presentations to this workshop, some of which also addressed security aspects, can be found under [3GPP and BBF 2010]. 3GPP and the BBF agreed to work in their respective organizations to address various aspects. These include basic connectivity, host-based mobility and network-based mobility for untrusted accesses, network discovery and selection functions, IP address allocation, authentication, policy, and quality of service (QoS). This work is assumed to be performed on top of the Release 9 baseline architecture that is specified in [TS23.402]. The corresponding security architecture is specified in [TS33.402] and is described in section 11.2 of this book. Work will start with applying the procedures for untrusted access (section 11.2.3) to fixed broadband access networks, while trusted access (section 11.2.2) to fixed broadband access networks will be considered at a later phase.

Security aspects discussed at the workshop included the observation that, in fixed broadband networks, typically only the access line is authenticated while, in 3GPP networks, it is the user who is authenticated – a fact that may create problems for a unified service delivery. Another presentation to the workshop raised the question whether 3GPP credentials (i.e. USIMs) could be also used for authentication in fixed broadband networks.

The work – especially the security work – is in its infancy at the time of completing this book, and the potential impact on 3GPP and BBF security specifications is therefore difficult to assess.

Security for voice over LTE

Chapter 12 describes three methods for providing voice services over LTE: IMS, CSFB and SRVCC. Improvements with at least some impact on security are considered in all three areas.

IMS over LTE

For SIP signalling security in IMS, no developments affecting IMS over LTE are discernible. For IMS media security, it seems likely that support for non-real-time media (e.g. messaging) will be added to the specifications. There are also discussions on security enhancements for conferencing and deferred delivery, such as when using voice mail boxes.

Circuit Switched Fallback (CSFB)

From a security perspective, only minor adjustments are discernible aiming to improve performance by measures such as reducing the frequency of authentication and TMSI reallocation.

Single Radio Voice Call Continuity (SRVCC)

In section 12.1.3, the types of handovers supported by SRVCC in 3GPP Release 9 are listed. They include handovers from LTE or HSPA to UTRAN or GERAN. At the time of completing this book, a study is under way in 3GPP on SRVCC support for handovers in the reverse direction, namely from UTRAN or GERAN to LTE or HSPA. If 3GPP decides to go ahead with this SRVCC enhancement, then security work analogous to that described in section 12.2.3 will be required; that is, the corresponding security context mappings will have to be defined.

Security for machine-type communication

Machine-Type Communication (MTC) is characterized by the fact that terminals are not attended by humans. It has attracted tremendous interest recently owing to the significant potential for growth inherent in the envisaged applications that include smart metering, fleet management, surveillance, remote control, patient monitoring in healthcare, and many others. It is expected that many of these MTC applications could greatly benefit from using cellular communication rather than, for example, fixed line communication.

On the other hand, the tremendous success of GSM and its successor systems has been almost exclusively based on mobile communication between humans. It is therefore understandable that the current cellular networks are optimized for human-to-human communication and less so for MTC. Consequently, 3GPP has started work on network improvements for MTC in order to prepare the ground for realizing the full growth potential of MTC over 3GPP networks. This work is intended to cover all 3GPP networks including LTE.

As security procedures represent an important part of the overall effort required in setting up a connection between a terminal and the network, it is quite natural to also study the possibility of optimizing security procedures for MTC. At the time of writing, the security work has, however, barely started, and it is therefore too early to predict which security features may be affected.

Security for home base stations

With the finalization of Release 9 work on Home eNodeBs (HeNBs), also all security features for the deployment of a small 'femto' base station in the home environment were provided. This solution technically allows any EPS-capable user equipment to camp on a HeNB, and then to communicate with the 'rest of the world' via the operator network.

On an architectural level, extensions of this basic architecture are discussed, not only to enhance the benefits for the user operating a HeNB within his home, but also to widen the usage scenarios for HeNBs into areas outside the deployment of single HeNBs for private homes or small enterprises. In addition, extensions planned for macro eNBs may also be applicable to HeNBs, but with the need for adaptation to this particular environment.

Many of these future features are discussed in the Femto Forum [FF], a non-profit organization of stakeholders in the 'femto cell' area – see the introduction to Chapter 13.

There have been no decisions on any of the possible extensions to the HeNB architecture described in the following paragraphs up to now, and the necessary new security features or security enhancements have not yet been discussed in standardization either. Consequently, the mentioned possible influences on security may change during future standardization work.

Enterprise femto

While the standardization in Release 9 allowed the sharing of one CSG by multiple HeNBs, the main deployment scenario seen was the home environment with one HeNB per CSG only. Usage of multiple HeNBs is seen as an important use case for small and medium enterprises. In this case all HeNBs belong to the same CSG and are managed by the enterprise as hosting party. Different services are envisaged ranging from using LIPA – see next paragraph – for access to the enterprise IP network to the integration of enterprise PBX functionality with the connectivity of the UEs being allowed to camp on the enterprise HeNBs. These features are foreseen for Releases 10 or 11.

Security aspects will relate, for example, to the access of UEs camping on the enterprise HeNBs to the enterprise network. Preliminary assessments showed that a future solution should encompass both (a) allowing access to the enterprise network based on CSG membership only, and (b) enforcing a separate access control by the enterprise. The integration of PBX and public telephony functions may raise security issues as well, depending on the solution(s) specified in future releases.

Local IP Access (LIPA) and Selective IP Traffic Offload (SIPTO)

LIPA is defined in [TS22.220] as follows:

> H(e)NB Local IP Access to the home based network provides access for a directly connected (i.e. using H(e)NB radio access) IP capable UE to other IP capable devices in the home.[2]

---

[2] Extract reproduced with permission from © 2010, 3GPP™.

Such IP traffic is diverted from the HeNB directly to the home network of the hosting party, without being tunnelled through the backhaul link. It is envisaged that all members of the HeNB Closed Subscriber Group (CSG) shall have access to the home network.

Apart from LIPA being already specified in stage 1 service requirements, all architectural specifications are expected for Releases 10 and 11. A possible security issue may be that the access control to the home network is coupled to the CSG membership. As the enforcement of CSG-based access control is performed by the operator, the hosting party would not have sole control over its home based network (i.e. the network in the customer premises). Another security issue may arise from the fact that traffic from a UE to the Internet is carried through the home based network in cleartext, so the hosting party of a HeNB operating in open or hybrid mode can eavesdrop on user data of any UE camping on this HeNB.

The Work Item Description in [3GPP 2009] states on SIPTO:

> Due to the fact that 3GPP radio access technologies enable data transfer at higher data rates, the 3GPP operator community shows strong interest to offload selected IP traffic not only for the Home (e)NodeB Subsystem but also for the macro layer network, i.e. offload selected IP traffic from the cellular infrastructure and save transmission costs.

It is too early to say at the time of writing what the security impact would be.

New cryptographic algorithms

It was explained in Chapter 10 that Release 8 includes two pairs of cryptographic algorithms for encryption and integrity protection: SNOW 3G-based UEA1 and UIA1, and AES-based UEA2 and UIA2. In addition to export control regulations [Wassenaar], there exist also import control requirements, especially in China. Mainly for this purpose, design work has been started for a third algorithm pair, UEA3 and UIA3. This work item is carried out by ETSI SAGE, together with Chinese cryptography experts.

14.2   Far-term Outlook

In the very beginning of this book it was explained that a new kind of cellular system and associated radio interface have been created approximately once every ten years. Based on this pattern it would be tempting to predict that another major redesign would happen around the year 2020. If we assume that 3GPP creates a new release of specifications every 18 months, then that would imply that Release 16 might, once again, contain specifications for a completely new system. In the previous section we listed many enhancements of LTE and EPS that are expected to happen in Releases 10 and 11. Some of these enhancements may well be still under specification even after Release 11; but even if that is the case there is still a gap of at least a couple of releases before the new revolutionary release would appear.

Since this book is about security, it is not worthwhile speculating what the exact extension features are that are going to appear after Release 11. What is certain is that most of them will need security of some sort. The 3GPP security features are usually specified in such a manner that they are future-proof at least to some extent, so there is a good chance they could be applied somewhat more widely than just to those particular features they are originally intended for. On the other hand, new features in mobile systems are typically intended to enable some new use cases. Then it tends to be so that, together with new use cases, new ways to misuse the system appear as well. Therefore, it is a safe bet to predict that each new release will also involve extensions to security specifications.

Key lengths of cryptographic algorithms are an area where speculative predictions and educated guesses are common. As explained in earlier chapters, EPS has been prepared for introducing 256-bit keys in all security mechanisms. According to some estimates [Smart 2009; Barker *et al.* 2007], the generic cryptographic strength provided by 128-bit keys will be adequate until around the year 2030. If we assume that LTE is going to be in use as long as GSM (i.e. definitely more than 20 years), this extension capability will be needed at some point but probably not very soon.

Cryptographic algorithms themselves constitute another area where advances may be needed. Algorithms sometimes get broken and it is relatively easy to introduce new algorithms into the EPS system. Hence, it is likely that, during the lifetime of LTE, new algorithms will be introduced even before the longer keys are needed. One constant source of speculation around cryptography is the potential effect of quantum computing. For secret key cryptography the effect of quantum computing would not be as drastic as for some of the most popular public-key algorithms. It has been estimated [Smart 2009] that 256-bit keys would provide protection also against attacks by quantum computing into the 'foreseeable future'.

Privacy has been a rising trend for several years, emphasized by huge amounts of data that is cumulatively collected in the Internet. Lots of this data is about ordinary people; a big part is even contributed by the people themselves via social networks and user-generated content. Mobile systems necessarily need to have lots of data about their users; the systems cannot operate unless whereabouts of the terminals are known. A certain amount of user-related data is also logged because of lawful interception. Mobile systems constitute a good platform for location-based services and other context-aware services. For these reasons, it is probable that some mechanisms to enhance protection of user data and other personally identifiable information would be introduced into mobile systems and these may have an effect also on LTE and EPS.

One area of privacy that has been discussed already several times in this book is the feature of identity confidentiality. As explained earlier, the current protection mechanism by temporary identities is vulnerable to active attacks. Protection against these would probably require introduction of public key technology into the access security. The cost of such mechanisms has so far prohibited their introduction but it is conceivable that during the lifetime of LTE the situation may change, partly due to new privacy requirements and partly due to increased processing power that makes it faster to carry out complex public-key operations.

Another factor on location and identity privacy is the fact that modern terminals support many different radio technologies, most of which are not defined by 3GPP. This implies that protection mechanisms that are applied to only a subset of these technologies have only a limited effect on identity privacy; users may still be tracked based on those technologies that do not have a good protection. Issues like this emphasize the need for further work on interworking with non-3GPP networks.

Let us now take another look at the prediction above that a new major system redesign will appear at some point in the future. It could be argued that the creation of a new radio interface does not necessarily imply that the whole system needs to be changed. The support of many different access technologies is already a core feature of EPS. Therefore, it may happen that a new radio interface is created, either in 3GPP or somewhere else, and EPC is simply adapted to support that technology as well.

Another line of study is cognitive radio. The leading idea in it is to optimize the use of radio frequencies and technologies dynamically and locally. The terminal senses its radio surroundings and uses the radio technology that is most suitable for both the environment and the current communication task. From a security point of view this raises some new challenges. Although all possible radio technologies have their own protection mechanisms, combining them in this dynamic manner is not a trivial task.

Convergence of Internet technologies and mobile communication technologies drives to a direction where a full-blown redesign of the cellular system is not anymore needed, at least not independently of the Internet. The future Internet would certainly contain mobility built in as a core property. One consequence could be that differences between the roles of mobile network operators and Internet service providers become more blurred. There are also potential efficiency gains around Cloud Computing; many tasks on the network side could be carried out wherever it is optimal to do so. Therefore, the functional split inside the network would become much more dynamic than is the case today. This kind of evolution provides also challenges to security, since more and more legacy security features have to be supported in a single system simultaneously. We have also a more heterogeneous set of terminals in the system, provisioned with many different kinds of credentials.

Some of the large-scale security issues with the Internet, such as distributed DoS attacks, botnets and spam, stem from the fact that sending data is easy and cheap while it is more costly to process the data on the receiving end. One possible architectural solution to the problem of unwanted traffic that plagues the Internet is to align more to the 'publish-and-subscribe' paradigm instead of the 'send-and-receive' paradigm. It is probably not an overstatement to claim that security and privacy issues will have a major impact on the shape of the future Internet.

We will certainly have an extremely heterogeneous terminal base when the vision of practically everything being connected to everything via the Internet comes about [ITU 2005]. This kind of system obviously provides lots of possibilities for attacks also. New security and privacy mechanisms will certainly be needed on the way.

Still another avenue stems from the fact that many communication needs are local and there is no need to be connected to the other side of the world. The wide-area coverage and connectivity to the Internet could be complemented by various ad-hoc types of network, such as for car-to-car communication or for communication needs at mass events. There are lots of security challenges in such settings; one possibility to solve them is to use the security features provided by wide-area mobile systems and extend them to the ad-hoc networks, but this may not be feasible in many cases owing to the lack of a centralized infrastructure.

We have listed several different avenues that the evolution of EPS networks could take. What is common to all these directions is that the concepts of security, trust and privacy have major roles to play. To be able to continue the success stories of mobile networks and communications, continuous evolution of the security concepts is a necessary requirement. Properties like flexibility, agility and usability provide key ingredients on the way towards this goal.

Abbreviations

3G	3rd Generation, often also used for the 3rd Generation System defined by 3GPP also known as UMTS
3GPP	3rd Generation Partnership Project
3GPP2	3rd Generation Partnership Project 2
A3	GSM authentication algorithm
A5	GSM encryption algorithm
A8	GSM key generation algorithm
AAA	Authentication, Authorization and Accounting
ACS	Auto-Configuration Server
AES	Advanced Encryption Standard
AK	Anonymity Key
AKA	Authentication and Key Agreement
AMF	Authentication and key Management Field
AMPS	Advanced Mobile Phone System
AN	Access Network
APN	Access Point Name
ARIB	Association of Radio Industries and Businesses
AS	Access Stratum
AS	Application Server
ASME	Access Security Management Entity
ATIS	Alliance for Telecommunications Industry Solutions
AuC	Authentication Centre
AUTN	Authentication Token
AV	Authentication Vector
BBF	Broadband Forum
BS	Base Station
BSC	Base Station Controller
BSF	Bootstrapping Server Function
BSS	Base Station Subsystem
BTS	Base Transceiver Station
CA	Certification Authority
CBC	Cipher-Block Chaining

CCSA	China Communications Standards Association
CFN	Connection Frame Number
CK	Ciphering Key in 3G
CKSN	GPRS CK Sequence Number
CLF	Connectivity session Location and repository Function
CM	Communication Management
CMAC	Cipher-based MAC
CMP	Certificate Management Protocol
CMS	Cryptographic Message Syntax
CPE	Customer Premises Equipment
CRL	Certificate Revocation List
CRMF	Certificate Request Message Format
C-RNTI	Cell Radio Network Temporary Identity
CS	Circuit Switched
CSCF	Call Session Control Function
CSFB	Circuit Switched Fallback
CSG	Closed Subscriber Group
CSG-ID	CSG Identification
CT	Core network and Terminals
DES	Data Encryption Standard
DHCP	Dynamic Host Configuration Protocol
DNS	Domain Name System
DoS	Denial of Service
DRB	Data Radio Bearer
DSL	Digital Subscriber Line
DSMIPv6	Dual Stack Mobile IPv6
DTLS	Datagram TLS
EAP	Extensible Authentication Protocol
EAPOL	EAP over Local Area Network
EARFCN-DL	E-UTRA Absolute Radio Frequency Channel Number-Down Link
ECB	Electronic Code Book
E-CSCF	Emergency CSCF
EDGE	Enhanced Data rates for GSM Evolution
EEA	EPS Encryption Algorithm
EIA	EPS Integrity Algorithm
eKSI	Key Set Identifier in EPS
EMSK	Extended Master Session Key
eNB	evolved NodeB
EPC	Evolved Packet Core
ePDG	evolved Packet Data Gateway
EPS	Evolved Packet System
ESP	Encapsulating Security Payload
ETSI	European Telecommunications Standards Institute
E-UTRAN	Evolved UTRAN
FA	Foreign Agent
FDMA	Frequency Division Multiple Access

FIPS	Federal Information Processing Standard
FQDN	Fully Qualified Domain Name
FSM	Finite State Machine
FTP	File Transfer Protocol
FTPS	FTP over TLS
GBA	Generic Bootstrapping Architecture
GEA	GPRS Encryption Algorithm
GERAN	GSM/Edge Radio Access Network
GGSN	Gateway GPRS Support Node
GIBA	GPRS-IMS-Bundled Authentication
GMSC	Gateway MSC
GNSS	Global Navigation Satellite System
GPRS	General Packet Radio Service
GPS	Global Positioning System
GSM	Global System for Mobile communications
GSMA	GSM Association
GTP	GPRS Tunnelling Protocol
GUMMEI	Globally Unique MMEI
GUTI	Globally Unique Temporary UE Identity
GW	Gateway
H(e)NB	HNB and/or HeNB
HA	Home Agent
HE	Home Environment
HeMS	HeNB Management System
HeNB	Home eNodeB
HeNB-GW	HeNB Gateway
HFN	Hyper Frame Number
HLR	Home Location Register
HMAC	Keyed-Hashing for Message Authentication (also Keyed-Hash Message Authentication Code)
HNB	Home NodeB
HP	Hosting Party
HPM	HP Module
HRPD	High Rate Packet Data
HS-GW	HRPD Serving Gateway
HSPA	High Speed Packet Access
HSS	Home Subscriber Server
HTTP	Hypertext Transfer Protocol
HTTPS	HTTP over TLS
ICS	IMS Centralized Services
ID	Identity
IDi	Identification - Initiator
IDr	Identification - Responder
IE	Information Element
IEC	International Electrotechnical Commission
IEEE	Institute of Electrical and Electronics Engineers

IETF	Internet Engineering Task Force
IK	Integrity Key in 3G
IKE	Internet Key Exchange
IMEI	International Mobile Equipment Identity
IMEISV	IMEI and Software Version Number
IMPI	IP Multimedia Private Identity
IMS	IP Multimedia Subsystem
IMS-ALG	IMS Application Level Gateway
IMSI	International Mobile Subscriber Identity
IMT	International Mobile Telecommunications
IP	Internet Protocol
IPsec	Security architecture for IP
ISIM	IP multimedia Services Identity Module
ISO	International Organization for Standardization
ISR	Idle state Signalling Reduction
ITU	International Telecommunication Union
K_{ASME}	Local Master Key in EPS
k_c	Ciphering Key in GSM
KD	Key Distributor
KDF	Key Derivation Function
K_{eNB}	Intermediate Key at eNB Level
KSI	Key Set Identifier used in 3G
LAI	Location Area Identity
LAN	Local Area Network
LFSR	Linear Feedback Shift Register
LI	Lawful Interception
LIPA	Local IP Access
LLC	Logical Link Control
LMA	Local Mobility Anchor
LTE	Long Term Evolution
LTE-A	LTE-Advanced
MAC	Medium Access Control
MAC	Message Authentication Code
MAC-I	Message Authentication Code for Integrity
MAG	Mobile Access Gateway
MAP	Mobile Application Part
MCC	Mobile Country Code
MD5	Message-Digest algorithm 5
ME	Mobile Equipment
MIPv4	Mobile IPv4
MM	Mobility Management
MME	MM Entity
MMEI	MME Identifier
MN	Mobile Node
MNC	Mobile Network Code
MS	Mobile Station

MSIN	Mobile Subscriber Identification Number
MSC	Mobile Switching Centre
MSK	Master Session Key
MT	Mobile Terminal
MTC	Machine-Type Communication
NAI	Network Access Identifier
NAPT	Network Address Port Translation
NAS	Non-Access Stratum
NAS-MAC	MAC for NAS for integrity
NASS	Network Attachment Sub-System
NAT	Network Address Translation
NBA	NASS-IMS-Bundled Authentication
NCC	Next hop Chaining Counter
NDS	Network Domain Security
NDS/AF	Network Domain Security/Authentication Framework
NDS/IP	Network Domain Security/IP network layer security
NE	Network Element
NH	Next Hop parameter in E-UTRAN
NIST	National Institute of Standards and Technology
NMT	Nordic Mobile Telephone
NTP	Network Time Protocol
O&M	Operations and Management
OCSP	Online Certificate Status Protocol
OFDMA	Orthogonal Frequency Division Multiple Access
OMA	Open Mobile Alliance
PBX	Private Branch Exchange
PCI	Physical Cell Id
P-CSCF	Proxy CSCF
PDC	Personal Digital Cellular
PDCP	Packet Data Convergence Protocol
PDG	Packet Data Gateway
PDN	Packet Data Network
PDN GW	PDN Gateway
PDU	Protocol Data Unit
PIN	Personal Identification Number
PKCS	Public Key Cryptography Standards
PKI	Public Key Infrastructure
PLMN	Public Land Mobile Network
PMIP	Proxy Mobile IP
PS	Packet Switched
PSAP	Public Safety Answering Point
PSTN	Public Switched Telephone Network
P-TMSI	Packet TMSI
QoS	Quality of Service
RA	Registration Authority
RAI	Routing Area Identity

RANAP	Radio Access Network Application Protocol
RAND	Random 128-bit string
RAT	Radio Access Technology
RAU	Routing Area Update
RFC	Request For Comments
RK	Root Key
RLC	Radio Link Control
RLC-SN	RLC Sequence Number
RN	Relay Node
RNC	Radio Network Controller
RRC	Radio Resource Control
RRM	Radio Resource Management
RTP	Real-Time Transport Protocol
SA	Service and System Aspects
SA	Security Association
SAGE	Special Algorithm Group of Experts
SCC	Service Centralization and Continuity
SC-FDMA	Single Carrier FDMA
S-CSCF	Serving CSCF
SDO	Standards Development Organisation
SDP	Session Description Protocol
SEG	Security Gateway (in NDS)
SeGW	Security Gateway (for HeNBs)
SGSN	Serving GPRS Support Node
S-GW	Serving Gateway
SHA	Secure Hash Algorithm
shortMAC-I	authentication token based on truncated MAC-I
SIM	Subscriber Identity Module
SIP	Session Initiation Protocol
SIPTO	Selective IP Traffic Offload
SK	Session Key
SKC	Session Keys Context
SMC	Security Mode Command
SMG	Special Mobile Group
SMSC	Short Message Service Centre
SN	Serving Network
SPI	Security Parameter Index
SQN	Sequence Number used in AKA
SRB	Signalling Radio Bearer
SRES	Signed Response
SRVCC	Single Radio Voice Call Continuity
SSL	Secure Sockets Layer
STUN	Simple Traversal of UDP through NATs
SW	Software
TAI	Tracking Area Identifier
TAU	Tracking Area Update

TCG	Trusted Computing Group
TCP	Transmission Control Protocol
TDMA	Time Division Multiple Access
TEK	Transient EAP Key
TIA	Telecommunications Industry Association
TID	Temporary ID
TIN	Temporary Identity used in Next update
TISPAN	Telecommunications and Internet converged Services and Protocols for Advanced Networking
TLinkID	Temporary Link ID
TLS	Transport Layer Security
TMSI	Temporary Mobile Subscriber Identity
TR	Technical Report
TrE	Trusted Environment
TS	Technical Specification
TSG	Technical Specification Groups
TTA	Telecommunications Technology Association
TTC	Telecommunication Technology Committee
UDP	User Datagram Protocol
UE	User Equipment
UEA	UMTS Encryption Algorithm
UIA	UMTS Integrity Algorithm
UICC	Universal Integrated Circuit Card
UMTS	Universal Mobile Telecommunications System
UP	User Plane
URI	Uniform Resource Identifier
URL	Uniform Resource Locator
USIM	Universal Subscriber Identity Module
UTRAN	Universal Terrestrial Radio Access Network
VCC	Voice Call Continuity
VLR	Visitor Location Register
VPN	Virtual Private Network
WCDMA	Wideband Code Division Multiple Access
WG	Working Group
WiMAX	Worldwide Interoperability for Microwave Access
WLAN	Wireless LAN
XMAC-I	Expected MAC-I
XOR	Exclusive or (operation)
XRES	Expected Response

References

All the IETF RFCs, 3GPP TRs and 3GPP TSs are listed as separate alphabetically ordered lists. Further references then follow.

3GPP Technical Reports and Technical Specifications

This book is based on versions of 3GPP specifications from March 2010. All 3GPP TSs and TRs are available at the 3GPP server, under the address http://www.3gpp.org/ftp/Specs/html-info/*xyabc*.htm, where *xyabc* corresponds to the number of the specification (e.g. TS33.401 in /ftp/Specs/html-info/33401.htm).

[TR21.801], 3GPP, Specification drafting rules.
[TR21.905], 3GPP, Vocabulary for 3GPP Specifications.
[TR23.830], 3GPP, Architecture aspects of Home Node B (HNB) / Home enhanced Node B (HeNB).
[TR31.900], 3GPP, SIM/USIM internal and external interworking aspects.
[TR33.820], 3GPP, Security of Home Node B (HNB) / Home evolved Node B (HeNB).
[TR33.821], 3GPP, Rationale and track of security decisions in Long Term Evolution (LTE) RAN / 3GPP System Architecture Evolution (SAE).
[TR33.901], 3GPP, Criteria for cryptographic Algorithm design process.
[TR33.908], 3GPP, 3G Security: General report on the design, specification and evaluation of 3GPP standard confidentiality and integrity algorithms.
[TR35.909], 3GPP, 3G Security: Specification of the MILENAGE algorithm set: an example algorithm set for the 3GPP authentication and key generation functions f1, f1*, f2, f3, f4, f5 and f5*; Document 5: Summary and results of design and evaluation.
[TR35.919], 3GPP, Specification of the 3GPP Confidentiality and Integrity Algorithms UEA2 and UIA2; Document 5: Design and evaluation report.
[TR36.912], 3GPP, Feasibility study for Further Advancements for E-UTRA (LTE-Advanced).
[TS21.133], 3GPP, 3G security; Security threats and requirements.
[TS22.101], 3GPP, Service aspects; Service principles.
[TS22.220], 3GPP, Service requirements for Home Node B (HNB) and Home eNode B (HeNB).
[TS22.278], 3GPP, Service requirements for the Evolved Packet System (EPS).
[TS23.002], 3GPP, Network architecture.
[TS23.003], 3GPP, Numbering, addressing and identification.
[TS23.060], 3GPP, General Packet Radio Service (GPRS); Service description; Stage 2.
[TS23.122], 3GPP, Non-Access-Stratum (NAS) functions related to Mobile Station (MS) in idle mode.
[TS23.167], 3GPP, IP Multimedia Subsystem (IMS) emergency sessions.
[TS23.216], 3GPP, Single Radio Voice Call Continuity (SRVCC); Stage 2.

[TS23.228], 3GPP, IP Multimedia Subsystem (IMS); Stage 2.

[TS23.234], 3GPP, 3GPP system to Wireless Local Area Network (WLAN) interworking; System description.

[TS23.237], 3GPP, IP Multimedia Subsystem (IMS) Service Continuity; Stage 2.

[TS23.272], 3GPP, Circuit Switched (CS) fallback in Evolved Packet System (EPS); Stage 2.

[TS23.292], 3GPP, IP Multimedia System (IMS) centralized services; Stage 2.

[TS23.334], 3GPP, IMS Application Level Gateway Control Function (ALGCF) – IMS Access Media Gateway (IMA-MGW); Iq Interface; Procedures description.

[TS23.401], 3GPP, General Packet Radio Service (GPRS) enhancements for Evolved Universal Terrestrial Radio Access Network (E-UTRAN) access.

[TS23.402], 3GPP, Architecture enhancements for non-3GPP accesses.

[TS24.229], 3GPP, Internet Protocol (IP) multimedia call control protocol based on Session Initiation Protocol (SIP) and Session Description Protocol (SDP); Stage 3.

[TS24.234], 3GPP, 3GPP system to Wireless Local Area Network (WLAN) interworking; WLAN User Equipment (WLAN UE) to network protocols; Stage 3.

[TS24.301], 3GPP, Non-Access-Stratum (NAS) protocol for Evolved Packet System (EPS); Stage 3.

[TS24.302], 3GPP, Access to the Evolved Packet Core (EPC) via non-3GPP access networks; Stage 3.

[TS25.467], 3GPP, UTRAN architecture for 3G Home Node B (HNB); Stage 2.

[TS29.060], 3GPP, General Packet Radio Service (GPRS); GPRS Tunnelling Protocol (GTP) across the Gn and Gp interface.

[TS29.228], 3GPP, IP Multimedia (IM) Subsystem Cx and Dx Interfaces; Signalling flows and message contents.

[TS29.229], 3GPP, Cx and Dx interfaces based on the Diameter protocol; Protocol details.

[TS29.234], 3GPP, 3GPP system to Wireless Local Area Network (WLAN) interworking; Stage 3.

[TS29.273], 3GPP, Evolved Packet System (EPS); 3GPP EPS AAA interfaces.

[TS31.101], 3GPP, UICC-terminal interface; Physical and logical characteristics.

[TS31.102], 3GPP, Characteristics of the Universal Subscriber Identity Module (USIM) application.

[TS31.103], 3GPP, Characteristics of the IP Multimedia Services Identity Module (ISIM) application.

[TS32.582], 3GPP, Telecommunications management; Home Node B (HNB) Operations, Administration, Maintenance and Provisioning (OAM&P); Information model for Type 1 interface HNB to HNB Management System (HMS).

[TS32.591], 3GPP, Telecommunication management; Concepts and requirements for Type 1 interface H(e)NB to H(e)NB Management System (H(e)MS).

[TS32.592], 3GPP, Telecommunications management; Home eNodeB (HeNB) Operations, Administration, Maintenance and Provisioning (OAM&P); Information model for Type 1 interface HeNB to HeNB Management System (HeMS).

[TS32.593], 3GPP, Telecommunication management; Procedure flows for Type 1 interface H(e)NB to H(e)NB Management System (H(e)MS).

[TS33.102], 3GPP, 3G security: Security architecture.

[TS33.106], 3GPP, Lawful interception requirements.

[TS33.107], 3GPP, 3G security: Lawful interception architecture and functions.

[TS33.108], 3GPP, 3G security: Handover interface for Lawful Interception (LI).

[TS33.120], 3GPP, Security Objectives and Principles.

[TS33.203], 3GPP, 3G security; Access security for IP-based services.

[TS33.210], 3GPP, 3G security; Network Domain Security (NDS); IP network layer security.

[TS33.220], 3GPP, Generic Authentication Architecture (GAA); Generic bootstrapping architecture.

[TS33.234], 3GPP, 3G security; Wireless Local Area Network (WLAN) interworking security.

[TS33.310], 3GPP, Network Domain Security (NDS); Authentication Framework (AF).

[TS33.320], 3GPP, Security of Home Node B (HNB) / Home evolved Node B (HeNB).

[TS33.328], 3GPP, Solutions for IMS media plane security.

[TS33.401], 3GPP, 3GPP System Architecture Evolution (SAE); Security architecture.

[TS33.402], 3GPP, 3GPP System Architecture Evolution (SAE); Security aspects of non-3GPP accesses.

[TS33.822], 3GPP, Security aspects for inter-access mobility between non 3GPP and 3GPP access network.

[TS35.201], 3GPP, Specification of the 3GPP confidentiality and integrity algorithms; Document 1: f8 and f9 specification.

[TS35.202], 3GPP, Specification of the 3GPP confidentiality and integrity algorithms; Document 2: Kasumi specification.

[TS35.203], 3GPP, Specification of the 3GPP confidentiality and integrity algorithms; Document 3: Implementors' test data.

[TS35.204], 3GPP, Specification of the 3GPP confidentiality and integrity algorithms; Document 4: Design conformance test data.

[TS35.205], 3GPP, 3G Security; Specification of the MILENAGE algorithm set: An example algorithm set for the 3GPP authentication and key generation functions f1, f1*, f2, f3, f4, f5 and f5*; Document 1: General.

[TS35.206], 3GPP, 3G Security; Specification of the MILENAGE algorithm set: An example algorithm set for the 3GPP authentication and key generation functions f1, f1*, f2, f3, f4, f5 and f5*; Document 2: Algorithm specification.

[TS35.207], 3GPP, 3G Security: Specification of the MILENAGE algorithm set: An example algorithm set for the 3GPP authentication and key generation functions f1, f1*, f2, f3, f4, f5 and f5*; Document 3: Implementors' test data.

[TS35.208], 3GPP, 3G Security; Specification of the MILENAGE algorithm set: An example algorithm set for the 3GPP authentication and key generation functions f1, f1*, f2, f3, f4, f5 and f5*; Document 4: Design conformance test data.

[TS35.215], 3GPP, Specification of the 3GPP Confidentiality and Integrity Algorithms UEA2 & UIA2; Document 1: UEA2 and UIA2 specifications.

[TS35.216], 3GPP, Specification of the 3GPP Confidentiality and Integrity Algorithms UEA2 & UIA2; Document 2: SNOW 3G specification.

[TS36.300], 3GPP, Evolved Universal Terrestrial Radio Access (E-UTRA) and Evolved Universal Terrestrial Radio Access Network (E-UTRAN); Overall description; Stage 2.

[TS36.323], 3GPP, Evolved Universal Terrestrial Radio Access (E-UTRA); Packet Data Convergence Protocol (PDCP) specification.

[TS36.331], 3GPP, Evolved Universal Terrestrial Radio Access (E-UTRA); Radio Resource Control (RRC); Protocol specification.

[TS36.806], 3GPP, Evolved Universal Terrestrial Radio Access (E-UTRA); Relay architectures for E-UTRA (LTE-Advanced).

[TS43.020], 3GPP, Security-related network functions.

[TS55.205], 3GPP, Specification of the GSM-MILENAGE algorithms: An example algorithm set for the GSM Authentication and Key Generation Functions A3 and A8.

[TS55.216], 3GPP, Specification of the A5/3 encryption algorithms for GSM and ECSD, and the GEA3 encryption algorithm for GPRS; Document 1: A5/3 and GEA3 specification.

IETF Requests For Comments

All IETF RFCs are available under the address http://www.ietf.org/rfc/rfc*xyzv*.txt, where *xyzv* corresponds the RFC number (e.g. RFC2131 in /rfc/rfc2131.txt).

[RFC1305], Mills, D., Network Time Protocol (Version 3) Specification, Implementation and Analysis.

[RFC1912], Barr, D., Common DNS Operational and Configuration Errors.

[RFC2104], Krawczyk, H., Bellare, M. and Canetti, R., HMAC: Keyed-Hashing for Message Authentication.

[RFC2131], Droms, R., Dynamic Host Configuration Protocol.

[RFC2246], Dierks, T. and Allen, C., The TLS Protocol Version 1.0.

[RFC2315], Kaliski, B., PKCS #7: Cryptographic Message Syntax Version 1.5.

[RFC2401], Kent, S. and Atkinson, R., Security Architecture for the Internet Protocol.

[RFC2406], Kent, S. and Atkinson, R., IP Encapsulating Security Payload (ESP).

[RFC2409], Harkins, D. and Carrel, D., The Internet Key Exchange (IKE).

[RFC2451], Pereira, R. and Adams, R., The ESP CBC-Mode Cipher Algorithms.

[RFC2560], Myers, M. *et al.*, X.509 Internet Public Key Infrastructure Online Certificate Status Protocol – OCSP.

[RFC2617], Franks, J. *et al.*, HTTP Authentication: Basic and Digest Access Authentication.

[RFC2818], Rescorla, E., HTTP Over TLS.

[RFC2903], Laat, C.D. *et al.*, Generic AAA Architecture.

[RFC2989], Aboba, B. *et al.*, Criteria for Evaluating AAA Protocols for Network Access.

[RFC3174], Eastlake, D. and Jones, P., US Secure Hash Algorithm 1 (SHA1).

[RFC3261], Rosenberg, J. *et al.*, SIP: Session Initiation Protocol.

[RFC3310], Niemi, A., Arkko, J. and Torvinen, V., Hypertext Transfer Protocol (HTTP) Digest Authentication Using Authentication and Key Agreement (AKA).

[RFC3329], Arkko, J. *et al.*, Security Mechanism Agreement for the Session Initiation Protocol (SIP).

[RFC3344], Perkins, C., IP Mobility Support for IPv4.

[RFC3489], Rosenberg, J. *et al.*, STUN - Simple Traversal of User Datagram Protocol (UDP) Through Network Address Translators (NATs).

[RFC3550], Schulzrinne, H. *et al.*, RTP: A Transport Protocol for Real-Time Applications.

[RFC3579], Aboba, B. and Calhoun, P., RADIUS (Remote Authentication Dial In User Service) Support For Extensible Authentication Protocol (EAP).

[RFC3602], Frankel, S., Glenn, R. and Kelly, S., The AES-CBC Cipher Algorithm and Its Use with IPsec.

[RFC3748], Aboba, B. *et al.*, Extensible Authentication Protocol (EAP).

[RFC3947], Kivinen, T. *et al.*, Negotiation of NAT-Traversal in the IKE.

[RFC3948], Huttunen, A. *et al.*, UDP Encapsulation of IPsec ESP Packets.

[RFC4067], Loughney, J. *et al.*, Context Transfer Protocol (CXTP).

[RFC4072], Eronen, P., Hiller, T. and Zorn, G., Diameter Extensible Authentication Protocol (EAP) Application.

[RFC4169], Torvinen, V., Arkko, J. and Naslund, M., Hypertext Transfer Protocol (HTTP) Digest Authentication Using Authentication and Key Agreement (AKA) Version 2.

[RFC4186], Haverinen, H. and Salowey, J., Extensible Authentication Protocol Method for Global System for Mobile Communications (GSM) Subscriber Identity Modules (EAP-SIM).

[RFC4187], Arkko, J. and Haverinen, H., Extensible Authentication Protocol Method for 3rd Generation Authentication and Key Agreement (EAP-AKA).

[RFC4210], Adams, C. *et al.*, Internet X.509 Public Key Infrastructure Certificate Management Protocol (CMP).

[RFC4211], Schaad, J., Internet X.509 Public Key Infrastructure Certificate Request Message Format (CRMF).

[RFC4217], Ford-Hutchinson, P., Securing FTP with TLS.

[RFC4282], Aboba, B. *et al.*, The Network Access Identifier.

[RFC4301], Kent, S. and Seo, K., Security Architecture for the Internet Protocol.

[RFC4303], Kent, S., IP Encapsulating Security Payload (ESP).

[RFC4305], Eastlake, D.3., Cryptographic Algorithm Implementation Requirements for Encapsulating Security Payload (ESP) and Authentication Header (AH).

[RFC4306], Kaufman, C., Internet Key Exchange (IKEv2) Protocol.

RFC4346], Dierks, T. and Rescorla, E., The Transport Layer Security (TLS) Protocol Version 1.1.

[RFC4347], Rescorla, E. and Modadugu, N., Datagram Transport Layer Security.

[RFC4366], Blake-Wilson, S. *et al.*, Transport Layer Security (TLS) Extensions.

[RFC4555], Eronen, P., IKEv2 Mobility and Multihoming Protocol (MOBIKE).

[RFC4566], Handley, M., Jacobson, V. and Perkins, C., SDP: Session Description Protocol.

[RFC4739], Eronen, P. and Korhonen, J., Multiple Authentication Exchanges in the Internet Key Exchange (IKEv2) Protocol.

[RFC4806], Myers, M. and Tschofenig, H., Online Certificate Status Protocol (OCSP) Extensions to IKEv2.

[RFC4949], Shirey, R., Internet Security Glossary, Version 2.

[RFC4962], Housley, R. and Aboba, B., Guidance for Authentication, Authorization, and Accounting (AAA) Key Management.

[RFC5213], Gundavelli, S. *et al.*, Proxy Mobile IPv6.

[RFC5246], Dierks, T. and Rescorla, E., The Transport Layer Security (TLS) Protocol Version 1.2.

[RFC5247], Aboba, B., Simon, D. and Eronen, P., Extensible Authentication Protocol (EAP) Key Management Framework.

[RFC5280], Cooper, D. *et al.*, Internet X.509 Public Key Infrastructure Certificate and Certificate Revocation List (CRL) Profile.

[RFC5448], Arkko, J., Lehtovirta, V. and Eronen, P., Improved Extensible Authentication Protocol Method for 3rd Generation Authentication and Key Agreement (EAP-AKA').

[RFC5555], Soliman, H., Mobile IPv6 Support for Dual Stack Hosts and Routers.

Further References

[3GPP and BBF, 2010], Joint 3GPP-BBF Workshop on Fixed/Mobile Convergence (Feb. 2010). Available at: http://www.3gpp.org/ftp/workshop/2010-02-18_FMC_BBF/

[3GPP, 2005], Review of recently published papers on GSM and UMTS security. *S3-050101, 3GPP TSG SA WG3 Security #37*, (Feb. 2005). Available at: ftp://ftp.3gpp.org/TSG_SA/WG3_Security/TSGS3_37_Sophia/

[3GPP, 2006], *UTRA-UTRAN Long Term Evolution (LTE) and 3GPP System Architecture Evolution (SAE)*. Available at: ftp://ftp.3gpp.org/Inbox/2008_web_files/LTA_Paper.pdf

[3GPP, 2009], Local IP Access and Selected IP Traffic Offload. SP-090761, 3GPP TSG SA#46, (Dec. 2009). Available at: ftp://ftp.3gpp.org/TSG_SA/TSG_SA/TSGS_46/Docs/

[3GPP], *3rd Generation Partnership Project*, URL: http://www.3gpp.org/

[3GPP2], *3rd Generation Partnership Project 2*, URL: http://www.3gpp2.org/

[ARIB], *Association of Radio Industries and Businesses*, URL: http://www.arib.or.jp/english/

[ATIS], *Alliance for Telecommunications Industry Solutions*, URL: http://www.atis.org/

[Aura, T. & Roe, M., 2005], Reducing Reauthentication Delay in Wireless Networks. In *Proceedings of the First International Conference on Security and Privacy for Emerging Areas in Communications Networks*. IEEE Computer Society, pp. 139–148. Available at: http://portal.acm.org/citation.cfm?id=1128478

[Barkan, E., Biham, E. and Keller, N., 2003], Instant Ciphertext-Only Cryptanalysis of GSM Encrypted Communication. *In Proceedings of Crypto 2003*. LNCS 2729. Santa Barbara, California, United States: Springer-Verlag, S. 600-616.

[Barker, E. *et al.*, 2007], Recommendation for Key Management. Available at: http://csrc.nist.gov/publications/nistpubs/800-57/sp800-57_PART3_key-management_Dec2009.pdf

[BBF TR-069], Broadband Forum, *BBF TR-069 CPE WAN Management Protocol v1.1 Issue 1 Amendment 2*, Dec. 2007. Available at: http://www.broadband-forum.org/technical/download/TR-069_Amendment-2.pdf

[BBF TR-098], Broadband Forum, *BBF TR-098 Internet Gateway Device Data Model for TR-069 Issue 1 Amendment 2*, Sep. 2008. Available at: http://www.broadband-forum.org/technical/download/TR-098_Amendment-2.pdf

[BBF TR-196], Broadband Forum, *BBF TR-196 Femto Access Point Service Data Model Issue 1*, Mar. 2009. Available at: http://www.broadband-forum.org/technical/download/TR-196.pdf

[BBF], *Broadband Forum (BBF)*, URL: http://www.broadband-forum.org/

[Bierbrauer, J. et al., 1993], On families of hash functions via geometric codes and concatenation. In *Proceedings of Crypto '93*. LNCS 773. Santa Barbara, California, United States: Springer-Verlag, S. 331-342.

[Bluetooth], *Bluetooth Special Interest Group*, URL: https://www.bluetooth.org

[C.S0024-A v2.0], 3GPP2, *cdma2000 High Rate Packet Data Air Interface Specification*, Oct. 2000. Available at: http://www.3gpp2.org/public_html/specs/index.cfm

[Camarillo, G. and García-Martín, M., 2008], *The 3G IP Multimedia Subsystem (IMS). Third Edition.*, Chichester: John Wiley & Sons.

[Carter, J.L. and Wegman, M.N., 1979], Universal Classes of Hash Functions. *J. Computer and System Sciences*, 18, 143–154.

[CCSA], *China Communications Standards Association*, URL: http://www.ccsa.org.cn/english/

[Diffie, W. and Hellman, M., 1976], New directions in cryptography. *IEEE Transactions on Information Theory*, 22(6), 644–654.

[draft-freier-ssl-version3-02], Freier, A.O., Karlton, P. and Kocher, P.C., *The SSL 3.0 Protocol*, Nov. 1996, Internet Engineering Task Force. Available at: http://www.mozilla.org/projects/security/pki/nss/ssl/draft302.txt

[draft-ietf-hokey-preauth-ps-09], Ohba, Y. and Zorn, G., *Extensible Authentication Protocol (EAP) Early Authentication Problem Statement*, Jul. 2009, IETF. Available at: http://www.ietf.org/internet-drafts/draft-ietf-hokey-preauth-ps-09.txt

[draft-ietf-ipsecme-eap-mutual-00], Eronen, P., Tschofenig, H. and Sheffer, Y., *An Extension for EAP-Only Authentication in IKEv2*, Feb. 2010, IETF. Available at: https://datatracker.ietf.org/doc/draft-ietf-ipsecme-eap-mutual/

[draft-ietf-pana-preauth-07], Ohba, Y. and Yegin, A., *Pre-authentication Support for PANA*, Oct. 2009, IETF. Available at: http://www.ietf.org/internet-drafts/draft-ietf-pana-preauth-07.txt

[draft-ietf-pkix-cmp-transport-protocols-07], Kapoor, A. et al., *Internet X.509 Public Key Infrastructure – Transport Protocols for CMP*, Oct. 2009, IETF. Available at: http://www.ietf.org/internet-drafts/draft-ietf-pkix-cmp-transport-protocols-07.txt

[draft-irtf-aaaarch-handoff-04], Arbaugh, W.A., *Handoff Extension to RADIUS*, Oct. 2003, Internet Engineering Task Force. Available at: http://www.watersprings.org/pub/id/draft-irtf-aaaarch-handoff-04.txt

[Dunkelmann, O. and Keller, N., 2008], An Improved Impossible Differential Attack on MISTY1. In *Proceedings of ASIACRYPT 2008*. LNCS 5350. Springer-Verlag, S. 441–454.

[Dunkelmann, O., Keller, N. and Shamir, A., 2010], A Practical-Time Attack on the A5/3 Cryptosystem Used in Third Generation GSM Telephony. Available at: http://eprint.iacr.org/2010/013

[Ekdahl, P. and Johansson, T., 2002], A new version of the stream cipher SNOW. In *Proceedings of SAC 2002*. LNCS 2595. Springer-Verlag, S. 47–61.

[ETSI ES 282 004], ETSI, *Telecommunications and Internet Converged Services and Protocols for Advanced Networking (TISPAN); NGN functional architecture; Network Attachment Sub-System (NASS)*, 2006.

[ETSI ES 283 035], ETSI, *Telecommunications and Internet Converged Services and Protocols for Advanced Networking (TISPAN); Network Attachment Sub-System (NASS); e2 interface based on the DIAMETER protocol*, Aug. 2008.

[ETSI TS 102 221], ETSI, *Smart Cards; UICC-Terminal interface; Physical and logical characteristics*, Feb. 2010.

[ETSI], *European Telecommunications Standards Institute*, URL: http://www.etsi.org/

[EUROSMART], *The Association representing the Smart Security Industry*, http://www.eurosmart.com/

[FF], *Femto Forum* (FF), URL: http://www.femtoforum.org/

[FIPS 140-2], NIST, *Security Requirements for Cryptographic Modules*, May. 2001. Available at: http://csrc.nist.gov/publications/fips/fips140-2/fips1402.pdf

[FIPS 180-2], NIST, *Secure Hash Standard*, Aug. 2002. Available at: http://csrc.nist.gov/publications/fips/fips180-2/fips180-2.pdf

[FIPS 197], NIST, *Advanced Encryption Standard (AES)*, Nov. 2001. Available at: http://csrc.nist.gov/publications/fips/fips197/fips-197.pdf

[Forsberg, D., 2007], Protected session keys context for distributed session key management. *Wireless Personal Commununications* 43(2), 665–676.

[Forsberg, D., 2010], LTE Key Management Analysis with Session Keys Context. *Computer Communications*, DOI: http://dx.doi.org/10.1016/j.comcom.2010.07.002

[GSMA, 2009], Press release 30 December 2009. Available at: http://www.gsmworld.com/newsroom/press-releases/2009/4490.htm

[GSMA], GSM Association, URL: http://www.gsmworld.com/

[Hillebrand, F. ed., 2001], *GSM and UMTS. The Creation of Global Mobile Communication.*, Chichester: John Wiley & Sons.

[Holtmanns, S. et al., 2008], *Cellular Authentication for Mobile and Internet Services - Overview and Application of the Generic Bootstrapping Architecture*, Chichester: John Wiley & Sons

[Horn, G. and Howard, P., 2000], Review of third generation mobile system security architecture. *Information Security Solutions Europe* (ISSE2000), Barcelona, Sept. 2000.

[IEEE 802.11], IEEE, *IEEE Standard for Information technology: Telecommunications and information exchange between systems, Local and metropolitan area networks, Specific requirements. Part 11: Wireless LAN Medium Access Control (MAC) and Physical Layer (PHY) Specifications*, March 2007.

[IEEE 802.11F], IEEE, *IEEE Trial-Use Recommended Practice for Multi-Vendor Access Point Interoperability via an Inter-Access Point Protocol Across Distribution Systems Supporting IEEE 802.11TM Operation*, July 2003, IEEE Computer Society.

[IEEE 802.11i], IEEE, *IEEE Standard for Information technology: Telecommunications and information exchange between systems, Local and metropolitan area networks, Specific requirements. Part 11: Wireless LAN Medium Access Control (MAC) and Physical Layer (PHY) Specifications Amendment 6: Medium Access Control (MAC) Security Enhancements*, July 2004.

[IEEE 802.16], IEEE, *IEEE Standard for Local and metropolitan area networks. Part 16: Air Interface for Fixed Broadband Wireless Access Systems*, June 2004.

[IEEE 802.1X], IEEE, *IEEE Standard for Local and Metropolitan Area Networks: Port-Based Network Access Control*, 2004, IEEE Standards Association.

[IEEE Std 1003.1], IEEE / The Open Group, *Portable Operating System Interface (POSIX) Base Specifications*, Issue 7, 2008. Available at: http://www.opengroup.org/onlinepubs/9699919799/toc.htm

[IETF], *Internet Engineering Task Force*, URL: http://www.ietf.org/

[ISO 7498-2], ISO, *ISO 7498-2: Information Processing Systems, Open Systems Interconnection, Basic Reference Model. Part 2: Security Architecture*, Jan. 1989, ISO (International Organization for Standardization).

[ISO/IEC 19790], ISO, *Information technology, Security techniques, Security requirements for cryptographic modules*, 2006, ISO (International Organization for Standardization). Available at: http://www.iso.org/iso/iso_catalogue/catalogue_tc/catalogue_detail.htm?csnumber=33928

[ITU X.509], ITU-T, *Information technology, Open Systems Interconnection, The Directory: Public-key and attribute certificate frameworks*, Nov. 2008. Available at: http://www.itu.int/rec/T-REC-X.509-200811-I/en

[ITU, R., 2005], The Internet of Things. In ITU Report. Available at: http://www.itu.int/osg/spu/ publications/internetofthings/InternetofThings_summary.pdf

[ITU], *International Telecommunication Union*, URL: http://www.itu.int/

[Kaaranen, H., 2005], *UMTS Networks*, Chichester: John Wiley & Sons.

[Kassab, M. *et al.*, 2005], Fast pre-authentication based on proactive key distribution for 802.11 infrastructure networks. In *Proceedings of the 1st ACM workshop on Wireless multimedia networking and performance modeling*. Montreal, Quebec, Canada: ACM, pp. 46–53. Available at: http://portal.acm.org/citation.cfm?id=1089737. 1089746

[Komarova, M. and Riguidel, M., 2007], Optimized ticket distribution scheme for fast re-authentication protocol (fap). In *Proceedings of the 3rd ACM workshop on QoS and security for wireless and mobile networks*. Chania, Crete Island, Greece: ACM, pp. 71–77. Available at: http://portal.acm.org/citation.cfm?id=1298239. 1298253

[Kühn, U., 2001], Cryptanalysis of Reduced Round MISTY. In *Proceedings of EUROCRYPT 2001*. LNCS 2045. Innsbruck, Austria: Springer-Verlag, pp. 325–339.

[Matsui, M., 1997], Block encryption algorithm MISTY. *Proceedings of Fast Software Encryption (FSE97)*, pp. 64–74.

[McGrew, D. and Viega, J., 2004], The security and performance of the Galois/Counter mode of operation. *IACR eprint 2004/193*. Available at: http://eprint.iacr.org/2004/193.pdf

[Menezes, A., Oorschot, P.V. and Vanstone, S., 1996], *Handbook of Applied Cryptography*, Boca Raton: CRC Press.

[Meyer, U. and Wetzel, S., 2004a], A Man-in-the-Middle Attack on UMTS. *Proceedings of ACM Workshop on Wireless Security* (WiSe 2004), ACM, 2004.

[Meyer, U. and Wetzel, S., 2004b], On the impact of GSM Encryption and Man-in-the-Middle Attacks on the Security of Interoperating GSM/UMTS Networks. *Proceedings of IEEE International Symposium on Personal, Indoor and Mobile Radio Communications* (PIMRC2004), IEEE, 2004.

[Miller, S.P. *et al.*, 1987], *Kerberos authentication and authorization system*. Available at: http://eprints. kfupm.edu.sa/47456/

[Mishra, A. *et al.*, 2003], *Proactive Key Distribution to support fast and secure roaming*, IEEE. Available at: http://www.ieee802.org/11/Documents/DocumentHolder

[Mishra, A. *et al.*, 2004], Proactive key distribution using neighbor graphs. *IEEE Wireless Communications* 11(1), 26–36.

[Neuman, B. and Ts'o, T., 1994], Kerberos: an authentication service for computer networks. *IEEE Communications Magazine* 32(9), 33–38.

[Niemi, V. and Nyberg, K., 2003], *UMTS Security*, Chichester: John Wiley & Sons.

[NIST], National Institute of Science and Technology; Information Technology Laboratory; Computer Security Resource Center, http://csrc.nist.gov/

[NIST800-38A, 2001], *Recommendations for Block Cipher Modes of Operation: Methods and Techniques*, NIST. Available at: http://csrc.nist.gov/publications/PubsSPs.html

[NIST800-38B, 2005], *Recommendations for Block Cipher Modes of Operation: The CMAC Mode for Authentication*, NIST. Available at: http://csrc.nist.gov/publications/PubsSPs.html

[Nohl, K. and Paget, C., 2009], GSM: SRSLY? *26th Chaos Communication Congress*. Available at: http://events.ccc.de/congress/2009/Fahrplan/attachments/1519_26C3.Karsten.Nohl.GSM.pdf

[Ohba, Y., Das, S. and Dutta, A., 2007], Kerberized handover keying: a media-independent handover key management architecture. In *Proceedings of 2nd ACM/IEEE international workshop on Mobility in the evolving internet architecture*. Kyoto, Japan: ACM, pp. 1–7. Available at: http://portal.acm.org/citation.cfm?id=1366932

[OMA], *Open Mobile Alliance*, URL: http://www.openmobilealliance.org/

[Pack, S. and Choi, Y., 2002a], Fast Inter-Ap Handoff Using Predictive Authentication Scheme in a Public Wireless LAN. In *Proceedings of IEEE Networks conference (confunction of IEEE ICN and IEEE ICWLHN*. Available at: http://citeseerx.ist.psu.edu/viewdoc/summary?doi=10.1.1.20.138

[Pack, S. and Choi, Y., 2002b], Pre-Authenticated Fast Handoff in a Public Wireless LAN Based on IEEE 802.1x Model. *IFIP TC6 Personal Wireless Communications*, pp. 175–182.

[Poikselkä, M. and Mayer, G., 2009], *The IMS. IP Multimedia Concepts and Services. Third Edition.*, Chichester: John Wiley & Sons.

[Sklavos, N., Denazis, S. and Koufopavlou, O., 2007], AAA and mobile networks: security aspects and architectural efficiency. In *Proceedings of the 3rd international conference on Mobile multimedia communications*. Nafpaktos, Greece: ICST (Institute for Computer Sciences, Social-Informatics and Telecommunications Engineering), pp. 1-4. Available at: http://portal.acm.org/citation.cfm?id=1385343

[Smart, N. ed., 2009], ECRYPT2 Yearly Report on Algorithms and Keysizes (2008–9). Available at: http://www.ecrypt.eu.org/documents/D.SPA.7.pdf

[Smetters, D.B. *et al.*, 2002], Talking To Strangers: Authentication in Ad-Hoc Wireless Networks. In *Proceedings of Network and Distributed System Security Symposium NDSS'02*. San Diego. Available at: http://citeseerx.ist.psu.edu/viewdoc/summary?doi=10.1.1.16.1408

[Stinson, D., 1992], Universal hashing and authentication codes. *Proceedings of Crypto '91*, LNCS 576. Santa Barbara, California, United States: Springer-Verlag, S. 74–85.

[TCG Mobile Phone Work Group, 2008], *TCG Mobile Trusted Module Specification* 1st ed., Version 1 rev. 1.0. Available at: http://www.trustedcomputinggroup.org/files/resource_files/87852F33-1D09-3519-AD0C0F141CC6B10D/Revision_6-tcg-mobile-trusted-module-1_0.pdf

[TIA], Telecommunications Industry Association, URL: http://www.tiaonline.org/

[TTA], Telecommunications Technology Association, URL: http://www.tta.or.kr/English/

[TTC], Telecommunication Technology Committee, URL: http://www.ttc.or.jp/e/

[Vedder, K., 2010], The UICC: Recent Work of SCP and Related Security Aspects. In *5th ETSI Security Workshop*, 20–22 January 2010. Available at: http://docbox.etsi.org/Workshop/2010/201001_ SECURITYWORK-SHOP/04INTERNATIONALSTANDARDIZATION/Vedder_GandD_UICC.pdf

[Wassenaar], Wassenaar Arrangement, URL: http://www.wassenaar.org/

[WiMAX], *Worldwide interoperability for Microwave Access*, URL: http://www.wimaxforum.org/

[X.S0042-0 v1.0], 3GPP2, *Voice Call Continuity between IMS and Circuit Switched System*, Oct. 2007. Available at: http://www.3gpp2.org/public_html/specs/index.cfm

[Zhang, M. and Fang, Y., 2005], Security Analysis and Enhancements of 3GPP Authentication and Key Agreement Protocol. *IEEE Transactions on Wireless Communications* 4(2), 734–742.

Index

3G Authentication and Key Agreement protocol
  (UMTS AKA), 37–42, 174
3G–WLAN Interworking, 63
3rd Generation Partnership Project (3GPP), 2, 5, 21,
  35
3GPP IP access, 64, 74–7
AAA Server, 67, 218, 221, 234, 237
Access Authorization for Home eNodeB, 236–7
Access control 59, 89, 220, 236–237, 254
Access Security Management Entity (ASME),
  79
Access Stratum (AS), 7, 80, 127
Access-independence, 91, 98, 202
Algorithm
  A3, 30, 32
  A5, 31–3, 55
  A8, 30, 32
  Advanced Encryption Standard (AES), 16, 47, 51,
    78, 167
  agility, 165
  AKA, 47, 51
  Data Encryption Standard (DES), 16, 47
  EPS Encryption Algorithm (EEA), 168
  EPS Integrity Algorithm (EIA), 169
  GPRS Encryption Algorithm (GEA), 30, 33
  KASUMI, 33, 47, 167
  MILENAGE, 51, 82, 117
  MISTY, 47
  null, 166
  Secure Hash Algorithm (SHA), 18, 51
  SNOW 3G, 49, 167
  type distinguisher, 120, 170
  UMTS Encryption Algorithm (UEA), 33, 48–50,
    55, 167
  UMTS Integrity Algorithm (UIA), 48–51
Alliance for Telecommunications Industry Solutions
  (ATIS), 21
AMF separation bit, 111, 114, 185

Association of Radio Industries and Businesses (ARIB),
  21
Attack, 12, 27, 32
  active, 19, 27, 32, 36
  algebraic, 48
  attack models, 20
  bidding down, 128, 129, 180
  birthday, 18
  chosen plaintext, 20, 33, 48
  ciphertext only, 20
  configuration, 219
  Denial of Service (DoS), 12, 87, 220, 262
  exhaustive search, 20, 28
  false base station, 38
  flooding, 87
  known plaintext, 20
  lunch time, 29
  man-in-the-middle, 44, 220
  on Home eNodeB, 219–220
  packet injection, 86, 162
  physical, 86, 101, 219
  pre-play, 42
  proxying, 252
  radio jamming, 39, 87
  related key, 20, 33, 48
  replay, 44
  tracking, 32, 46, 86
  side channel, 20
  SIM cloning, 29
Authentication, 12, 16, 88
  access-network bundled, 208
  data, 79, 115–16
  failure, 114–15
  failure reporting, 115
  for base station enrolment, 139
  of Home eNodeB, 220, 222–3, 228, 234, 247
  request, 111
  response, 113